AF780744

Antimicrobial Activity of Plant Volatiles in Vapor Phase

Antimicrobial Activity of Plant Volatiles in Vapor Phase

Editor

Ladislav Kokoska

MDPI

Basel • Beijing • Wuhan • Barcelona • Belgrade • Novi Sad • Cluj • Manchester

Editor
Ladislav Kokoska
Department of Crop Sciences
and Agroforestry, Faculty of
Tropical AgriSciences
Czech University of Life
Sciences Prague
Prague
Czech Republic

Editorial Office
MDPI
St. Alban-Anlage 66
4052 Basel, Switzerland

This is a reprint of articles from the Special Issue published online in the open access journal *Molecules* (ISSN 1420-3049) (available at: www.mdpi.com/journal/molecules/special_issues/Ant_Vap).

For citation purposes, cite each article independently as indicated on the article page online and as indicated below:

Lastname, A.A.; Lastname, B.B. Article Title. *Journal Name* **Year**, *Volume Number*, Page Range.

ISBN 978-3-0365-8751-6 (Hbk)
ISBN 978-3-0365-8750-9 (PDF)
doi.org/10.3390/books978-3-0365-8750-9

© 2023 by the authors. Articles in this book are Open Access and distributed under the Creative Commons Attribution (CC BY) license. The book as a whole is distributed by MDPI under the terms and conditions of the Creative Commons Attribution-NonCommercial-NoDerivs (CC BY-NC-ND) license.

Contents

About the Editor

Ladislav Kokoska

Ladislav Kokoska works at the Department of Crop Sciences and Agroforestry of the Faculty of Tropical AgriSciences of the Czech University of Life Sciences Prague as full professor since 2013. He teaches courses on Economic Botany (Tropical Ethnobotany) and Special Crops (Secondary Metabolite-Bearing Crops). The Laboratory of Ethnobotany and Ethnopharmacology, of which he is in charge, has a rich tradition in the research and development activities focused on biologically active natural products derived from underutilized tropical crops, especially from species used by native peoples in traditional folk medicinal systems. Prof. Kokoska is author or co-author of 1 patent, 1 monograph, 2 chapters in monographs, 116 scientific papers, more than 200 contributions presented on international symposiums, 16 works popularizing science (2 books and 14 articles) and 1 documentary movie. As a principal investigator, co-investigator or co-worker, he has participated in 20 research projects. He is a member of several advisory boards, including those of scientific journals (e.g., *Journal of Advanced Research, Frontiers in Pharmacology, Molecules*).

Preface

Plant-derived volatile products (essential oils, volatile extracts and their constituents) are commercially important as pharmaceuticals, agrochemicals, food flavours and fragrances. Because of antimicrobial properties and high volatility, these products have strong potential to be used in novel agricultural, food and pharmaceutical applications such as inhalation therapies and agents used in active food packaging and controlled atmosphere storage of agricultural products. However, lack of knowledge on chemistry and antimicrobial activity of their vapours is the main limiting factor for their introduction to the practice. Therefore, the aim of the Special Issue is to present the recent advances and developments in the area of antimicrobial activity of plant-derived volatile agents in vapour phase. The papers provide information on various aspects of antibacterial/antifungal effects and chemical properties of plant volatiles, with special emphasis on innovative bioassays and modern sampling methods for testing and analysing their vapours.

Recent progress in the development of novel methods for testing of antimicrobial activity of volatile agents in vapour phase and advances in the analysis of volatile substances in gaseous state were the main reasons motivating me to edit this Special Issue. It includes 11 full research papers addressing various aspect of antimicrobial activity and chemical composition of plant-derived volatile agents such as development and validation of new methods for the assessment of essential oils and their constituents, antibacterial and antifungal effects of volatiles in vapour phase against human, food and plant pathogens (including their combinatory and anti-biofilm effects), and innovative approaches in sampling and analysis of bioactive vapours.

This Special Issue should be useful for medicinal, pharmaceutical, food and agricultural experts working in the areas related to the management of infectious diseases, food preservation, and protection of agriculture products. It may also stimulate interest of pharmaceutical, food and agriculture industries in research and development of new antibacterial and antifungal agents of natural origin. The issue therefore appeals to communities of industrial stakeholders, pharmacists, physicians, food experts, agriculturists and researchers in related areas such as pharmacology, medicinal chemistry, microbiology, natural product chemistry, food preservation, plant protection, etc.

Finally, I would like to thank all authors for publishing their research papers in this Special Issue. A special thanks goes to the research teams from institutions contributing by two or more papers to the issue. It has been a great pleasure collaborating with all the enthusiastic researchers around the world working in the quickly developing field of antimicrobial plant-derived volatile agents.

Ladislav Kokoska
Editor

molecules

MDPI

Article

Validation of Broth Macrodilution Volatilization Method for Testing of Essential Oils in Liquid and Vapor Phase: Chemical Composition, Cytotoxicity, and Antibacterial Effect of Indian Medicinal Plants against Pneumonia-Causing Pathogens

Aishwarya Chaure [1], Marketa Houdkova [1], Julien Antih [1], Klara Urbanova [2], Ivo Doskocil [3], Mukund Lal Naik [4], Khageshwar Singh Patel [5] and Ladislav Kokoska [1,*]

1 Department of Crop Sciences and Agroforestry, Faculty of Tropical AgriSciences, Czech University of Life Sciences Prague, 16500 Prague, Czech Republic; chaure@ftz.czu.cz (A.C.); houdkovam@ftz.czu.cz (M.H.); antih@ftz.czu.cz (J.A.)
2 Department of Sustainable Technologies, Faculty of Tropical AgriSciences, Czech University of Life Sciences Prague, 16500 Prague, Czech Republic; urbanovak@ftz.czu.cz
3 Department of Microbiology, Nutrition and Dietetics, Faculty of Agrobiology, Food and Natural Resources, Czech University of Life Sciences Prague, 16500 Prague, Czech Republic; doskocil@af.czu.cz
4 National Center for Natural Resources, Pt. Ravishankar Shukla University, Raipur 492010, India; mlnaik1943@gmail.com
5 Department of Applied Sciences, Amity University, Manth (Kharora), State Highway 9, Raipur 493225, India; patelkhageshwarsingh@gmail.com
* Correspondence: kokoska@ftz.czu.cz; Tel.: +420-224-382-180

Abstract: Essential oils (EOs) have great potential in inhalation therapy for the treatment of respiratory infections. However, innovative methods for evaluation of antimicrobial activity of their vapors are still needed. The current study reports validation of the broth macrodilution volatilization method for assessment of the antibacterial properties of EOs and shows the growth-inhibitory effect of Indian medicinal plants against pneumonia-causing bacteria in liquid and vapor phase. Among all samples tested, *Trachyspermum ammi* EO exhibits the strongest antibacterial effect against *Haemophilus influenzae*, with minimum inhibitory concentrations of 128 and 256 μg/mL in the liquid and vapor phases, respectively. Furthermore, *Cyperus scariosus* EO is found to be nontoxic to normal lung fibroblasts assessed by modified thiazolyl blue tetrazolium bromide assay. Chemical analysis performed using gas chromatography–mass spectrometry identified α-citral, cyperotundone, and thymol as the main constituents of *Cymbopogon citratus*, *C. scariosus*, and *T. ammi* EOs, respectively. In addition, β-cymene is identified as the major compound of *T. ammi* EO vapors when analyzed using solid-phase microextraction and gas-tight syringe sampling techniques. This study demonstrates the validity of the broth macrodilution volatilization method for antimicrobial screening of volatile compounds in the vapor phase and suggests the therapeutic potential of Indian medicinal plants in inhalation therapy.

Keywords: antimicrobial activity; *Cymbopogon citratus*; *Cyperus scariosus*; GC/MS; headspace analysis; macrodilution; MTT assay; respiratory infections; *Trachyspermum ammi*; vapor phase; volatiles

Citation: Chaure, A.; Houdkova, M.; Antih, J.; Urbanova, K.; Doskocil, I.; Naik, M.L.; Patel, K.S.; Kokoska, L. Validation of Broth Macrodilution Volatilization Method for Testing of Essential Oils in Liquid and Vapor Phase: Chemical Composition, Cytotoxicity, and Antibacterial Effect of Indian Medicinal Plants against Pneumonia-Causing Pathogens. *Molecules* **2023**, *28*, 4625. https://doi.org/10.3390/molecules28124625

Academic Editor: Daniela Rigano

Received: 10 May 2023
Revised: 2 June 2023
Accepted: 5 June 2023
Published: 7 June 2023

Copyright: © 2023 by the authors. Licensee MDPI, Basel, Switzerland. This article is an open access article distributed under the terms and conditions of the Creative Commons Attribution (CC BY) license (https://creativecommons.org/licenses/by/4.0/).

1. Introduction

Pneumonia is defined as acute infection of the lung parenchyma [1], typically caused by bacterial species including gram-positive *Staphylococcus aureus*, *Streptococcus pneumoniae*, *Streptococcus pyogenes*, and gram-negative *Haemophilus influenzae* [2]. It is a major health problem associated with high morbidity and mortality, especially in developing countries, where vulnerable populations such as children of <5 years of age and older adults with prior chronic conditions are at great risk [3]. In the year 2020, India alone contributed

23% of the global pneumonia burden with case fatality rates between 14 and 30% [4]. The general approach in the treatment of pneumonia includes the timely administration of appropriate antibiotic therapy. Recently, nebulized antibiotics have become preferred over intravenous and oral therapies, as they allow high pulmonary efficacy and minimal systemic side effects [5]. Antibiotics currently marketed for inhalation treatment of chronic *Pseudomonas aeruginosa* infections include tobramycin, colistin, and aztreonam. Although both the U.S. Food and Drug Administration (FDA) and European Medicines Agency have approved their use for cystic fibrosis, they have not been approved in other disease areas because of a lack of supportive clinical trial evidence [6]. Moreover, other problems are associated with the distribution and deposition of aerosol particles, whose size plays a major role in efficient delivery. For example, larger particles preferably accumulate in the oropharyngeal area, while smaller particles deposit in the lower airways, which alters drug delivery, as small particles carry fewer active substances. In addition, patient-related factors such as age, physical capability, disease severity, and the cognitive ability of the patient to perform specific inhalation affect drug delivery [7].

Essential oils (EOs) can be promising sources for the development of new inhalation preparations, as the physical property of being volatile at room temperature enables them to freely distribute among the lung tissues [8]. Therefore, in contrast with nebulized antibiotics, EO vapors allow easy inhalation and uniform distribution of active substances in the upper and lower parts of the respiratory tract [9]. Furthermore, due to their antimicrobial and anti-inflammatory potency in the vapor phase, they offer effective treatment for various respiratory infections. In the last few years, several inhalation devices using a combination of different EOs have been developed and patented. For example, adhesive inhalation antiviral patches (Axogen Inc., Alachua, FL, USA), which contain safe and effective amounts of EOs obtained from plant species such as *Cinnamomum verum* J. Presl, *Citrus limon* (L.) Osbeck, *Gaultheria procumbens* L., *Matricaria recutita* L., *Mentha* × *piperita* L., *Salvia sclarea* L., *Syzygium aromaticum* (L.) Merr. & L.M. Perry, and *Zingiber officinale* Roscoe are usually placed near the nasal pathway with the application of an appropriate mask, which prevents the entry of various respiratory-infection-causing pathogens [10]. In addition, various herbal products based on EOs, e.g., GeloMyrtol (G. Pohl-Boskamp, Hohenlockstedt, Germany) and Bronchipret (Bionorica, Neumarkt, Germany), are recommended for the treatment of acute and chronic bronchitis [11]. EOs are very complex natural mixtures of compounds that work in synergy to provide various medicinal properties. They are mainly composed of aromatic and aliphatic compounds, hydrocarbon terpenes (isoprenes), and terpenoids (terpene hydrocarbons such as phenols, alcohols, ketones, aldehydes, acids, esters, and ethers) [12]. Especially the phenolic monoterpenoids, such as carvacrol and thymol, have been reported to exhibit a strong antibacterial effect in the vapor phase against pathogens that cause pneumonia [13]. The bioactivity of EOs depends greatly on their chemical composition, which makes the chemical characterization of volatile components an important work [14]. Nowadays, static headspace extraction, coupled with the gas chromatography–mass spectrometry (S-HS-GC/MS) technique, is commonly used to investigate EO vapors. It is a simple, rapid, and solventless technique that requires a small amount of sample and allows the analysis of highly volatile compounds [15]. Correspondingly, several studies have explored the chemical composition of EO vapors with the aim to find the most prominent volatile compounds responsible for antibacterial activity in the vapor phase using S-HS-GC/MS. For example, a study conducted by Schweitzer et al. [16] has explored the chemical composition of *C. citratus* EO vapors using solid-phase microextraction followed by GC/MS. Using this technique, citral, α-terpineol, γ-cadinene, and calamenene have been found to be the major components of EO vapors that are attributed to their antibacterial and antifungal activities. In another study, chemical investigation of *T. ammi* EO has been conducted using headspace–solid phase microextraction (HS-SPME) coupled with GC/MS, where γ-terpinene, *p*-cymene, and thymol were found to be major compounds [17].

In terms of plant diversity, 1500 species have been reported to have medicinal uses in India [18], many of which have not been phytochemically and pharmacologically explored [19]. Indian Ayurvedic medicine has a long tradition in the management of various respiratory diseases using poly-herbal preparations and multicomponent therapeutics. Several pharmaceutical companies are manufacturing and marketing different Ayurvedic formulations [20]. For example, Bresol-NS (Himalaya, Bengaluru, India), a nasal spray with a combination of *Coleus aromaticus* Benth., *Eucalyptus globulus* Labill., and *Glycyrrhiza glabra* L. extracts, helps relieve nasal congestion caused by upper respiratory tract infections. Administration of drugs through the nasal route is an integral part of Ayurvedic practices such as *dhumpana* (medicinal smoking), *nasya* (nasal administration of therapeutic oil), and *dhupanartha* (herbal fumigation). In addition, ethnobotanical evidence of using vapor-based medicines has been well-documented in northeast India. For example, in Manipur, the people of the Meitei community use leaves of *Phlogacanthus thyrsiflorus* Nees for steam inhalation in the treatment of pneumonia. Similarly, the stems of *Clerodendrum indicum* (L.) Kuntze and *Rotheca serrata* (L.) Steane & Mabb. are inhaled via smoke for treatment of acute bronchitis [21]. Among various plant species used in Indian traditional medicine for the treatment of respiratory infections, *Cymbopogon citratus* (DC.) Stapf and *Trachyspermum ammi* (L.) Sprague and certain species of genera *Cyperus* (e.g., *Cyperus rotundus* L.) are common ingredients for vapor-based medicines [22–24]. In correspondence with their traditional use, the antibacterial potential of EO vapors of the above-mentioned species has been previously reported in several in vitro experiments. For example, a study performed by Inouye et al. [25] assessed the growth-inhibitory effects of *C. citratus* EO vapors against respiratory tract pathogens such as *H. influenzae*, *S. aureus*, *S. pneumoniae*, and *S. pyogenes* using the airtight box disc volatilization method. In another study, *T. ammi* EO vapors exhibited a growth-inhibitory effect against bovine respiratory bacterial pathogens using an agar plug vapor-phase assay [26]. However, the antibacterial potential of *T. ammi* and *Cyperus scariosus* R.Br. EO vapors against pneumonia-causing bacteria has not been fully explored yet. Moreover, the methods used for examining the antimicrobial effect of the above-mentioned EO vapors provided mainly qualitative results, or in the case of quantitative methods, the results were expressed in different ways, e.g., as a concentration of volatile agent per volume of air [25], and interpretation of results, including minimum inhibitory concentration (MIC) values, was varying greatly [27]. Although EOs and their compounds are usually considered as safe, and they are registered as generally recognized as safe (GRAS) products by the U.S. FDA, toxicological evaluations are still necessary to know the risks associated with direct inhalation of EOs or their constituents on lung tissues [28]. There are several studies describing the low cytotoxicity of *C. citratus* EO when tested in vitro on human non-cancer fibroblasts [29]; however, there are no data on toxicity of *T. ammi* and *C. scariosus* EOs to lung cells.

The current study reports an assessment of the antibacterial properties of selected EOs from Indian medicinal plants, namely *C. citratus*, *C. scariosus*, and *T. ammi*, with the aim to validate the newly developed broth macrodilution volatilization method for simultaneous testing of EOs in liquid and vapor phase against the bacteria causing pneumonia. Additionally, their major chemical constituents responsible for growth-inhibitory effects were identified using GC/MS, and the chemical profile of EO vapors was determined by a time series of headspace analyses including both HS-SPME and headspace–gas-tight syringe (HS-GTS) sampling techniques. Moreover, the cytotoxicity of EOs against normal lung fibroblasts using a modified thiazolyl blue tetrazolium bromide (MTT) assay was evaluated in this study.

2. Results

2.1. Antibacterial Activity and Cytotoxicity

The results showed that all EOs produced a certain level of in vitro growth-inhibitory activity against the pneumonia-causing bacterial strains included in this study. The effectiveness of EOs varied substantially in the MIC values ranging from 128 to 1024 μg/mL in the liquid phase and being equal or greater than 256 µg/mL in the vapor phase. Among all samples tested, *T. ammi* EO was found to be the most active with respective MICs of 128 and 256 µg/mL in the liquid and vapor phases against *H. influenzae,* followed by *C. citratus* EO with growth-inhibitory effect against *H. influenzae* with MIC 256 µg/mL in both phases. The *C. scariosus* EO exhibited weak antibacterial effect with MIC 1024 µg/mL in the liquid phase against all bacteria tested, while vapors of this EO were active against *H. influenzae* (MIC 1024 µg/mL) only. In general, a higher activity was observed in the liquid than in the vapor phase, and *H. influenzae* was the most susceptible bacterium to the EOs assayed. Complete results on the antibacterial activity of Indian EOs against respiratory pathogens in both liquid and vapor phases are shown in Table 1.

Table 1. In vitro growth-inhibitory effect of Indian essential oils in the liquid and vapor phases against respiratory pathogens.

Sample	**Bacterium/Growth Medium/Minimum Inhibitory Concentration (µg/mL)**								
Essential Oil	***Haemophilus influenzae***		***Staphylococcus aureus***		***Streptococcus pneumoniae***		***Streptococcus pyogenes***		**x̄-MIC**
	Broth	**Agar**	**Broth**	**Agar**	**Broth**	**Agar**	**Broth**	**Agar**	
Cymbopogon citratus	256	256	512	1024	512	1024	512	1024	448
Cyperus scariosus.	1024	1024	1024	>1024	1024	>1024	1024	>1024	1024
Trachyspermum ammi	128	256	512	512	512	1024	512	1024	416
Positive antibiotic control	1 [a]	n.d.	0.5 [b]	n.d.	0.25 [c]	n.d.	0.25 [d]	n.d.	-

x̄-MIC: mean value of minimal inhibitory concentrations in broth medium; positive antibiotic control: [a] ampicillin, [b] oxacillin, [c] amoxicillin, [d] tetracycline; n.d.: not detected.

The results of a cytotoxicity assay showed that the values of half maximal inhibitory concentration (IC_{50}) varied substantially for all EOs tested. *C. scariosus* EO was evaluated as nontoxic ($IC_{50} > 258$ µg/mL), whereas both *C. citratus* and *T. ammi* EOs were found to be moderately toxic with respective IC_{50} values of 19.63 and 82.04 µg/mL. Similarly, the highest 80% inhibitory concentration of proliferation (IC_{80}) values were calculated for *C. scariosus* ($IC_{80} \geq 258$ µg/mL). According to the WHO [30], the EOs were classified as nontoxic (*C. scariosus*) and moderately toxic (*C. citratus* and *T. ammi*). The therapeutic index (TI) calculated for comparison of the antibacterial and cytotoxic effects indicates safety of the *C. scariosus* EO (TI > 0.252). The detailed results of the cytotoxicity of EOs to lung fibroblasts are shown in Table 2 with a graphical presentation in Figure 1.

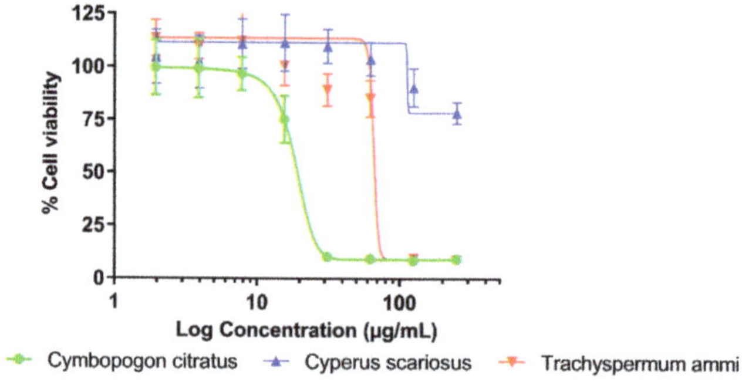

Figure 1. Cytotoxic activity of eight twofold serially diluted concentrations (2–256 µg/mL) of *Cymbopogon citratus*, *Cyperus scariosus* and *Trachyspermum ammi* EOs to lung fibroblast cells MRC-5 tested by MTT assay performed in microtiter plates sealed with vapor barrier EVA Capmat.

Table 2. Cytotoxicity of Indian essential oils to the normal lung fibroblast cell line MRC-5.

Sample	IC_{50} ± SD (µg/mL)	IC_{80} ± SD (µg/mL)	TI
Essential oil			
Cymbopogon citratus	19.63 ± 1.02	29.54 ± 2.18	0.065
Cyperus scariosus	>258	>258	>0.252
Trachyspermum ammi	82.04 ± 3.39	156.57 ± 13.88	0.376
Positive control			
vinorelbine	0.54 ± 0.26	>10	n.a.

IC_{50}: half maximal inhibitory concentration of proliferation in µg/mL; IC_{80}: 80% inhibitory concentration of proliferation in µg/mL; SD: standard deviation; TI: therapeutic index (TI = $IC_{80}/\bar{x}$-MIC); n.a.: not applicable.

2.2. Chemical Composition

Based on GC/MS analysis of EOs from *C. citratus* (aerial part), *C. scariosus* (rhizomes), and *T. ammi* (seeds), a total of 17, 28, and 9 components were identified using HP-5MS column, which represents 99.62, 91.48, and 99.47% of their total respective contents. Similarly, using the DB-HeavyWAX column, 15, 36, and 13 components were determined, which constituted 68.77, 77.17, and 84.26% of the EOs, respectively. In total, 21 compounds were identified in the *C. citratus* EO, 40 compounds in *C. scariosus* EO, and 15 in *T. ammi* EO. The analysis showed that in *C. citratus* and *T. ammi* EOs, monoterpene hydrocarbons and oxygenated monoterpenoids were the predominated groups of chemicals, whereas sesquiterpene hydrocarbons and oxygenated sesquiterpenoids were major classes of compounds in *C. scariosus* EO. For both the columns, in *C. citratus* EO, α- and β-citral were the most abundant compounds with peak areas of (HP-5MS = 48.9 and 35.8%) and (DB-HeavyWAX = 24.3 and 33.2%), respectively. Caryophyllene oxide was the third abundant compound in the EO (~3%). In the case of *C. scariosus* EO, cyperotundone was the major component with peak areas of (29.1 and 28.9%), followed by caryophyllene oxide (19.8 and 17.54%) and cyperene with (9.9 and 8.5%), when measured using HP-5MS/DB-HeavyWAX columns, respectively. Thymol was the most predominant substance reported for both the columns with 51.2 and 45.8% in *T. ammi* EO, followed by β-cymene (22.6 and 17.1%) and γ-terpinene (21.5 and 17.6%) for HP-5MS/DB-HeavyWAX columns, subsequently. The complete chemical profiles of *C. citratus*, *C. scariosus*, and *T. ammi* EOs are provided in Tables 3–5, respectively.

Table 3. Chemical composition of *Cymbopogon citratus* essential oil.

RI [a]		Compound	Cl. [b]	Column			
				Content [c] (%)		Identification [f]	
Obs.	Lit.			HP-5MS	DB-WAX	HP-5MS	DB-WAX
915	926	Tricyclene	MH	0.26 ± 0.12	tr. [d]	RI, MS, Std,	MS
927	939	α-Pinene	MH	0.20 ± 0.04	- [e]	RI, MS, Std	-
942	953	Camphene	MH	2.42 ± 0.79	1.62 ± 0.34	RI, MS, Std	MS
979	985	Sulcatone	MO	0.35 ± 0.18	-	RI, MS	-
1023	1029	*D*-Limonene	MH	0.23 ± 0.03	-	RI, MS	-
1029	1050	*trans*-β-Ocimene	MH	0.31 ± 0.09	-	RI, MS	-
1039	1011	3-Carene	MH	0.13 ± 0.02	-	RI, MS, Std	-
1063	1030	4-Nonanone	MO	1.41 ± 0.38	1.55 ± 0.97	RI, MS	MS
1098	1098	Linalool	MO	0.46 ± 0.24	0.31 ± 0.07	RI, MS, Std	MS
1178	1184.7	Isogeranial	MO	0.70 ± 0.01	0.55 ± 0.05	RI, MS	MS
1190	1189	α-Terpineol	MO	0.39 ± 0.03	-	RI, MS	-
1238	1240	β-Citral	MO	35.8 ± 0.61	24.3 ± 8.82	RI, MS, Std	MS
1268	1270	α-Citral	MO	48.9 ± 0.55	33.2 ± 11.2	RI, MS, Std	MS
1376	1381	Geranyl acetate	MO	2.02 ± 0.20	tr.	RI, MS	MS
1412	1418	Caryophyllene	SH	0.45 ± 0.31	0.75 ± 0.05	RI, MS	MS

Table 3. *Cont.*

RI [a]		Compound	Cl. [b]	Column			
				Content [c] (%)		Identification [f]	
Obs.	Lit.			HP-5MS	DB-WAX	HP-5MS	DB-WAX
1510	1513	γ-Cadinene	SH	1.19 ± 0.33	1.20 ± 0.23	RI, MS	MS
g	1797	Geraniol	MO	-	0.77 ± 0.07	-	MS
1582	1581	Caryophyllene oxide	SH	3.0 ± 1.34	2.88 ± 1.40	RI, MS, Std	MS
g	1430	α-Cyclocitral	MO	-	0.45 ± 0.04	-	MS
g	NA	Isoneral	MO	-	0.36 ± 0.02	-	MS
g	1669	Isoborneol	MO	-	0.83 ± 0.01	-	MS
		Total content (%)		**99.62**	**68.77**		

[a] RI = retention indices; Obs. = retention indices determined relative to a homologous series of *n*-alkanes (C_8–C_{40}) on an HP-5MS column; Lit. = literature RI values [31,32], NA = RI values not available in the literature; [b] Cl = class; MH—monoterpene hydrocarbons, MO—oxygenated monoterpenoids, SH—sesquiterpene hydrocarbons; [c] relative peak area percentage as mean of three measurements ± standard deviation; [d] tr. = traces, relative peak area < 0.05%; [e] - = not detected; [f] identification method: MS = mass spectrum was identical to that of the National Institute of Standards and Technology Library (ver. 2.0.f), RI = the retention index was matching literature database; Std = constituent identity confirmed by co-injection of authentic standards; [g] retention indices were not calculated for compounds determined by DB-WAX column.

Table 4. Chemical composition of *Cyperus scariosus* essential oil.

RI [a]		Compound	Cl. [b]	Column			
				Content [c] (%)		Identification [f]	
Obs.	Lit.			HP-5MS	DB-WAX	HP-5MS	DB-WAX
927	939	α-Pinene	MH	1.34 ± 0.10	0.50 ± 0.01	RI, MS, Std	MS
970	980	β-Pinene	MH	1.88 ± 0.33	0.06 ± 0.00	RI, MS, Std	MS
1025	1032	Eucalyptol	MO	0.24 ± 0.16	- [d]	RI, MS	-
1137	1137	Pinocarveol	MO	1.85 ± 0.23	0.60 ± 0.01	RI, MS	MS
1158	1165	Pinocarvone	MO	0.37 ± 0.32	0.09 ± 0.00	RI, MS	MS
1168	1193	Myrtenal	MO	0.41 ± 0.27	0.13 ± 0.30	RI, MS	MS
1314	1327	Cyprotene	SH	0.16 ± 0.05	tr. [e]	RI, MS	MS
1344	1349	α-Terpinyl acetate	MO	1.65 ± 0.18	tr.	RI, MS	MS
1371	1376	Copaene	SH	1.46 ± 0.47	tr.	RI, MS	MS
1394	1398	Cyperene	SH	9.87 ± 0.59	8.5 ± 0.04	RI, MS	MS
1446	1477	α-Muurolene	SH	0.16 ± 0.04	tr.	RI, MS	MS
1456	1461	Rotundene	SH	1.94 ± 0.07	1.25 ± 0.01	RI, MS	MS
1483	1473.7	γ-Patchoulene	SH	0.19 ± 0.05	tr.	RI, MS	MS
1489	1491	Valencene	SH	0.63 ± 0.08	0.57 ± 0.60	RI, MS	MS
1518	1518	β-Cadinene	SH	0.31 ± 0.18	0.08 ± 0.40	RI, MS	MS
1528	1532	Cyperene epoxide	SO	2.65 ± 0.26	1.50 ± 0.00	RI, MS	MS
1541	1542	α-Calacorene	SH	0.13 ± 0.05	-	RI, MS	-
1565	1579	Isoaromadendrene epoxide	SO	0.62 ± 0.07	1.03 ± 0.00	RI, MS	MS
1572	1627	Longiverbenone	SO	1.33 ± 0.15	1.20 ± 0.08	RI, MS	MS
1582	1581	Caryophyllene oxide	SH	19.79 ± 0.58	17.54 ± 0.12	RI, MS, Std	MS
1591	NA	β-Santalol	SO	0.38 ± 0.12	-	RI, MS	-
1609	1608	Humulene epoxide 2	SO	1.69 ± 0.25	2.60 ± 0.10	RI, MS	MS
1656	1604	Globulol	SO	0.23 ± 0.04	0.39 ± 0.03	RI, MS	MS
1664	1663	Patchouli alcohol	SO	0.50 ± 0.05	-	RI, MS	-
1677	1676	Mustakone	SO	6.26 ± 0.26	3.67 ± 0.30	RI, MS	MS
1697	1694	Cyperotundone	SO	29.1 ± 1.11	28.91 ± 0.72	RI, MS	MS
1750	1752	Aristolone	SO	3.17 ± 0.77	3.73 ± 0.10	RI, MS	MS
1808	1807	Nootkatone	SO	2.17 ± 0.49	2.03 ± 0.40	RI, MS	MS
g	NA	β-Pinone	MO	-	tr.	-	MS
g	1586	β-Elemene	SH	-	tr.	-	MS
g	1652	*cis*-Verbenol	MO	-	tr.	-	MS

Table 4. *Cont.*

RI [a]		Compound	Cl. [b]	Column			
				Content [c] (%)		Identification [f]	
Obs.	Lit.			HP-5MS	DB-WAX	HP-5MS	DB-WAX
g	NA	Aristolochene	SH	-	tr.	-	MS
g	1680	α-Terpineol			tr.	-	MS
g	NA	α-Maaliene	SH	-	tr.	-	MS
g	1784	Myrtenol	MO	-	tr.	-	MS
g	2063	Cubenol	SO	-	0.49 ± 0.00	-	MS
g	1978	α-Cedrene epoxide	SO	-	0.40 ± 0.10	-	MS
g	NA	Aromadendrene oxide-(1)	SO	-	1.35 ± 0.40	-	MS
g	NA	Calarene epoxide	SO	-	0.52 ± 0.02	-	MS
g	NA	Diepicedrene-1-oxide	SO	-	0.05 ± 0.00	-	MS
		Total content (%)		**91.48**	**77.17**		

[a] RI = retention indices; Obs. = retention indices determined relative to a homologous series of *n*-alkanes (C_8–C_{40}) on an HP-5MS column; Lit. = literature RI values [31,32], NA = RI values not available in the literature; [b] Cl = class; MH—monoterpene hydrocarbons, MO—oxygenated monoterpenoids, SH—sesquiterpene hydrocarbons, SO—oxygenated sesquiterpenoids; [c] relative peak area percentage as mean of three measurements ± standard deviation; [d] - = not detected; [e] tr. = traces, relative peak area < 0.05%; [f] identification method: MS = mass spectrum was identical to that of the National Institute of Standards and Technology Library (ver. 2.0.f), RI = the retention index was matching literature database; Std = constituent identity confirmed by co-injection of authentic standards; [g] retention indices were not calculated for compounds determined by DB-WAX column.

Table 5. Chemical composition of *Trachyspermum ammi* essential oil.

RI [a]		Compound	Cl. [b]	Column			
				Content [c] (%)		Identification [e]	
Obs.	Lit.			HP-5MS	DB-WAX	HP-5MS	DB-WAX
920	917	β-Thujene	MH	0.30 ± 0.03	0.17 ± 0.02	RI, MS	MS
964	925	α-Pinene	MH	0.19 ± 0.01	0.42 ± 0.13	RI, MS, Std	MS
981	971	β-Pinene	MH	1.87 ± 0.25	2.13 ± 0.62	RI, MS, Std	MS
994	984	β-Myrcene	MH	0.21 ± 0.06	0.38 ± 0.03	RI, MS	MS
1007	974	2-Carene	MH	0.20 ± 0.02	- [d]	RI, MS	MS
1050	1031	β-Cymene	MH	22.6 ± 0.89	17.1 ± 3.99	RI, MS, Std	MS
1079	1065	γ-Terpinene	MH	21.5 ± 0.86	17.6 ± 0.99	RI, MS, Std	MS
1175	1086	Isoterpinolene	MH	0.08 ± 0.06	-	RI, MS	MS
1315	1290	Thymol	MO	51.2 ± 1.25	45.8 ± 4.41	RI, MS, Std	MS
f	1172	α-Terpinene	MH	-	0.12 ± 0.01	-	MS
f	1244	*D*-Limonene	MH	-	0.10 ± 0.01	-	MS
f	1195	β-Phellandrene	MH	-	0.08 ± 0.03	-	MS
f	NA	*trans*-2-Caren-4-ol	MO	-	0.18 ± 0.02	-	MS
f	1680	Terpineol	MO	-	0.06 ± 0.01	-	MS
f	1635	Terpinen-4-ol	MO	-	0.20 ± 0.04	-	MS
		Total content (%)		**99.47**	**84.26**		

[a] RI = retention indices; Obs. = retention indices determined relative to a homologous series of *n*-alkanes (C_8–C_{40}) on an HP-5MS column; Lit. = literature RI values [31,32], NA = RI values not available in the literature; [b] Cl = class; MH—monoterpene hydrocarbons, MO—oxygenated monoterpenoids, SH—sesquiterpene hydrocarbons; [c] relative peak area percentage as mean of three measurements ± standard deviation; [d] - = not detected; [e] identification method: MS = mass spectrum was identical to that of the National Institute of Standards and Technology Library (ver. 2.0.f), RI = the retention index was matching literature database; Std = constituent identity confirmed by co-injection of authentic standards; [f] retention indices were not calculated for compounds determined by DB-WAX column.

2.3. *Chemical Composition of T. ammi EO Vapors*

In the current study, the composition of headspace above the mixture of *T. ammi* EO and Mueller–Hinton broth (MHB) has been carried out using HS-SPME and HS-GTS in a time series every 3 h during a 12 h period using the HP-5MS column. Using HS-SPME extraction, a total of six volatile compounds were identified in the EO sample, which represented 97.25% of their respective total constituents. When using the HS-GTS extraction method, a lower number (five) of compounds was detected, which accounted for 93.4% of their total content. Regardless of the extraction method used, monoterpenes were the most predominant chemical groups of volatile compounds identified in the headspace. Using the HS-SPME extraction method, β-cymene was the most abundant constituent of the headspace of *T. ammi* EO. The content of β-cymene gradually decreased during the whole experiment with the peak area value ranging from 49% (time—0 h) to 43% (time—12 h). Similarly, a slight decrease in the concentration of γ-terpinene, the second-most abounding compound in the sample, was observed during overtime incubation from 39 to 31%. In contrast, the concentration of the thymol increased in the vapor over the time from 4.9 to 12%. In the case of HS-GTS extraction, β-cymene, γ-terpinene, and β-pinene were detected as the predominant compounds. For β-cymene and γ-terpinene, a steady decrease in the concentrations was observed during the experiment, ranging from 52 to 45% and 35 to 28%, respectively. Although the concentrations of both most-abundant compounds are nearly similar to those obtained using the HS-SPME method, the third-most abounding compound differed for the HS-GTS extraction method, β-pinene being detected with peak area values ranging from 2 to 6.5%. Interestingly, the chemical analysis showed that content of thymol detected by HS-SPME extraction was much higher in comparison with HS-GTS extraction (nearly 10%). Apart from this, no discrepancies for either sampling method were observed, and there were no significant changes in the chemical composition in the vapor of *T. ammi* EO over time. A complete chemical profile of *T. ammi* EO vapors is provided in Table 6.

Table 6. Chemical composition of a headspace above a *Trachyspermum ammi* essential oil dissolved in Mueller–Hinton broth at a concentration of 256 μg/mL.

RI [a]		Compound	Extraction Method/Time (h)/Content [b] (%)										Ident. [e]
			Solid Phase Microextraction					Gas Tight Syringe Extraction					
Obs.	Lit.		0	3	6	9	12	0	3	6	9	12	
925	939	α-Pinene	1.27 ± 0.10	1.46 ± 0.00	1.42 ± 0.01	1.41 ± 0.00	1.25 ± 0.01	tr. [c]	tr.	tr.	tr.	tr.	RI, MS
961	1011	2-Carene	0.64 ± 0.00	0.44 ± 0.00	0.62 ± 0.01	0.65 ± 0.02	0.72. ±0.01	- [d]	-	-	-	-	RI, MS
971	980	β-Pinene	1.91 ± 0.01	2.80 ± 0.05	2.36 ± 0.00	2.31 ± 0.01	1.79 ± 0.20	6.57 ± 0.20	4.93 ± 0.40	3.89 ± 0.10	4.11 ± 0.01	2.02 ± 0.30	RI, MS
1031	1030	β-Cymene	49.14 ± 1.00	48.00 ± 1.10	46.57 ± 0.80	45.97 ± 1.80	43.17 ± 0.90	52.00 ± 3.50	49.18 ± 2.90	48.67 ± 2.50	46.32 ± 1.40	45.60 ± 0.60	RI, MS
1065	1062	γ-Terpinene	39.36 ± 0.70	35.41 ± 0.50	35.11 ± 0.90	33.26 ± 0.60	31.01 ± 0.20	35.00 ± 2.50	32.23 ± 0.09	31.60 ± 0.50	32.22 ± 1.30	28.20 ± 2.90	RI, MS
1306	1290	Thymol	4.96 ± 0.01	5.66 ± 0.40	9.82 ± 0.08	11.90 ± 0.93	12.10 ± 0.80	tr.	1.23 ± 0.10	2.21 ± 0.20	2.07 ± 0.03	tr.	RI, MS
Total content (%)			**97.25**	**94.21**	**95.9**	**95.5**	**90.66**	**93.40**	**88.57**	**87.98**	**85.72**	**79.05**	

[a] RI = retention indices; Obs. = retention indices determined relative to a homologous series of *n*-alkanes (C_8–C_{40}) on an HP-5MS column; Lit. = literature RI values [31,32]; [b] relative peak area percentage as mean of three measurements ± standard deviation; [c] tr. = traces, relative peak area < 0.05%; [d] - = not detected; [e] identification method: MS = mass spectrum was identical to that of the National Institute of Standards and Technology Library (ver. 2.0.f), RI = the retention index was matching literature database.

3. Discussion

T. ammi is a medicinal plant highly valued in traditional Ayurvedic medicine [23]. Its EO has been reported to have pivotal antibacterial properties against foodborne and spoilage bacteria; however, only a few studies have reported its effect against pneumonia-causing pathogens [33,34]. The current study suggests the significant antibacterial activity of *T. ammi* EO against the targeted bacteria in both liquid and vapor phases. The observed endpoints for the liquid phase correspond well with a previously published study reporting the MIC of *T. ammi* EO against *S. aureus* at 500 μg/mL [35] and within a three-dilution MIC range for *S. pneumoniae* (MIC = 250 μg/mL) [36]. According to our best knowledge, the antibacterial activity of *T. ammi* EO vapors was described for the first time in this study. Considering the toxicity, the current study reports that *T. ammi* EO is moderately cytotoxic to lung fibroblasts. This result corresponds with an in vivo study published by Vazirian et al. [37], who observed mild oral toxicity of the EO in a rat model. Regarding inhalation toxicity, the sources of information on the safety of EO vapors are scarce. However, data on its predominant compound, thymol, might suggest its possible inhalation safety. The European Chemicals Agency reported thymol to be nontoxic to mice when they are exposed to the chemical via inhalation of its vapor for 2 h with a lethal dose 50% (LD_{50}) of 7.57 mg/L [38]. Moreover, another study published by Xie et al. [39] reported no evidence of chronic toxicity of thymol through inhalation in a mouse model. Nevertheless, further experiments on their in vivo inhalation toxicity are necessary to determine the safety of *T. ammi* EO. The biological properties of *T. ammi* EO are attributed to its chemical composition, which has already been studied. The chromatographic profile obtained in the current study corresponds with previously published reports identifying thymol as the main component (ranging from 54.32 to 67.4%), β-cymene as the second-most abundant compound (ranging from 17.9 to 21.74%, respectively), and γ-terpinene as third-most abundant component (ranging from 11.3 to 19.38%) [35,40]. Moreover, research led by Modareskia et al. [41] investigated the bioactive potential of different populations of *T. ammi* EO in liquid phase and reported that the antibacterial effects are mainly due to the presence of phenolic and hydrocarbon monoterpenes such as thymol, *p*-cymene, and γ-terpinene. β-Cymene, γ-terpinene, and thymol were identified as the key constituents of vapors of *T. ammi* EO in the sample obtained using the HS-SPME technique. This is well-corresponding with the results of Liu et al. [17], who reported γ-terpinene (26.21%), *p*-cymene (23.58%), and thymol (20.02%) as the major components of *T. ammi* EO vapor analysis by HS-SPME. Because the EO samples for headspace analysis performed in this study were prepared by dissolving them in microbiological growth medium, variances in the concentration of γ-terpinene can be attributed to the different experimental conditions [42]. Our paper identifying β-cymene, γ-terpinene, and β-pinene as predominant components of *T. ammi* EO vapors is the first report on its analysis using the HS-GTS method. When comparing both sampling techniques, the present study reports disparities with thymol concentration. In comparison with the HS-SPME method, the analysis of the sample collected by the HS-GTS technique showed a lower amount of thymol. The possible explanation may be that it was due to the sensitivity and affinity of SPME fibers toward certain volatile components [43]. In contrast, the HS-GTS technique is less selective but provides a precise and, perhaps, closer assessment of the real distribution of volatile compounds in vials [44]. These findings are consistent with the data published by Antih et al. [42], who performed *Thymus vulgaris* L. EO headspace analysis using similar experimental conditions. The antibacterial potential of volatile monoterpene hydrocarbons (*p*-cymene and γ-terpinene) and phenolic monoterpene (thymol) are well-known; for instance, the bioactivity of thymol in the vapor phase against respiratory tract bacteria such as *H. influenzae*, *S. aureus*, and *S. pneumoniae* has been previously published [13]. As mentioned before, the antibacterial activity of *T. ammi* EO and its vapors is mostly due to its composition; particularly, the presence of hydrocarbon and phenolic monoterpenes in high concentrations are related to the growth-inhibitory effect. Moreover, the current study also reports higher activity in the broth medium as compared to the vapor phase. This can be related to data obtained from the headspace analysis, where thymol, which is regarded as the main antimicrobial constituent, was detected in lower amounts in the EO vapors. The hydrophobic nature of thymol worsening its solubility in water-based media and, subsequently, reducing its volatility, can be a possible reason of lower antimicrobial activity [45].

C. citratus EO is extensively used in Ayurvedic medicine as a folk remedy for coughs, flu, and pneumonia [46]. Previously published studies have reported the antibacterial and anti-fungal properties of *C. citratus* EO [47,48]. As a result of the modified agar dilution method, Inouye et al. [48] observed a moderate growth-inhibitory effect of *C. citratus* EO against *S. pyogenes* (MIC = 400 μg/mL), *H. influenzae*, and *S. pneumoniae* (MIC = 800 μg/mL). In addition, *C. citratus* EO vapors were reported to be more efficient against *H. influenzae* as compared to *S. aureus*, *S. pneumoniae*, and *S. pyogenes*. Despite certain variances in the MIC values caused probably by different bacterial strains and antimicrobial assays used, these results correspond well with our findings. Regarding cytotoxicity, the current study reports that *C. citratus* EO is moderately toxic (IC_{50} = 19.63 μg/mL) to the lung fibroblasts. A previous study reported a similar mild cytotoxic effect on human fibroblast cell line WI38 with an IC_{50} value of 49.39 μg/mL [29]. The minor variance in the IC_{50} values can be attributed to the modification of the cytotoxicity assay previously recommended for evaluation of biological properties of the volatile agents. The use of EVA Capmat protects microtiter plates against vapor transition and provides more reliable results [49]. The safety of *C. citratus* EO for potential inhalation use can be supported by a previously published study reporting a nontoxic effect of vapors of its predominant compound citral on Sprague–Dawley rats at concentrations up to 34 ppm [50]. Furthermore, other published data on the in vivo acute oral toxicity of *C. citratus* EO in mice and rabbit models have been reported to be nontoxic with an LD_{50} value > 2000 mg/kg [51]. Although *C. citratus* EO is classified as GRAS, further toxicological evaluation is necessary to confirm its nontoxicity for practical application in inhalation therapy. GC/MS analysis showed that the major constituents of *C. citratus* EO were α- and β-citral. These findings are consistent with those of several previously published studies. For instance, El-Kased and El-Kersh [52] and Hanaa et al. [53] reported α-citral (36.35 and 34.98%) and β-citral (34.99 and 40.72%) as the main components of hydrodistilled EO from this species. Citral has also been identified as the main antimicrobial compound of *C. citratus* EO [54].

The EO from rhizomes of *C. scariosus* is used as an ingredient in several Ayurvedic formulations as an anti-infective agent [55]. Previous studies have reported the growth-inhibitory effect of *C. scariosus* EO against carbapenem-resistant *Klebsiella pneumoniae* and methicillin-resistant *S. aureus* in liquid media [56]. In our study, we have observed only weak antibacterial activity of the EO. Considering cytotoxicity, the present research reports that *C. scariosus* EO is nontoxic to the human lung cells. According to our best knowledge, there are no studies reporting the toxicity of the EO from this species. In correspondence with results of our study, cyperene (20.1%) and cyperotundone (10.30%) have previously been identified as abundant components of *C. scariosus* EO [57]. Since a significant variability of chemical composition has been reported for *C. scariosus* EO extracted from plants grown in different geographical locations in India (e.g., content of caryophyllene oxide varied from 2.42 to 10.38%) [58], the geographical origin of plant material analyzed in this study can explain minor differences in the concentrations of detected compounds.

Despite the recent progress in evaluation of biological properties of volatile agents, development of new methodologies suitable for determination of antimicrobial effects in the vapor phase remains a challenge [27]. Recently, a broth macrodilution volatilization assay was developed by our team for evaluation of antimicrobial activity of volatile agents in liquid and vapor phases [59], which combines the principles of broth microdilution volatilization [13] and standard macrodilution methods [60]. The method is performed in commercially available microtubes, which can be tightly closed with a snap cap preventing loss of the active agents by evaporation. Another advantage is that appropriate amounts of media suitable for the cultivation of a broad spectrum of microorganisms (including slow growing fungi) can be applied in microtubes and their caps. However, our previous study assessed only several representatives of volatile phytochemicals. The results of experiments described in this paper clearly demonstrate the validity of a new assay for testing the susceptibility of bacterial pathogens causing respiratory infections to EOs and their vapors.

4. Materials and Methods

4.1. Chemicals

In biological assays, dimethyl sulfoxide (DMSO, CAS 67-68-5) and MTT dye (CAS 298-93-1) were used as solvent and indicator of cell viability, respectively. Amoxicillin (90%, CAS 26787-78-0), ampicillin (84.5%, CAS 69-52-3), oxacillin (86.3%, CAS 7240-38-2), and tetracycline (98–102%, CAS 60-54-8) were assayed as positive controls. Chemical standards, namely 3-carene (99%, CAS: 498-15-7), camphene (97.5%, CAS: 79-92-5), caryophyllene oxide (99%, CAS: 1139-30-6), citral (95%, CAS: 5392-40-5), linalool (97%, CAS: 78-70-6), *m*-cymene (99%, CAS: 535-77-3), thymol (99%, CAS: 89-83-8), α-pinene (99%, CAS: 7785-70-8), β-pinene (99.0%, CAS: 18172-67-3), γ-terpinene (97%, CAS: 99-85-4), and *n*-alkanes were used for GC-MS analyses. With the exception of *n*-hexane (CAS: 110-54-3) obtained from (Merck KGaA, Darmstadt, Germany), all other chemicals were purchased from Sigma-Aldrich (Prague, Czech Republic).

4.2. Plant Material and Sample Preparation

Plant materials were purchased from Bhagyashree Herbal Farms (Chhattisgarh, India). Dried aerial parts of *C. citratus*, rhizomes of *C. scariosus*, and seeds of *T. ammi* were homogenized using a Grindomix apparatus (GM 100 Retsch, Haan, Germany). The residual moisture contents of both samples were determined gravimetrically at 130 °C for 1 h by a Scaltec SMO 01 analyzer (Scaltec Instruments, Gottingen, Germany) in triplicate according to the Official Methods of Analysis of the Association of Official Agricultural Chemists [61]. EOs were extracted by hydrodistillation of 100 g of ground plant materials in 1 L of distilled water for 3 h using a Clevenger-type apparatus (Merci, Brno, Czech Republic) according to the procedure described in the European Pharmacopeia (2013) [62] and stored in sealed glass vials at +4 °C.

4.3. Bacterial Strains and Culture Media

The following four bacterial standard strains from the American Type Culture Collection (ATCC, Manassas, VA, USA) were used: *H. influenzae* ATCC 49247, *S. aureus* ATCC 29213, *S. pneumoniae* ATCC 49619, and *S. pyogenes* ATCC 19615. Cultivation and assay media (broth/agar) were Mueller–Hinton broth (MHB) complemented by Haemophilus test medium (*H. influenzae*), MHB (*S. aureus*), and brain–heart infusion (*S. pneumoniae* and *S. pyogenes*). The pH of the broths was adjusted to a final value of 7.6 using Trizma base (Sigma-Aldrich). All microbial strains and cultivation media were purchased from Oxoid (Basingstoke, UK). Stock cultures of bacterial strains were cultivated in broth medium at 37 °C for 24 h prior to testing in the incubator (Memmert GmbH & Co. KG, Buchenbach, Germany). For the preparation of inoculum, the turbidity of the bacterial suspension was adjusted to 0.5 McFarland standard using a Densi-La-Meter II (Lachema, Brno, Czech Republic) to obtain a final concentration of 10^8 colony-forming units per mL.

4.4. Antimicrobial Assay

EOs were tested for their in vitro antibacterial activity using the broth macrodilution volatilization method at various concentrations in the liquid and vapor phases [59] in standard 2 mL microtubes with snap covers (Eppendorf, Hamburg, Germany). Each EO was dissolved in DMSO, which was subsequently diluted in the appropriate broth medium. To make sufficient stock solutions and to prevent the loss of active chemicals by evaporation, six twofold serially diluted concentrations of samples were created in 15 mL test tubes (Gama Group, Ceske Budejovice, Czech Republic). The starting concentration of tested EOs was 1024 µg/mL. In the subsequent phase, 90 mL of melted agar was pipetted into the rims of the caps, and when the agar had solidified, 5 mL of bacterial suspension was added. Then, with a final volume of 1500 μL, the proper concentrations of each sample that had been prepared in test tubes were pipetted into microtubes. The microtubes were then sealed properly after being injected with 10 μL of bacterial solution. As growth and purity controls, microtubes with inoculated and non-inoculated medium were prepared, respectively. The experiment was incubated at 37 °C for 24 h. Thereafter, the MICs were determined by visual assessment of bacterial growth after coloring the metabolically active bacterial colonies with MTT dye. The respective volumes of 30 and 375 μL of MTT at a concentration of 600 μg/mL were pipetted into the caps and in the microtubes when the interface of color change from yellow to purple (relative to that of colors in control wells) was recorded in broth and agar. The MIC values were determined as the lowest concentrations that inhibited bacterial growth compared with the compound-free control, and they were expressed in μg/mL. DMSO, assayed as the negative control, did not inhibit any of the strains at the concentrations used in the assay ($\leq$1%). The respective susceptibilities of *H. influenzae*, *S. aureus*, *S. pneumoniae*, and *S. pyogenes* to ampicillin, oxacillin, amoxicillin, and tetracycline were taken as positive antibiotic controls [63]. All tests were performed as three independent experiments, each carried out in triplicate, and the results were presented as median/modal values. According to the widely accepted norm in MIC testing, the mode and median were used for the final value calculation when triplicate endpoints were within the two- and three-dilution range, respectively.

4.5. Cell Cultures

Lung fibroblast cell line MRC-5, obtained from ATCC (Manassas, VA, USA), was propagated in Eagle's minimum essential medium (EMEM) supplemented with 10% fetal bovine serum (FBS), 2 mM glutamine, 10 μL/mL nonessential amino acids, and 1% penicillin–streptomycin solution (10,000 units/mL of penicillin and 10 mg/mL of streptomycin); all these components were purchased from Sigma-Aldrich. The cells were pre-incubated in 96-well microtiter plates at a density of 2.5×10^3 cells per well for 24 h at 37 °C in a humidified incubator (Sanyo Electric Co, Ltd., Osaka, Japan) in an atmosphere of 5% CO_2 in air. Vinorelbine ($\geq$98%, CAS 125317-39-7), used as positive control, was also obtained from Sigma-Aldrich.

4.6. Cytotoxicity Assay

The modified MTT test, which is based on the metabolism of MTT to blue formazan by mitochondrial dehydrogenases in living lung cells, was used in the current investigation [64]. The evaluated EOs samples were applied to MRC-5 lung fibroblast cells grown in 96-well plates for 72 h. EOs were diluted in EMEM medium supplemented with 10% FBS after being dissolved in DMSO at a maximum concentration of 1%. The samples were then serially diluted eight times, yielding concentrations ranging from 256 to 2 μg/mL. The microtiter plates were incubated for 72 h at 37 °C in a humidified atmosphere of 5% CO_2 in air with ethylene vinyl acetate EVA Capmats (USA Scientific Inc., Ocala, FL, USA). After that, MTT reagent (1 mg/mL) in EMEM solution was added to each well and the plates were incubated for an additional 2 h at 37 °C in a humidified atmosphere of 5% CO_2 in air. Once the medium had been removed from the incubation, the intracellular formazan product had been dissolved in 100 μL of DMSO. The viability of the lung cells at the tested concentration (1%) was unaffected by the solvent utilized. The absorbance was measured at 555 nm using a Tecan Infinite M200 spectrometer (Tecan Group, Mannedorf, Switzerland), and the viability was computed in comparison to the untreated control. For each test, three independent experiments, each with two replicates, were conducted. Using the GraphPad Prism program (GraphPad Software, La Jolla, CA, USA), the results of the cytotoxicity effect were calculated and expressed as average IC_{50} with standard deviation in μg/mL. According to the Special Programme for Research and Training in Tropical Diseases (WHO-Tropical Diseases) [30], the levels of cytotoxic effects were categorized as cytotoxic (IC_{50} < 2 μg/mL), moderately cytotoxic (IC_{50} 2–89 μg/mL), and nontoxic (IC_{50} > 90 μg/mL). Furthermore, for the purpose of comparing microbiological and toxicological data, IC_{80} was determined to be equal to the MIC endpoint [65]. To compare the quantity of toxic antibacterial agents with the amount of effective antibacterial agents, therapeutic indices (TIs) were defined as the ratio of IC_{80} and $\bar{x}$-MIC values [66].

4.7. Chemical Analysis of EOs' Liquid Phase

The dual-column/dual-detector gas chromatograph Agilent GC-7890B system (Agilent Technologies, Santa Clara, CA, USA) equipped with autosampler Agilent 7693, two columns, a fused-silica HP-5MS (30 m × 0.25 mm, film thick-ness 0.25 μm, Agilent 19091s-433), a DB-Heavy WAX (30 m × 0.25 mm, film thickness 0.25 μm, Agilent 122–7132), and a flame ionization detector (FID) coupled with single quadrupole mass selective detector Agilent MSD-5977B were used to characterize the chemical composition of the targeted EOs. Helium was used as the carrier gas at a rate of 1 mL/min, and the injector temperature for both columns was 250 °C. After 3 min, the oven temperature for both columns increased from 50 to 280 °C. Initially, the heating velocity was 3 °C/min until the system reached a temperature of 120 °C. Subsequently, the velocity increased to 5 °C/min until a temperature of 250 °C, and after 5 min holding time, the heating speed reached 15 °C/min until obtaining a temperature of 280 °C. Heating was followed by an isothermic period of 20 min. The EO samples were diluted in *n*-hexane for GC/MS at a concentration of 20 μL/mL. One microliter of the solution was injected in split mode in a split ratio of 1:30. The mass detector was set to the following conditions: ionization energy 70 eV, ion source temperature 230 °C, scan time 1 s, and mass range 40–600 *m*/*z*. Identification of constituents was based on a comparison of their retention indices (RI) and retention times (RT) and spectra with the National Institute of Standards and Technology Library ver. 2.0.f (NIST, Gaithersburg, MD, USA) [33], as well as with authentic standards. The RIs were calculated for compounds separated by the HP-5MS column using the retention times of *n*-alkanes series ranging from C_8 to C_{40}. For each EO analyzed, the final number of compounds was calculated as the sum of components simultaneously identified using both columns and the remaining constituents identified by individual columns only. Quantitative data were expressed as relative percentage content of constituents determined by the FID.

4.8. Chemical Analysis of EOs' Vapor Phase

For the chemical analysis of EOs' vapors, *T. ammi* was chosen, as it exhibited the highest antimicrobial potential. Similarly, like Antih et al. [43], two distinct sampling methods—HS-SPME and HS-GTS—were used to sample the headspace above a solution of MHB and EO at a concentration of 256 μg/mL, which was observed as the lowest MIC value of this EO in vapor phase. Five samples were made for each experiment (one for every three hours during the 12 h incubation period), and a volume of 2 mL of the combination was added to a 4 mL glass vial. Except for the first sample (t = 0 h), all the samples were incubated in the same way as the bacterial cultures at a temperature of 37 °C until they were analyzed at 3, 6, and 12 h. For the HS-SPME, a fiber assembly coated with a 50/30 m mixed layer of divinylbenzene/carboxen/polydimehylsiloxane (DVB/CAR/PDMS—SUPELCO, Bellefonte, PA, USA) was used. The coated fiber was exposed to the headspace for 15 min to allow the adsorption of the volatile chemicals above the mixture (EO and MHB) until the headspace had reached equilibrium. The fiber was then retained in the injector for the duration of the analysis until the next measurement, when the needle was removed and put into the GC injector port. The injection method was set at splitless mode and injector temperature was set at 250 °C. For the HS-GTS sampling technique, a 2.5 mL SampleLock gas-tight syringe (Hamilton Bonaduz AG, Bonaduz, Switzerland) was used, which has a twist valve lock and a positive rear plunger stop to prevent sample loss. At equilibrium, the needle was passed through the vial septum and inserted until reaching the middle of the headspace, and after collecting the headspace, the valve of the syringe was closed. Afterward, the syringe was removed from the vial and inserted into the GC injector at a similar temperature of 250 °C, but the injection mode was set as spitless mode. For both sampling methods, measurements were repeated every 3 h during a 12 h incubation period.

Furthermore, analysis was performed on the HP-5MS column with similar operational parameters as described earlier for GC/MS analysis.

The chemical analysis of the all EO samples was performed in triplicate, including the chromatographic analysis of its liquid phase and the headspace analysis using both extraction techniques (HS-SPME and HS-GTS). Relative peak area percentages were expressed as mean average of these three measurements ± standard deviation.

5. Conclusions

In summary, this study reports validation of the broth macrodilution volatilization method for assessment of antibacterial properties of EOs and shows in vitro growth-inhibitory effect of EOs hydrodistilled from three Indian medicinal plants, namely *C. citratus*, *C. scariosus*, and *T. ammi*, against pneumonia-causing bacteria *H. influenzae*, *S. aureus*, *S. pneumoniae*, and *S. pyogenes* in liquid and vapor phase. Among all EOs tested, the strongest antibacterial effect was observed for *T. ammi* EO against *H. influenzae* with MIC 128 and 256 μg/mL in the liquid and vapor phases. The results of the cytotoxicity testing showed no toxicity of *C. scariosus* EO to the normal lung fibroblasts. The GC/MS analysis identified α-citral, cyperotundone, and thymol as the main constituents of *C. citratus*, *C. scariosus*, and *T. ammi* EOs, respectively. In a series of headspace experiments, β-cymene (52–49%), γ-terpinene (39–35%), thymol (12–2%), and β-pinene (6–2%) were identified as the major compounds of *T. ammi* EO vapors when analyzed using HS-GTS and HS-SPME sampling techniques. These compounds can be associated with imparting an antibacterial effect in the vapor phase. Furthermore, HS-SPME and HS-GTS have been proven to be complementary methods suitable for studying the qualitative and quantitative aspects of the volatile profiles of EO, subsequently. Overall, the broth macrodilution volatilization method can be recommended for testing the susceptibility of bacteria to EOs. In addition, *T. ammi* EO seems to be a promising antibacterial agent for further research in the area of inhalation therapies. However, further experiments focused on the safety and in vivo efficacy of EOs analyzed in this study are required to verify their practical applicability.

Author Contributions: Conceptualization, L.K.; formal analysis, A.C.; investigation, A.C.; plant material collection, M.L.N. and K.S.P.; performance of the cytotoxicity testing I.D.; supervision for chemical analysis L.K., K.U. and J.A.; supervision for microbiology experiments, M.H. and L.K.; writing—original draft, A.C.; writing—review and editing, M.H., J.A. and L.K. All authors have read and agreed to the published version of the manuscript.

Funding: This work was supported by the Czech University of Life Sciences Prague (project IGA.20233109) and the METROFOOD-CZ research infrastructure project (MEYS Grant No: LM2023064), including access to its facilities.

Institutional Review Board Statement: Not applicable.

Informed Consent Statement: Not applicable.

Data Availability Statement: Not applicable.

Conflicts of Interest: The authors declare no conflict of interest.

Sample Availability: Not available.

References

1. Grossman, T.H.; Fyfe, C.; O'Brien, W.; Hackel, M.; Minyard, M.B.; Waites, K.B.; Dubois, J.; Murphy, T.M.; Slee, A.M.; Weiss, W.J.; et al. Fluorocycline TP-271 is potent against complicated community-acquired bacterial pneumonia pathogens. *mSphere* **2017**, *2*, e00004-17. [CrossRef]
2. Baron, S. *Medical Microbiology*, 4th ed.; University of Texas Medical Branch at Galveston: Galveston, TX, USA, 1996.
3. Torres, A.; Cilloniz, C.; Niederman, M.S.; Menendez, R.; Chalmers, J.D.; Wunderink, R.G.; van der Poll, T. Pneumonia. *Nat. Rev. Dis. Primers* **2021**, *7*, 25. [CrossRef] [PubMed]
4. Eshwara, V.K.; Mukhopadhyay, C.; Rello, J. Community-acquired bacterial pneumonia in adults: An update. *Indian J. Med. Res.* **2020**, *151*, 288. [CrossRef]

5. Borghardt, J.M.; Kloft, C.; Sharma, A. Inhaled therapy in respiratory disease: The complex interplay of pulmonary kinetic processes. *Can. Respir. J.* **2018**, *2018*, 2732017. [CrossRef]
6. Quon, B.S.; Goss, C.H.; Ramsey, B.W. Inhaled antibiotics for lower airway infections. *Ann. Am. Thorac. Soc.* **2014**, *11*, 425–427. [CrossRef]
7. Tiddens, H.A.W.M.; Bos, A.C.; Mouton, J.W.; Devadason, S.; Janssens, H.M. Inhaled antibiotics: Dry or wet? *Eur. Clin. Respir. J.* **2014**, *44*, 1308–1318. [CrossRef]
8. Netopilova, M.; Houdkova, M.; Urbanova, K.; Rondevaldova, J.; Damme, P.; Kokoska, L. In vitro antimicrobial combinatory effect of *Cinnamomum cassia* essential oil with 8-hydroxyquinoline against *Staphylococcus aureus* in liquid and vapour phase. *J. Appl. Microbiol.* **2020**, *129*, 906–915. [CrossRef] [PubMed]
9. Baser, K.H.C.; Buchbauer, G. *Handbook of Essential Oils: Science, Technology, and Applications*, 3rd ed.; CRC Press: Boca Raton, FL, USA, 2020.
10. Horvath, G.; Acs, K. Essential oils in the treatment of respiratory tract diseases highlighting their role in bacterial infections and their anti-inflammatory action: A review. *Flavour Fragr. J.* **2015**, *30*, 339–340. [CrossRef]
11. Kokoska, L.; Kloucek, P.; Leuner, O.; Novy, P. Plant-derived products as antibacterial and antifungal agents in human health care. *Curr. Med. Chem.* **2019**, *26*, 5501–5541. [CrossRef] [PubMed]
12. Hashemi, S.M.B.; Khaneghah, A.M.; Sant'Ana, A.S. *Essential Oils in Food Processing: Chemistry, Safety and Applications*, 1st ed.; John Wiley & Sons Ltd.: Hoboken, NJ, USA, 2018.
13. Houdkova, M.; Rondevaldova, J.; Doskocil, I.; Kokoska, L. Evaluation of antibacterial potential and toxicity of plant volatile compounds using new broth microdilution volatilization method and modified MTT assay. *Fitoterapia* **2017**, *118*, 56–62. [CrossRef] [PubMed]
14. Jugreet, B.S.; Suroowan, S.; Rengasamy, R.R.K.; Mahomoodally, M.F. Chemistry, bioactivities, mode of action and industrial applications of essential oils. *Trends Food Sci. Technol.* **2020**, *101*, 89–105. [CrossRef]
15. Ross, C.F. Headspace analysis. In *Comprehensive Sampling and Sample Preparation*; Elsevier: Amsterdam, The Netherlands, 2012; pp. 27–50.
16. Schweitzer, B.; Balazs, V.L.; Molnar, S.; Szogi-Tatar, B.; Boszormenyi, A.; Palkovics, T.; Horvath, G.; Schneider, G. Antibacterial effect of lemongrass (*Cymbopogon citratus*) against the aetiological agents of pitted keratolyis. *Molecules* **2022**, *27*, 1423. [CrossRef] [PubMed]
17. Liu, L.; Wang, Q.; Tang, J.; Zhang, Z. Component analysis of volatile oil from fruits of *Trachyspermum ammi* by headspace solid-phase microextraction GC-MS. *Chin. J. Pharm. Anal.* **2013**, *33*, 607–610.
18. Upasani, S.V.; Beldar, V.G.; Tatiya, A.U.; Upasani, M.S.; Surana, S.J.; Patil, D.S. Ethnomedicinal plants used for snakebite in India: A brief overview. *Integr. Med. Res.* **2017**, *6*, 114–115. [CrossRef]
19. Bhattacharyya, R.; Bhattacharya, S.; Chaudhuri, S. Conservation and documentation of the medicinal plant resources of India. *Biodivers. Conserv.* **2006**, *15*, 2705–2717. [CrossRef]
20. Vaidya, V.N.; Tatiya, A.U.; Elango, A.; Kukkupuni, S.K.; Vishnuprasad, C.N. Need for comprehensive standardization strategies for marketed Ayurveda formulations. *J. Ayurveda Integr. Med.* **2018**, *9*, 312–315. [CrossRef]
21. Ningthoujam, S.S.; Talukdar, A.D.; Potsangbam, K.S.; Choudhury, M.D. Traditional uses of herbal vapour therapy in Manipur, Northeast India: An ethnobotanical survey. *J. Ethnopharmacol.* **2013**, *147*, 136–147. [CrossRef]
22. Shah, G.; Shri, R.; Panchal, V.; Sharma, N.; Singh, B.; Mann, A.S. Scientific basis for the therapeutic use of *Cymbopogon citratus*, stapf (*Lemon grass*). *J. Adv. Pharm. Technol. Res.* **2011**, *2*, 87. [CrossRef] [PubMed]
23. Bairwa, R.; Sodha, R.S.; Rajawat, B.S. *Trachyspermum ammi*. *Pharmacogn. Rev.* **2012**, *6*, 56–60. [CrossRef]
24. Singh, V.; Ali, M.; Negi, A.; Sultana, S. Analysis and antimicrobial activity of the essential oil of *Cyperus rotundus* L. rhizomes. *J. Med. Plants Stud.* **2018**, *6*, 101–105.
25. Inouye, S.; Takizawa, T.; Yamaguchi, H. Antibacterial activity of essential oils and their major constituents against respiratory tract pathogens by gaseous contact. *J. Antimicrob. Chemother.* **2001**, *47*, 565–573. [CrossRef] [PubMed]
26. Amat, S.; Baines, D.; Alexander, T.W. A Vapour phase assay for evaluating the antimicrobial activities of essential oils against bovine respiratory bacterial pathogens. *Lett. Appl. Microbiol.* **2017**, *65*, 489–495. [CrossRef] [PubMed]
27. Houdkova, M.; Kokoska, L. Volatile antimicrobial agents and in vitro methods for evaluating their activity in the vapour phase: A review. *Planta Med.* **2020**, *86*, 822–857. [CrossRef]
28. Cimino, C.; Maurel, O.M.; Musumeci, T.; Bonaccorso, A.; Drago, F.; Souto, E.M.B.; Pignatello, R.; Carbone, C. Essential oils: Pharmaceutical applications and encapsulation strategies into lipid-based delivery systems. *Pharmaceutics* **2021**, *13*, 327. [CrossRef]
29. Toukourou, H.; Gbaguidi, F.; Quetin-Leclercq, J. Phytochemical composition, antibacterial activity against sore throat pathogens and toxicological evaluation of *Cymbopogon citratus* essential oil from Benin. *J. Pharmacogn. Phytochem.* **2019**, *8*, 3258–3263.
30. Special Programme for Research and Training in Tropical Diseases. Available online: http://www.who.int/tdr/grants/workplans/en/cytotoxicity_invitro.pdf (accessed on 21 December 2022).
31. Adams, R.P. *Identification of Essential Oil Components by Gas Chromatography/Mass Spectorscopy*, 4th ed.; Allured Publishing Corp: Carol Stream, IL, USA, 2007.
32. National Institute of Standards and Technology. Available online: https://www.nist.gov/ (accessed on 27 March 2023).
33. Paul, S.; Dubey, R.C.; Maheswari, D.K.; Kang, S.C. *Trachyspermum ammi* (L.) fruit essential oil influencing on membrane permeability and surface characteristics in inhibiting food-borne pathogens. *Food Control* **2011**, *22*, 725–731. [CrossRef]

34. Jebelli Javan, A.; Salimiraad, S.; Khorshidpour, B. Combined effect of *Trachyspermum ammi* essential oil and propolis ethanolic extract on some foodborne pathogenic bacteria. *Vet. Res. Forum* **2019**, *10*, 235–240. [CrossRef] [PubMed]
35. Vitali, L.A.; Beghelli, D.; Nya, P.C.B.; Bistoni, O.; Cappellacci, L.; Damiano, S.; Lupidi, G.; Maggi, F.; Orsomando, G.; Papa, F.; et al. Diverse biological effects of the essential oil from Iranian *Trachyspermum ammi*. *Arab. J. Chem.* **2016**, *9*, 775–786. [CrossRef]
36. Gardener, A.C.; Trifan, A.; Spac, A.; Brebu, M.; Miron, A.; Aprotosoaie, A.C. Antibacterial activity of traditional spices against lower respiratory tract pathogens: Combinatorial effects of *Trachyspermum ammi* essential oil with conventional antibiotics. *Lett. Appl. Microbiol.* **2018**, *67*, 449–457. [CrossRef]
37. Vazirian, M.; Hekmati, D.; Ostad, S.; Manayi, A. Toxicity evaluation of essential oil of *Trachyspermum ammi* in acute and sub-chronic toxicity experiments. *J. Med. Plants* **2019**, *18*, 70–77.
38. ECHA. European Chemicals Agency. Available online: https://echa.europa.eu/ (accessed on 26 February 2023).
39. Xie, K.; Tashkin, D.P.; Luo, M.Z.; Zhang, J.Y. Chronic toxicity of inhaled thymol in lungs and respiratory tracts in mouse model. *Pharmacol. Res. Perspect.* **2019**, *7*, e00516. [CrossRef]
40. Howyzeh, M.S.; Noori, S.A.S.; Shariati, J.V.; Niazian, M. Essential oil chemotype of Iranian Ajowan (*Trachyspermum ammi* L.). *J. Essent. Oil-Bear. Plants* **2018**, *21*, 273–276. [CrossRef]
41. Modareskia, M.; Fattahi, M.; Mirjalili, M.H. Thymol screening, phenolic contents, antioxidant and antibacterial activities of Iranian populations of *Trachyspermum ammi* (L.) Sprague (Apiaceae). *Sci. Rep.* **2022**, *12*, 15645. [CrossRef]
42. Antih, J.; Houdkova, M.; Urbanova, K.; Kokoska, L. Antibacterial activity of *Thymus vulgaris* L. essential oil vapours and their GC/MS analysis using solid-phase microextraction and syringe headspace sampling techniques. *Molecules* **2021**, *26*, 6553. [CrossRef]
43. Perestrelo, R.; Silva, C.L.; Rodrigues, F.; Caldeira, M.; Câmara, J.S. A powerful approach to explore the potential of medicinal plants as a natural source of odor and antioxidant compounds. *J. Food Sci. Technol.* **2016**, *53*, 132–144. [CrossRef] [PubMed]
44. Chialva, F.; Gabri, G.; Liddle, P.A.P.; Ulian, F. Qualitative evaluation of aromatic herbs by direct headspace GC analysis. Applications of the method and comparison with the traditional analysis of essential oils. *J. High. Resolut. Chromatogr.* **1982**, *5*, 182–188. [CrossRef]
45. Laird, K.; Phillips, C. Vapour phase: A potential future use for essential oils as antimicrobials? *Lett. Appl. Microbiol.* **2012**, *54*, 169–174. [CrossRef]
46. Manvitha, K.; Bidya, B. Review on pharmacological activity of *Cymbopogon citratus*. *Int. J. Herb. Med.* **2014**, *6*, 5–7.
47. Valkova, V.; Duranova, H.; Galovicova, L.; Borotova, P.; Vukovic, N.L.; Vukic, M.; Kacaniova, M. *Cymbopogon citratus* essential oil: Its application as an antimicrobial agent in food preservation. *Agronomy* **2022**, *12*, 155. [CrossRef]
48. Inouye, S.; Yamaguchi, H.; Takizawa, T. Screening of the antibacterial effects of a variety of essential oils on respiratory tract pathogens, using a modified dilution assay method. *J. Infect. Chemother.* **2001**, *7*, 251–254. [CrossRef] [PubMed]
49. Houdkova, M.; Albarico, G.; Doskocil, I.; Tauchen, J.; Urbanova, K.; Tulin, E.E.; Kokoska, L. Vapors of volatile plant-derived products significantly affect the results of antimicrobial, antioxidative and cytotoxicity microplate-based assays. *Molecules* **2020**, *25*, 6004. [CrossRef] [PubMed]
50. Gaworski, C.L.; Vollmuth, T.A.; York, R.G.; Heck, J.D.; Aranyi, C. Developmental toxicity evaluation of inhaled citral in Sprague-Dawley Rats. *Food Chem. Toxicol.* **1992**, *30*, 269–275. [CrossRef]
51. Lulekal, E.; Tesfaye, S.; Gebrechristos, S.; Dires, K.; Zenebe, T.; Zegeye, N.; Feleke, G.; Kassahun, A.; Shiferaw, Y.; Mekonnen, A. Phytochemical analysis and evaluation of skin irritation, acute and sub-acute toxicity of *Cymbopogon citratus* essential oil in mice and rabbits. *Toxicol. Rep.* **2019**, *6*, 1289–1294. [CrossRef]
52. El-Kased, R.F.; El-Kersh, D.M. GC–MS profiling of naturally extracted essential oils: Antimicrobial and beverage preservative actions. *Life* **2022**, *12*, 1587. [CrossRef]
53. Mohamed Hanaa, A.R.; Sallam, Y.I.; El-Leithy, A.S.; Aly, S.E. Lemongrass (*Cymbopogon citratus*) essential oil as affected by drying methods. *Ann. Agric. Sci.* **2012**, *57*, 113–116. [CrossRef]
54. Gao, S.; Liu, G.; Li, J.; Chen, J.; Li, L.; Li, Z.; Zhang, X.; Zhang, S.; Thorne, R.F.; Zhang, S. Antimicrobial activity of lemongrass essential oil (*Cymbopogon flexuosus*) and its active component citral against dual-species biofilms of *Staphylococcus aureus* and *Candida* species. *Front. Cell Infect. Microbiol.* **2020**, *10*, 603858. [CrossRef] [PubMed]
55. Bhawna, K.; Sharma, S.K.; Singh, L.; Mohapatra, S.; Singh, T. *Cyperus scariosus*: A potential herb. *Int. Res. J. Pharm.* **2013**, *4*, 17–20. [CrossRef]
56. Jha, V.; Patel, R.; Devkar, S.; Shaikh, M.A.; Rai, D.; Walunj, S.; Koli, J.; Jain, T.; Jadhav, N.; Shruti Narvekar, S.; et al. Chemical composition, bioactive potential, and thermal behaviour of *Cyperus scariosus* essential oil. *Chem. Sci. Int. J.* **2022**, *31*, 1–14. [CrossRef]
57. Clery, R.A.; Cason, J.R.L.; Zelenay, V. Constituents of cypriol oil (*Cyperus scariosus* R. Br.): N-containing molecules and key aroma components. *J. Agric. Food Chem.* **2016**, *64*, 4566–4573. [CrossRef]
58. Kumar, A.; Niranjan, A.; Lehri, A.; Srivastava, R.K.; Tewari, S.K. Effect of geographical climatic conditions on yield, chemical composition and carbon isotope composition of nagarmotha (*Cyperus scariosus* R. Br.) essential oil. *J. Essent. Oil-Bear. Plants* **2016**, *19*, 368–373. [CrossRef]
59. Houdkova, M.; Chaure, A.; Doskocil, I.; Havlik, J.; Kokoska, L. New broth macrodilution volatilization method for antibacterial susceptibility testing of volatile agents and evaluation of their toxicity using modified MTT assay in vitro. *Molecules* **2021**, *26*, 4179. [CrossRef] [PubMed]

60. Clinical and Laboratory Standards Institute (CLSI). *Performance Standards for Antimicrobial Disk Susceptibility Tests*, 11th ed.; Approved Standard, CLSI Document M02-A11; CLSI: Wayne, PA, USA, 2012; p. 32. ISBN 1-56238-782-0.
61. AOAC International. *Official Methods of Analysis, Official Method 925.10*; Association of Official Analytical Chemists: Gaithersburg, MD, USA, 2012; ISBN 0935584838.
62. European Pharmacopoeia. *Published in Accordance with the Convention on the Elaboration of a European Pharmacopoeia*, 7th ed.; European Treaty Series No. 50; Council of Europe: Strasbourg, France, 2013.
63. Clinical and Laboratory Standards Institute (CLSI). *Performance Standards for Antimicrobial Susceptibility Testing*; 25th Informational Supplement M100-S25; CLSI: Wayne, PA, USA, 2015; ISBN 1-56238-989-0.
64. Mosmann, T. Rapid colorimetric assay for cellular growth and survival: Application to proliferation and cytotoxicity assays. *J. Immunol. Methods* **1983**, *65*, 56–63. [CrossRef] [PubMed]
65. Kokjohn, K.; Bradley, M.; Griffiths, B.; Ghannoum, M. Evaluation of in vitro activity of ciclopirox olamine, butenafine HCl and econazole nitrate against dermatophytes, yeasts and bacteria. *Int. J. Dermatol.* **2003**, *42*, 11–17. [CrossRef] [PubMed]
66. Trevor, A.J.; Katzung, B.G.; Kruidering-Hall, M. *Katzung and Trevor's Pharmacology Examination and Board Review*, 11th ed.; McGraw-Hill Education: New York, NY, USA, 2015; p. 20. ISBN 978-0-07-182639-6.

Disclaimer/Publisher's Note: The statements, opinions and data contained in all publications are solely those of the individual author(s) and contributor(s) and not of MDPI and/or the editor(s). MDPI and/or the editor(s) disclaim responsibility for any injury to people or property resulting from any ideas, methods, instructions or products referred to in the content.

molecules

Article

Enhancing the Antimicrobial Effect of Ozone with *Mentha piperita* Essential Oil

Alin-Daniel Floare [1,2], Ramona Dumitrescu [1,2], Vlad Tiberiu Alexa [1,2], Octavia Balean [1,2], Camelia Szuhanek [2,3], Diana Obistioiu [4], Ileana Cocan [5], Alina-Georgeta Neacsu [6,*], Iuliana Popescu [6], Aurora Doris Fratila [7] and Atena Galuscan [1,2]

1 Translational and Experimental Clinical Research Centre in Oral Health, University of Medicine and Pharmacy "Victor Babes", Eftimie Murgu Sq. no 2, 300041 Timisoara, Romania
2 Department of Preventive, Community Dentistry and Oral Health, University of Medicine and Pharmacy "Victor Babes", Eftimie Murgu Sq. no 2, 300041 Timisoara, Romania
3 Orthodontic Research Center (ORTHO-CENTER), Faculty of Dental Medicine, "Victor Babes" University of Medicine and Pharmacy, Eftimie Murgu Sq. No 2, 300041 Timisoara, Romania
4 Faculty of Veterinary Medicine, University of Life Sciences "King Mihai I" from Timisoara, Calea Aradului No. 119, 300641 Timisoara, Romania
5 Faculty of Food Engineering, University of Life Sciences "King Mihai I" from Timisoara, Calea Aradului No. 119, 300645 Timisoara, Romania
6 Faculty of Agriculture, University of Life Sciences "King Mihai I" from Timisoara, Calea Aradului No. 119, 300641 Timisoara, Romania
7 Faculty of Dental Medicine, Ludwig-Maximilian University Munich, Goethestr. 70, 80336 Munich, Germany
* Correspondence: alinaneacsu@usab-tm.ro

Abstract: This study aimed to obtain and analyse *Mentha piperita* essential oil (MpEO) for the prospect of being used as an enhancement agent for the antimicrobial potential of ozone against gram-positive and gram-negative bacteria and fungi. The research was done for different exposure times, and it gained time–dose relationships and time–effect correlations. *Mentha piperita* (Mp) essential oil (MpEO) was obtained via hydrodistillation and further analysed by using GC-MS. The broth microdilution assay was used to determine the strain inhibition/strain mass growth by using spectrophotometric optical density reading (OD). The bacterial/mycelium growth rates (BGR/MGR) and the bacterial/mycelium inhibition rates (BIR/MIR) after ozone treatment in the presence and absence of MpEO on the ATTC strains were calculated; the minimum inhibition concentration (MIC) and statistical interpretations of the time–dose relationship and specific *t*-test correlations were determined. The effect of ozone on the following tested strains at maximum efficiency was observed after 55 s of single ozone exposure, in order of effect strength: *S. aureus* > *P. aeruginosa* > *E. coli* > *C. albicans* > *S. mutans*. For ozone with the addition of 2% MpEO (MIC), maximum efficacy was recorded at 5 s for these strains, in order of effect strength: *C. albicans* > *E. coli* > *P. aeruginosa* > *S. aureus* > *S. mutans*. The results suggest a new development and affinity regarding the cell membrane of the different microorganisms tested. In conclusion, the use of ozone, combined with MpEO, is sustained as an alternative therapy in plaque biofilm and suggested as helpful in controlling oral disease-causing microorganisms in medicine.

Keywords: *Mentha piperita*; essential oil; GC-MS; ozonation; gram-positive; gram-negative bacteria

Citation: Floare, A.-D.; Dumitrescu, R.; Alexa, V.T.; Balean, O.; Szuhanek, C.; Obistioiu, D.; Cocan, I.; Neacsu, A.-G.; Popescu, I.; Fratila, A.D.; et al. Enhancing the Antimicrobial Effect of Ozone with *Mentha piperita* Essential Oil. *Molecules* **2023**, *28*, 2032. https://doi.org/10.3390/molecules28052032

Academic Editors: Antoni Szumny and Monica Rosa Loizzo

Received: 27 November 2022
Revised: 8 February 2023
Accepted: 17 February 2023
Published: 21 February 2023

Copyright: © 2023 by the authors. Licensee MDPI, Basel, Switzerland. This article is an open access article distributed under the terms and conditions of the Creative Commons Attribution (CC BY) license (https://creativecommons.org/licenses/by/4.0/).

1. Introduction

Ozone is a chemical compound consisting of three oxygen atoms (O_3, triatomic oxygen) in an energy form larger than normal atmospheric oxygen (O_2). It is one of the most potent oxidants with a high oxidation capacity [1] and also with bactericidal, fungicidal, and virus-inactivating effects. Ozone is well-known for its use in water purification and drinking water treatment as well as for being used in heavily infected wounds such as burns, diabetic feet, and ulcus cruris, lesions infected with germs resistant to most antibiotics, such as MRSA (e.g., methicillin-resistant *Staphylococcus aureus*) [2].

Ozone allows microbial flora to pass from acidogenic and acidic microorganisms to normal oral commensals. In addition, ozone has a significant environmental advantage represented by rapid degradation, a process that clinically causes low cytotoxicity after contact with organic compounds. All of these features suggest that ozone could be widely used soon in restorative and preventive dentistry [1–5].

Due to its high oxidation power, ozone can oxidise the bacterial cell wall, causing its lysis, and the pyruvic acid produced by bacteria can then turn it into acetic acid and carbon dioxide.

In recent years, in the search for new natural therapeutic solutions, the use of essential oils (EOs) in alternative medicine and dentistry represents modern trends with a positive impact on the patient [6,7]. *Mentha piperitha* essential oil (MpEO) has multiple uses in medicine and dentistry. In dentistry, it was used as part of a mouthwash formula, positively influencing the management of oral mucositis. It was demonstrated that the MpEO, when loaded in chitosan polymeric nanogel, was influential in inhibiting *Streptococcus mutans*, an important pathogen of dental caries in humans. In addition, the MpEO-nanogel inhibited the formation of biofilms on the dental surface [8]. In particular, the antimicrobial effect of MpEO was studied against gram-positive and gram-negative bacteria [9,10]. The combination of conventional therapy with natural compounds such as EOs leads to synergistic/antagonistic effects. The interaction of MpEO with amphotericin B as a conventional antimicrobial agent revealed an antagonistic effect against *Candida albicans* and a synergistic effect against *Escherichia coli* and *Candida albicans*. However, moderate influence against *Staphylococcus aureus* was observed [11]. The inhibitory effect of MpEO against multidrug-resistant clinical pathogenic bacteria, including *Streptococcus pyogenes*, *Enterococcus faecalis*, and methicillin-resistant *Staphylococcus aureus*, was also reported [12].

The use of EOs as antimicrobial agents in water disinfection was recently studied. EOs of ginger, turmeric, lavender, and tulsi enhanced the disinfection rate up to five times when an innovative hybrid hydrodynamic cavitation process was applied [13]. MpEO's efficiency against *Pseudomonas aeruginosa* and *Staphylococcus aureus* using two different cavitating reactors and hydrodynamic cavitation was higher than 99%. The advantage of these procedures refers to enhancing the disinfection rate concurrently with significant lowering of the oil dose [14].

The use of ozonated vegetal oils was previously evaluated in terms of their antibacterial and antifungal activity with application in skin treatments [15–17] and dentistry [18–20]. In oral medicine, the control of periodontal and infectious diseases with oral microorganisms based on ozonated oils, such as sunflower, olive, and groundnut, were reported [20,21]. In addition, commercial formulations, such as ozone gel with olive oil in its composition, were patented to control the formation of dental plaque [22,23]. Indeed, ozonated sunflower oil (Oleozon®) was tested against pathogens such as *S. aureus*, *Enterococcus faecalis*, *S. pyogenes*, *Escherichia coli*, *Pseudomonas aeruginosa*, and different species of *Mycobacterium T.*, and the results confirmed the antimicrobial potential of this natural formulation. The MICs varied between 2.37 and 9.95 mg/mL for mycobacteria and between 1.18 and 9.5 mg/mL for all other bacteria [24].

Even if studies in the literature indicate a wide use of ozonated vegetable oils in medicine, the possibility of ozonation of essential oils from medicinal plants, or the association of the two elements, to our knowledge, was only briefly reported in the specialised literature. Some studies have been carried out on the influence of ozonation on the chemical composition of essential oils, emphasising that the presence of ozone does not change their chemical composition [25,26]. In addition, some commercial products are based on vegetal oils with essential oils added, but they do not present scientific support for antimicrobial effects [22].

This study aims to analyse the antimicrobial potential of ozone alone and ozone enhanced with MpEO against the following organisms at different times of exposure: bacteria *Escherichia coli* (*E. coli*), *Staphylococcus aureus* (*S. aureus*), *Pseudomonas aeruginosa* (*P. aeruginosa*), *Streptococcus mutans* (*S. mutans*), and fungi *Candida albicans* (*C. albicans*).

In this regard, two indicators, the bacterial/mycelium growth rate (BGR/MGR) and the bacterial/mycelium inhibition rate (BIR/MIR), were calculated after ozone treatment in the presence and absence of MpEO on the ATTC strains, and the minimum inhibition concentrations (MIC) of ozone in single and combined forms were determined.

To our knowledge, this is the first study that intends to highlight the antagonistic/synergistic effect obtained by combining ozone with MpEO against pathogenic bacteria and fungi.

2. Results

2.1. The Obtaining and Characterisation of EOs

The chemical composition of MpEO is presented in Table 1.

Table 1. Chemical composition of MpEO.

Compounds	Type	LRI [a]	% of Total Compounds
Linalool	MO	1533	0.752
α-pinene	MH	1021	1.476
β-pinene	MH	1106	1.179
Sabinene	MH	1135	0.508
Limonene	MH	1196	3.137
β-trans-ocimene	MH	1199	0.233
γ-terpinene	MH	1202	0.399
Eucalyptol	MO	1204	5.639
p-cymene	MH	1284	0.454
Menthofurane	MO	1474	2.960
Linalol acetate	MO	1541	0.558
Menthyl-acetate iso	MO	1546	0.375
Menthyl acetateone racemic	MO	1548	10.322
p-menthan-3-one	MO	1552	23.643
Menthol, acetate, iso-	MO	1562	1.882
Isomenthone	MO	1582	4.046
Terpinen-4-ol	MO	1593	1.191
Menthol	MO	1634	32.648
Germacrene D	SH	1708	5.410
γ-Elemene	SH	1717	0.244
p-menth-1-en-8-ol	MO	1724	0.433
Pulegone	MO	1730	1.636
Piperitone	MO	1750	0.521
Caryophyllene oxide	SO	1889	0.355
Total of compounds			100.000
Monoterpene hydrocarbonates	MH		7.386
Monoterpene oxygenate	MO		86.606

Table 1. *Cont.*

Compounds	Type	LRI [a]	% of Total Compounds
Sesquiterpene hydrocarbonates	SH		5.653
Sesquiterpene oxygenates	SO		0.355

[a] Linear retention indices (LRI) calculated according to n-alkanes (C8–C27) for AT WAX 30 m × 0.32 mm × 1 μm column [24].

The extraction yield of MpEO from the aerial parts of *Mentha piperita* prelevated was small (3.50%). In MpEO, 95.169% of terpene compounds in percentages over 1% were identified, of which the majority were oxygenated monoterpenes (83.967%), followed by monoterpene hydrocarbonates (5.792%) and sesquiterpene hydrocarbonates (5.410%). The main chemical compounds found in MpEO were menthol (32.658%), menthyl acetate (23.643%), and menthone (10.322%).

2.2. *Evaluation of MpEO Antimicrobial Activity and Ozone Potential as Single and Enhanced with MpEO Antimicrobial Agent*

Figures 1–6 show the bacterial inhibition rates (BIR%)/mycelial inhibition rate (MIR%) against analysed strains of MpEO, ozone, and MpEO after ozone application, which were calculated according to Formula (3), and the bacterial growth rates in relation to the time of treatment application, which was calculated according to Formula (2).

Concerning MpEO's efficacy against the strains tested, the results are presented in Figure 1. The results suggest that the efficacy of MpEO starts at the first concentration tested, meaning that the minimum concentration tested was 2%. *S. mutans* and *S. aureus* achieved a mean efficacy of 30% BIR at the first concentration tested, with their maximum efficacies reaching 50.13% and 67.59% BIR, respectively. Regarding *P. aeruginosa*, antibacterial activity evolved with a positive correlation to the concentration tested, with BIR% values ranging from 48.32% to 67.59%. The antimicrobial activity against *E. coli* was also recorded with BIR% values ranging from 62.08% to 74.06%. *C. albicans* was the only strain tested that showed a negative correlation; efficacy decreased alongside the increase in concentration. The BIR% value recorded for 2 μL was 67.54%, and for 32 μL, the BIR% was 51.07%, a decrease in value by almost 25%. However, even if the trend was negative, MIC was reached at the 2% concentration tested.

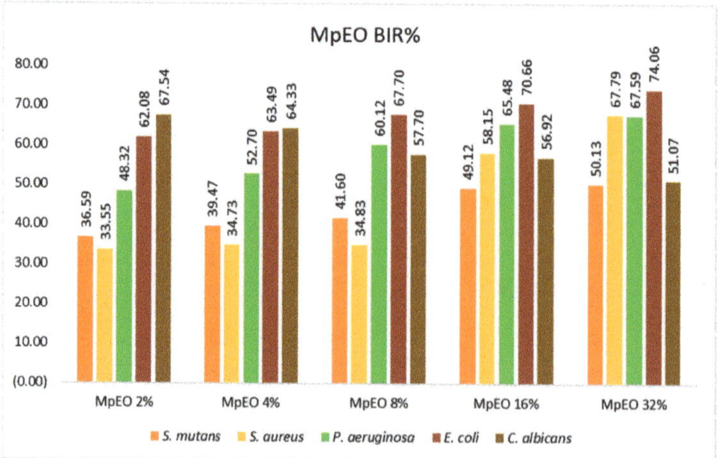

Figure 1. MpEO antimicrobial activity on the tested strains at different concentrations expressed as BIR%.

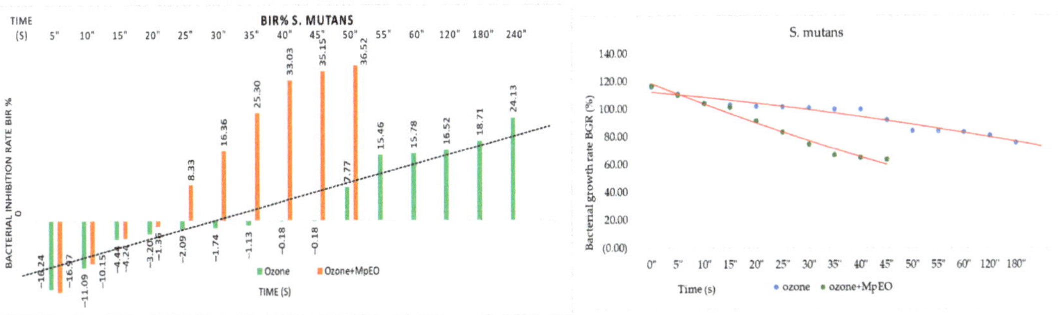

(**a**) (**b**)

Figure 2. Ozone and ozone+MpEO antimicrobial activity on *S. mutans* at different time exposures expressed in seconds (″). (**a**) Results expressed as BIR%; (**b**) results expressed as BGR%.

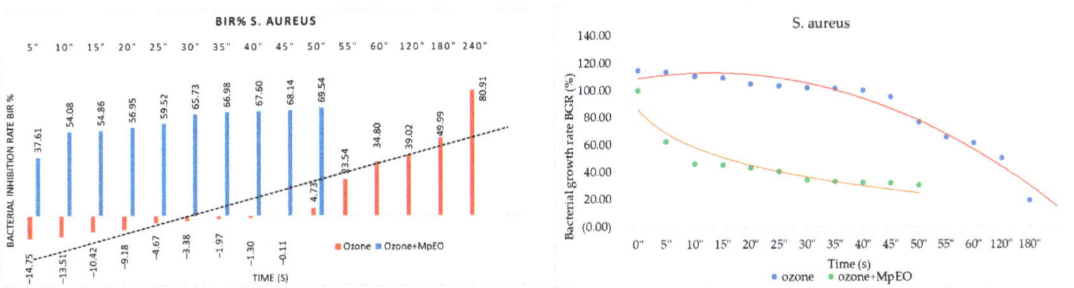

(**a**) (**b**)

Figure 3. Ozone and ozone+MpEO antimicrobial activity on *S. aureus* at different time exposures expressed in seconds (″). (**a**) Results expressed as BIR%; (**b**) results expressed as BGR%.

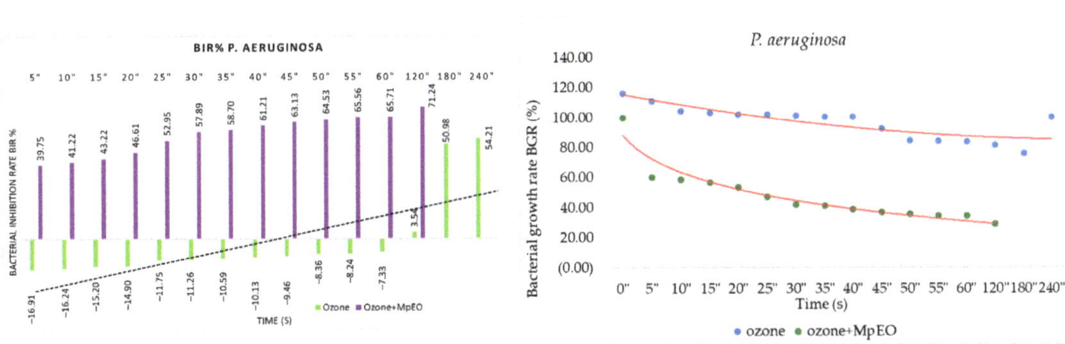

(**a**) (**b**)

Figure 4. Ozone and ozone+MpEO antimicrobial activity on *P. aeruginosa* at different time exposures expressed in seconds (″). (**a**) Results expressed as BIR%; (**b**) results expressed as BGR%.

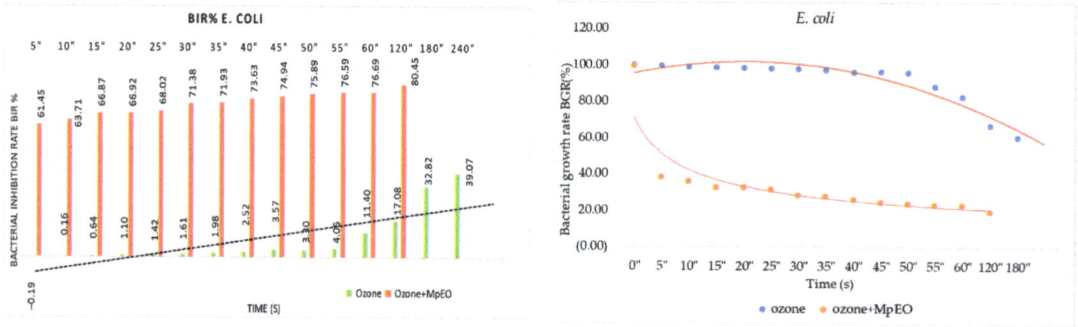

(a) (b)

Figure 5. Ozone and ozone+MpEO antimicrobial activity on *E. coli* at different time exposures expressed in seconds (″). (**a**) Results expressed as BIR%; (**b**) results expressed as BGR%.

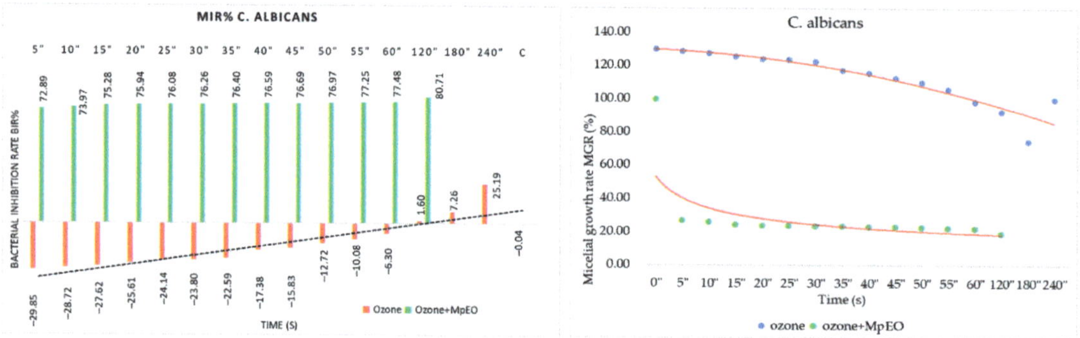

(a) (b)

Figure 6. Ozone and ozone+MpEO antimicrobial activity tested on *C. albicans* at different time exposures expressed in seconds (″). (**a**) Results expressed as MIR%; (**b**) results expressed as BGR%.

Comparing the BIR percentages concerning the ozone effect on *S. mutans*, the first effective duration was 50 s, with a value of 7.77% compared to the control (Figure 2). All of the values obtained in the previous tests were negative. The evolution trend was positively correlated with the duration of ozonation. The maximum duration tested, 240 s, reached a BIR% of 24.13% compared to the negative control, which was 0%. The negative values of BIR showed synergistic activity of ozone with the *S. mutans* strain tested; the values obtained proved a strain-boosting effect demonstrated by a bacterial mass growth of 2% compared to the positive control in the duration interval between 5 s and 45 s, with values ranging from −16.24% to −0.18% (Figure 2a).

All of the results changed when the ozone activity was enhanced by MpEO and used at the lowest MIC (2%). As a result, the first effective duration became half of the duration needed by ozone alone, 25 s, with a BIR% of 8.33%. Even if the starting point was similar in the first duration tested (16.24% compared to 16.97%), evolution improved compared to the inhibition caused by ozone alone. All of the subsequent durations used in the research showed higher BIR% than any initial durations, proving a BIR% enhancement rate of 15% on average. The inhibitory effects of the ozone and ozone enhanced with *M. piperitha* treatments were also revealed from the bacterial growth rate–time relationship, as shown in Figure 2b. Bacterial growth tended to decrease with ozone treatment time. The time–inhibition relationship clearly shows the inhibiting effect against *S. mutans* when ozone is applied in association with MpEO compared to simple ozone, which justifies this therapeutic approach.

Figure 3a summarises the ozone's antimicrobial activity on *S. aureus*, both alone and enhanced with MpEO. Ozone was effective against *S. aureus* with BIR values between −14.75% and 80.91%, obtained from all of the periods tested. Concerning the growth-boosting effect, it was present from 5 s to 45 s, just as for *S. mutans*, and the inhibitory values were obtained after 50 s. These results are not superior to the ones recorded in the test. Positive BIR values started at 50 s with a value of 4.73%; the value registered as the MIC and increased up to 80.91%, the value reached after 240 s. The BIR values after 55 s were higher against the *S. aureus* and *S. mutans* strains.

Regarding the antimicrobial effect on *S. mutans*, the MIC was reached at 25 s; concerning *S. aureus*, inhibition was proven at the first duration tested, 5 s, with a BIR% value of 37.61%. All of the subsequent durations tested showed higher inhibition efficacy for ozone enhanced by MpEO, with an average enhancement rate of 65%.

The time–bacterial growth relationship demonstrates the inhibitory effect against *S. aureus* (Figure 3b). A decrease in bacterial growth over time can be observed both in the treatment with ozone only and in the treatment with ozone and MpEO, but higher bacterial inhibition can be noticed in the combined application of ozone with MpEO.

Concerning the inhibition of *P. aeruginosa*, ozone tested alone proved less effective until after 120 s, when the MIC for *P. aeruginosa* was recorded at a BIR value of 3.54%. All of the data were collected between 5–60 s, proving a positive strain-boosting effect. The effects correlated with the duration of the test, and negative inhibitory values are presented in Figure 4a.

Regarding ozone/MpEO efficacy, the effect was similar to the activity against *S. mutans*. MIC proved to be effective at 5 s with a difference from the ozone alone of 56.66%. The trend was evident, as inhibitory efficacy grew over time. The ozone alone reached the MIC at 120 s, the ozone enhanced with MpEO demonstrated the MIC at 5 s, and the difference at 120 s was a 67.73% improvement of BIR%. The decreasing trend of bacterial development over time was also observed, as shown in Figure 4b; it is worth mentioning that the growth rate was lower when supplementing the treatment with MpEO. Bacterial growth rate BGR (%) with the use of ozone–MpEO was lower (between 22–60%) than when the treatment was done only with ozone (between 78–120%) (Figure 4b).

Figure 5a presents the results regarding the 15 time durations tested on *E. coli*. The results show a positive antimicrobial effect recorded against *E. coli* (with the BIR% ranging from 0.16% to 39.07%) for the ozone-alone treatment tested, starting from 10 s.

The only negative value was presented at 5 s. Even if time had a relatively weak influence on the value of BIR recorded, with 0.64% at 15 s and 4.05% at 55 s, the effect was still inhibitory considering that the value obtained was reported to be 0% growth of the control strain. The last four durations tested proved good inhibitory results ranging from 11.39% to 39.07% BIR. The values obtained for the ozone enhanced with MpEO show the inhibitory pattern improved over what was presented in the previous test. For ozone alone, the BIR% was −0.19, and for ozone enhanced with MpEO, the BIR% was 61.45%. The values obtained present an improvement of BIR% after MpEO enhancement by approximately 70%. Figure 5b shows the decrease of bacterial growth over time, especially for ozone treatment enhancement with MpEO.

Regarding the inhibition of *C. albicans*, there was a positive correlation between the antimicrobial effect with ozonation duration, as seen in Figure 6a. The 120 s test was the first active, with a MIR of 1.6%, while the final test at 240 s showed a mycelial inhibitory percentage of 25.19%.

The values obtained for the enhanced ozone with MpEO show a clear improvement in results. For the single ozone tested at 5 s, the MIR% was −29.84%, and for the ozone enhanced with MpEO, the value reached 72.89%; all of the other durations tested confirmed an average rise in efficacy by 50%. The inhibitory activity against *C. albicans* was also demonstrated by BGR (%), as presented in Figure 6b. The bacterial growth rate was higher for the use of ozone alone (between 78–130%) than when the treatment of ozone and MpEO was applied (20%).

Figures 7 and 8 present the inhibition percentages of all of the strains tested at different periods selected in the research. The difference in the figures regarding the strains and times of exposure to ozone is due to the fact that the data presented in Figure 8 are complementary to the data presented in Figure 7. The values presented in Figure 8 are results obtained from the first part of the analysis, from which we selected only the times necessary for obtaining the MIC for each strain, and all of the extra periods of time outside of the necessary timeframe were untested.

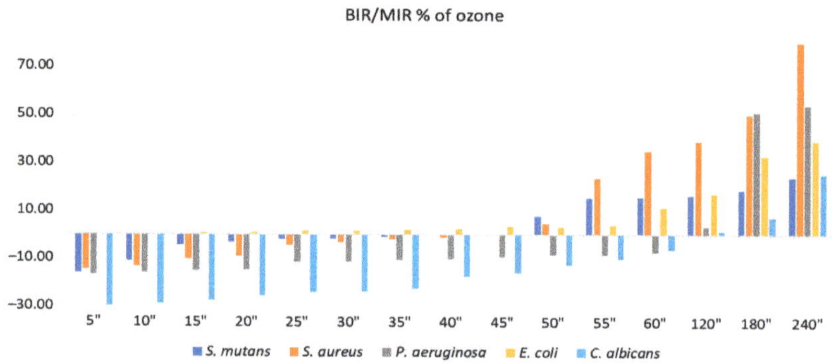

Figure 7. Antimicrobial activity (expressed as BIR/MIR%) of ozone tested on ATCC strains at different exposure times represented in seconds (″).

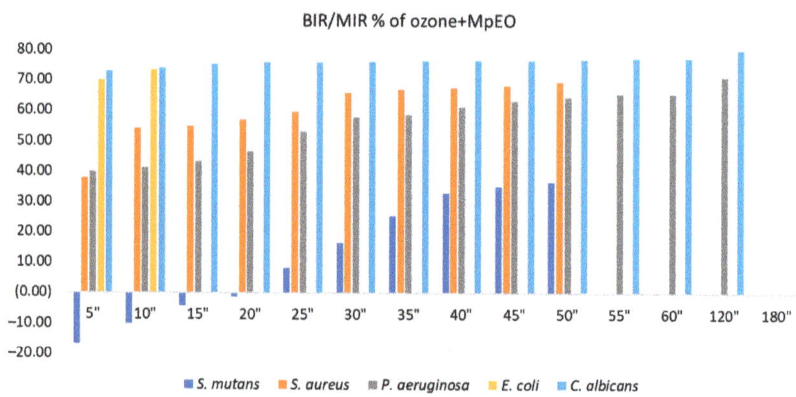

Figure 8. Antimicrobial activity (expressed as BIR/MIR%) of ozone+MpEO tested on ATCC strains at different exposure times expressed in seconds (″).

Concerning the ozone tested as a single element, the best results were on *S. aureus*, with the inhibition value reaching 80.91% at 240 s. The second strain most affected by ozone alone was *P. aeruginosa*, with a maximum inhibitory response of 54.21%, followed by *E. coli* at 39.07%. *S. mutans* and *C. albicans* showed similar values at maximum exposure: 24.13% and 25.18%, respectively. The results suggest the best effect was recorded for *C. albicans* with BIR% up to 80.71%, followed by *E. coli* (73.38%), *P. aeruginosa* (71.24%), *S. aureus* (69.54%), and finally, *S. mutans* (36.52%).

Regarding the BIR/MIR (%) values of MpEO (Figure 8), all bacterial strains were inhibited. For each strain, the trend was positive, with BIR% values ranging from 36.59% to 74.06%. There was a difference for *C. albicans*, where inhibition was present with MIR% values ranging from 67.54% to 51.07%, the difference being made by the negative trend of inhibition; the inhibitory values decreased with increases in EO concentration. For *C. albicans*, the increase in MpEO concentration negatively correlated with inhibition values.

The minimum inhibition concentrations (MIC) tested on the ATTC strains is presented in Table 2. MIC was determined as being the lowest time expressed in seconds of ozone exposure that inhibited the visible growth of cells.

Table 2. The MIC values of ozone (a), ozone+MpEO (b), and MpEO (c) on ATTC strains at different exposure times. The red background colour highlights the time duration or concentration where the MIC was determined for each strain. The effect was maintained together with an increase in the period/concentration. The ozonations with positive inhibitory responses are marked with grey background.

MIC Values of Ozone					MIC Values of Ozone+MpEO				
S. mutans	*S. aureus*	*P. aeruginosa*	*E. coli*	*C. albicans*	*S. mutans*	*S. aureus*	*P. aeruginosa*	*E. coli*	*C. albicans*
5″	5″	5″	5″	5″	5″	5″	5″	5″	5″
10″	10″	10″	10″	10″	10″	10″	10″	10″	10″
15″	15″	15″	15″	15″	15″	15″	15″	15″	15″
20″	20″	20″	20″	20″	20″	20″	20″	20″	20″
25″	25″	25″	25″	25″	25″	25″	25″	25″	25″
30″	30″	30″	30″	30″	30″	30″	30″	30″	30″
35″	35″	35″	35″	35″	35″	35″	35″	35″	35″
40″	40″	40″	40″	40″	40″	40″	40″	40″	40″
45″	45″	45″	45″	45″	45″	45″	45″	45″	45″
50″	50″	50″	50″	50″	50″	50″	50″	50″	50″
55″	55″	55″	55″	55″	55″	55″	55″	55″	55″
60″	60″	60″	60″	60″	60″	60″	60″	60″	60″
120″	120″	120″	120″	120″	120″	120″	120″	120″	120″
180″	180″	180″	180″	180″	180″	180″	180″	180″	180″
240″	240″	240″	240″	240″	240″	240″	240″	240″	240″

	S. mutans	*S. aureus*	*P. aeruginosa*	*E. coli*	*C. albicans*
MpEO%	2	2	2	2	2
MpEO%	4	4	4	4	4
MpEO%	8	8	8	8	8
MpEO%	16	16	16	16	16
MpEO%	32	32	32	32	32

Following statistical analysis with a *t*-test, the values obtained are recorded in the table presented as a Supplementary File, wherein significant differences ($p < 0.05$) are recorded between the ozone-treated and ozone-MpEO-treated samples.

3. Discussion

3.1. The Obtaining and Characterisation of EOs

The extraction yield of MpEO from the aerial parts of *Mentha piperita* prelevated from spontaneous flora was small (3.50%) but similar to the results reported in the literature (3.22%) [27].

Similar compositions have been reported in other studies regarding the chemotypes identified in MpEOs. Oxygenated monoterpenes predominated: 60.826% in MpEOs obtained from the aerial parts of *Mentha piperita* originating from the spontaneous flora of Romania [28], and 73.9–94.8% obtained from Mentha species from China [29]. Yield rates for menthol include the following: 36.02% [30], 45.05% [31], 32.14% [32], 10.6–66.6% (depending on cultivar and harvest stage), and 11.57% [33]; for menthone, 20.14% [34] and 5–35% [35] were identified in the majority of EOs originating from *Mentha piperita* species cultivated in different geographical areas.

3.2. The Ozone Potential as Single and Enhanced with MpEO Antimicrobial Agent

The results obtained regarding the antimicrobial effect of ozone on the tested strains show that, in terms of the effect related to the effective exposure time, it decreased in the following order: *E. coli* > *S. aureus* = *S. mutans* > *P. aeruginosa* = *C. albicans*. Regarding the intensity of the effect of ozone on the tested strains, maximum efficiencies were observed, after 240 s of ozone exposure, in decreasing order of strength: *S. aureus* > *P. aeruginosa* > *E. coli* > *C. albicans* > *S. mutans*.

All of the results obtained using the OD values and, subsequently, the BIR values demonstrate the synergistic effect of the ozone and MpEO mixture. This affirmation is sustained by the fact that the inhibitory rates obtained in the mixture increased compared to the inhibitory rates obtained for ozone and MpEO tested as single compounds. This fact is possibly a result of a dual effect: the oxidation reaction of the bacterial wall produced by the ozone, potentiated with the MpEO mechanism of weakening the cell walls of resistant bacteria, damages or kills these cells.

The effects of ozone on various bacterial and fungal strains and their mechanisms of action were previously reported in specialised studies [36]. Giuliani et al., 2018 explained that ozone attacks numerous cellular constituents such as proteins, unsaturated lipids, enzymes, and nucleic acids in the cytoplasm [36]. When ozone interacts with the cell wall, an oxidative burst occurs, creating a tiny hole in the cell wall; thus, the bacterium loses its shape, and after multiple collisions with ozone, the cell dies in only a few seconds [36]. Another explanation regarding the antimicrobial activity was given by Srinivasan SR, Amaechi BT, which highlighted the solid oxidising potential of ozone that degrades the cytoplasmatic membranes of unsaturated fatty acids [37].

Our data agree with previous studies that confirm the antimicrobial potential of ozone. In our study, the maximum effective rate of ozone inhibition on *S. aureus* was 80.91% after 240 s of exposure. Similar results were reported for different strains of *Staphylococcus* (*Staphylococcus aureus*, 94.0%; *Staphylococcus epidermidis*, 88.6%; *Enterococcus faecalis*, 79.7%) [38]. In 2011, Wilczińska-Borawska et al. confirmed ozone's bactericidal activity in bacterial strains most frequently isolated from the oral cavity. In addition, two gas application models on the infected medium were compared [39]. In agreement with our study, they found a statistically significant difference regarding inhibited bacterial growth on all media depending on the time of ozone action.

In 2000, Baysan et al. assessed the antimicrobial impact of ozone against *S. mutans* and *Streptococcus sobrinus* on primary root caries and observed an important inhibition effect for either 10 or 20 s after ozone introduction [40]. The killing rate was 78% after 6 s and 47% after twofold 18 s of exposure [40], but no completely eradicated strain was observed after applying ozone against *S. aureus*, *E. faecalis*, *Enterobacter cloacae*, and *C. albicans* [41]. Our results are in agreement with those obtained by Baysan et al. in 2000. The inhibition effect on *S. aureus* was proved at the first duration tested, meaning 5 s, with a BIR% value of 37.61% and a higher MIC value at 25 s. Against *S. mutans*, ozone significantly affected the bacterial load in dentin. Significant differences were found between the control and treated groups after 4 and 8 weeks after treatment, but no significant differences were found between 4 and 8 weeks [42].

The antibacterial effect of ozone on cariogenic bacterial species with and without the presence of saliva and a possible effect on the salivary proteins was studied by Johansson et al. in 2009 [43]. The results showed that in salt buffer, 92%, 73%, and 64% of the initial numbers of *A. naeslundii*, *S. mutans*, and *L. casei* were killed after 10 s of ozone exposure, and approximately 99.9% of the bacteria were dead after 60 s; however, in saliva, compared to the salt buffer, *S. mutans* and *L. casei* were less efficiently killed. In addition, saliva proteins were degraded by ozone after 60 s of exposure [43].

Huth KC et al., 2009 studied the influence of ozone aggregation conditions on the antimicrobial effect of treating different strains with not only either gaseous ozone or aqueous ozone, but also with 0.2%, 1%, and 2% chlorhexidine [44]. After 60 s of exposure, gaseous ozone at a concentration of 4 g m-3 was as effective as chlorhexidine. However, the highest concentration of aqueous ozone (20 μg/mL) proved even more effective than chlorhexidine in killing oral pathogens [44].

Although our experiments differed from those made by Huth et al. in 2009 regarding the time of exposure and ozone concentration, the results of this study agree with their results regarding the dose- and strain-dependence of ozone antimicrobial activity [44].

In another study, the effect of ozone on cultivatable microflora compared to chlorhexidine highlighted that the total reduction in the group without excavation was 7% after ozone treatment and 36% after chlorhexidine treatment. In comparison, 19% was the inhibition rate for ozone and 41% for chlorhexidine in the group with excavation [45]. The efficacy of ozone on cariogenic microorganisms (*S. mutans*, *S. salivarius*, *S. epidermidis*, *S. mitis*, *Lactobacillus*, and different kinds of *Staphylococcus*) exceeded the efficacy of 3% hydrogen peroxide and 0.2% chlorhexidine digluconate significantly [46].

In recent years, there has been an increasing emphasis on the association of ozonation with other natural preparations. The association of ozone with other natural preparations, such as essential oils, enhances its efficiency. The killing rates of ozonated oil against *S. aureus* were almost 100% [47]. In addition, ozonated water (4 mg/L) was discovered to be viable for treating gram-positive and gram-negative oral bacteria and oral *C. albicans* in plaque biofilms and helpful in controlling oral disease-causing microorganisms in teeth [48].

Montevecchi M. et al. in 2013 observed the antibacterial activity of ozonated sunflower oil against *Porphyromonas gingivalis* after 72 h of anaerobic incubation. Although ozonated oil showed a significantly greater zone of inhibition than 0.2% chlorhexidine against *S. aureus* and *P. gingivalis* [49], the evaluation of the antibacterial activity of ozonised olive oil against oral and periodontal pathogens compared with the other two gel agents, which were based on chlorhexidine, proved the moderate antiseptic potential of ozonated oil [22]. A previous study reported the effectiveness of ozonated oil on *Candida albicans*. Ozonated olive and sunflower oils demonstrated similar antimicrobial activity, with low MICs ranging from 0.53 to 0.2 μg/mL [23].

Other approaches have referred to the intensity and concentration of ozone applied [50]. Schneider H. in 2004 reported that ozone concentrations ranging between 300 and 800 ppm have bactericidal effects, with the maximal effect recorded at 525 ppm [51].

The efficiency of our method of using ozone mixed with MpEO as an antimicrobial agent is superior to methods reported by other authors who used only ozone. In the recent study of Rangel and al. in 2022, an inhibition rate of 17% against *P. aeruginosa* was reported after 60 s of exposure, and an inhibition effect of 99.99% against *S. aureus* was reported only after 40 min of exposure [52]. Our data highlighted the advantage of enhancing ozone with MpEO by showing the reduction of exposure time at 60 s with an inhibition rate of 65.71% against *P. aeruginosa* and 69.54% after 50 s against *S. aureus*.

In our study, the MIC values of MpEO varied. The MIC value of MpEO, reported by other authors, varied from 0.625 μg/mL against fungi such as *C. albicans* and *C. parapsilosis*, gram-positive bacteria such as *S. mutans*, *S. pyogenes* to 1.14–6.25 μg/mL for *E. coli* or 5 μg/mL for *P. aeruginosa* [34,35,53]. The MIC values obtained in our study for MpEO against *S. mutans*, *S. aureus*, *P. aeruginosa*, *E. coli*, and *C. albicans* were 2% lower than those reported by Sechi et al., who obtained MICs values against *S. aureus*, *Enterococcus faecalis*, *S. pyogenes*, *E. coli*, and *P. aeruginosa* that ranged between 1.18 and 9.5% [24].

Heydari in 2018 highlighted that MpEO possesses stronger antimicrobial activity with greater activity on the bacilli (*Bacillus subtilis* and *Bacillus cereus*) rather than on the cocci (*Staphylococcus aureus* and *Staphylococcus epidermidis*) [54]. Other authors reported only a moderate antibacterial effect against some tested bacteria [55].

The main chemical compounds of our MpEO are menthol (32.658%), menthyl acetate (23.643%), and menthone (10.322%). The MICs demonstrate that menthol is considerably toxic against *S. aureus* and the most toxic against *E. coli* [56]. The inhibition potential of these compounds against different bacteria and fungi is due to the toxic effects on their membrane structures and functions [57]. Monoterpenes have a lipophilic character responsible for the antimicrobial effects of menthol and menthone. The mechanism of action of menthol and menthone is a perturbation of the lipid fraction of a microorganism's plasma membrane, resulting in membrane permeability alterations and leakage of intracellular materials [56].

The use of the ozone–MpEO mixture increased inhibition effects, as proved in this study, and this increase is based on the combination of the two effects and mechanisms of action of the materials as explained previously. The obtained results strengthen the applicability of these mixtures in medical and dental techniques.

4. Materials and Methods

4.1. The Obtaining and Characterisation of EOs

The aerial parts of the investigated medicinal plant (Mp) were collected during the flowering period from the flora located in the outskirts of Timisoara, Romania (45°47′00.1″ N 21°12′37.2″ E) in 2020. A total of ~700 g fresh material was used. Additionally, a voucher specimen was botanically identified and deposited in the temperature-controlled herbarium (22–25 °C and 30–40% relative humidity) of the Faculty of Agriculture-Botany Department, University of Life Sciences "King Mihai I" Timisoara, Romania; the plant's name (accepted name) was identified as *Mentha x piperita* L *(M. aquatica x M. spicata)* (Voucher Specimen Number Herbarium-Botany Department, VSNH.ULST-BD65).

The plant material was dried under natural conditions. The MpEO was extracted from 300 g of plant material using Clevenger-type equipment for 2 h, and the essential oil was separated via decantation and stored at 2–4 °C until used in GC/MS and microbiological analyses. The extraction yield of MpEO was calculated using the following formula:

$$\text{Yield } (\%) = [\text{amount of EO (g)}/\text{amount of dry plant (g)}] \times 100 \quad (1)$$

The GC-MS characterisation of MpEO was done using Shimadzu QP 2010 Plus equipment (Shimadzu Corporation, Columbia, SC, USA), which was equipped with a capillary column AT WAX 30 m × 0.32 mm × 1 μm. The carrier gas used was helium, with a flow rate of 1 mL/min, and the injector and ion source temperatures were 250 °C and 220 °C, respectively. The gradient temperature was used for compound separation with an initial oven temperature of 40 °C maintained for 1 min; then, the temperature was raised to 210 °C at a rate of 5 °C/min and maintained for 5 min at this temperature. The sample injection volume was 1 μL, and a split ratio of 1:50 was used. Volatile compounds of MpEO were identified using the NIST5 Wiley 275 libraries database. The linear retention indexes (LRI) were calculated using n-alkane C8–C27 standards using the same experimental conditions. The results are presented as percentages from total compounds [24]. The chromatogram of the chemical compounds is presented in the Supplementary File.

4.2. Microbiological Assay

4.2.1. Microbial Strains

All of the strains used belong to the culture collection of the Microbiology Laboratory of the Interdisciplinary Research Platform within the University of Life Sciences "King Michael I" from Timișoara. The used microbial strains were: *Streptococcusmutans* (ATCC 19615), *Staphylococcus aureus* (ATCC 25923), *Escherichia coli* (ATCC 25922), *Pseudomonas aeruginosa* (ATCC 27853), and *Candida albicans* (ATCC 10231). All ATCC strains are maintained at −50 °C.

4.2.2. Microbial Culture Preparation and Essential Oil Efficacy Assessment

Antimicrobial activity was tested according to the method described by Hulea et al. [58] and Obistioiu et al. [59], and the CLSI method was used for microdilution for the antimicrobial susceptibility test [60]. Different concentrations of MpEO ranging from 2 μL–32 μL were tested using the broth microdilution method. Considering that the quantity of oil tested was immersed in 100 μL of Brain Heart Infusion Broth, the concentration reached is expressed as a percentage (2–32%). Subsequently, the minimum inhibitory concentration (MIC) was determined. The MIC is defined as the lowest compound concentration that yields no visible microorganism growth, and it is based on mass loss by measurement of optical density (OD) by using the spectrophotometric method at 540 nm. For ozone, MIC was determined as the lowest time expressed in seconds of the ozone or enhanced ozone exposure that inhibited the visible growth of cells. The values obtained are expressed as OD reading values (values presented as mean ± standard deviation in Supplementary Material). The simple strain multiplied in BHI was used as a negative control. The control group for each strain was represented by the ATCC strain multiplied in BHI after 24 h without any interference. The mean OD values reached for each experiment were used in Formulas (2) and (3) and were considered to be 100% BGR and 0% BIR for each strain.

Two indicators were calculated, MGR/BGR and MIR/BIR, to interpret the values obtained by using the following formulas:

$$\text{MGR/BGR} = \frac{OD_{sample}}{OD_{negativecontrol}} \times 100(\%) \tag{2}$$

$$\text{MIR/BIR} = 100 - \text{MGR/BGR}\ (\%) \tag{3}$$

where:

OD sample—optical density at 540 nm as the mean value of triplicate samples in the presence of the selected fungi/bacteria;

OD negative control—optical density at 540 nm as the mean value of triplicate readings for the selected fungi/bacteria in BHI.

4.2.3. Ozonation Procedure

Various authors researched the effect of ozone in either clinical or microbiological studies, all stating as the conclusion that the duration of action can be an important consideration in ozone's antibacterial effect. Noites et al. [61] used gaseous ozone applied with an ozone generator for 24, 60, 120, and 180 s and assessed the antimicrobial impact of ozone against *S. mutans* and *Streptococcus sobrinus* for either 10 or 20 s, whereas Baysan et al. [40,62] assessed that an ozone application of 10–20 s eliminated more than 99% of the microorganisms found in dental caries and associated biofilms, and a 40 s treatment time covered all eventualities. Starting from these conclusions, we selected a wider range of time periods to investigate the efficacy of ozone on the microorganisms tested to ensure we obtained the necessary MIC.

The ATCC strains were successively subjected to ozonation with gaseous O_3 produced with HealOzone X4 (KaVo Dental & Co., Biberach, Germany), which released O_3 at a fixed concentration of 2100 ppm with a flow rate of 615 cm^2/min for 5″ (s), 10″, 15″, 20″, 25″, 30″, 35″, 40″, 45″, 50″, 55″, 60″, 120″, 180″, and 240″. Each period selected in our research was measured according to the times suggested by the manufacturer and in research articles.

The ozonation device comprises an air filter, vacuum pump, an ozone generator, a handpiece fitted with a sealing silicone cup, and a flexible hose. Once the procedure was started, the hose end was submersed into the ATCC strain within the 96-microdilution plate, and therefore, no O_3 was lost from the dose injected in the BHI broth. According to the specifications, the device produced ozone at a concentration of 32 g/m^3 at an exposure of 60 s.

This analysis method was also used for broth microdilution, and all of the previous steps presented before in the microbiological methodology (Sections 4.2.1 and 4.2.2) were followed once the ozonation procedure was finished.

4.3. Statistical Analysis

Differences between means were analysed with a one-way ANOVA, followed by multiple comparisons using a *t*-test (two-sample assuming equal variances) using GraphPad Prism 8 (version 8.0.2; GraphPad Software Inc., La Jolla, CA, USA). Differences were considered significant when p-values < 0.05. Time–inhibition relationships were established with the Microsoft Excel 365 software.

5. Conclusions

The results underline that ozone in combination with MpEO exerts synergistic effects of potentiating antimicrobial activity. The antibacterial effects obtained depend on contact time and the strain type. The effect of ozone on the following tested strains at maximum efficiency was observed after 55 s of single ozone exposure, listed in descending order of effect strength: *S. aureus* > *P. aeruginosa* > *E. coli* > *C. albicans* > *S. mutans*. Exposure to the ozone–MpEO mixture decreased contact time with the bacteria, which is of practical and economic importance. For ozone with the addition of 2% MpEO, the maximum efficacy was recorded at 5 s in the following strains, listed in descending order of effect strength: *C. albicans* > *E. coli* > *P. aeruginosa* > *S. aureus* > *S. mutans*.

The results demonstrate that the combination of ozone with MpEO leads to an increase in efficiency and a decrease in exposure time. All bacterial strains were inhibited, with BIR% values ranging from 36.59% to 74.06%, and the exposure time was reduced from 120 s to an optimal 55 s.

In the search for alternative antimicrobial solutions to synthetic antibiotics, the use of ozone in combination with essential oils can represent innovative solutions with applications in medicine and dentistry.

Supplementary Materials: The following supporting information can be downloaded at: https://www.mdpi.com/article/10.3390/molecules28052032/s1, Table S1: Optical density (OD) values at 540 nm, Table S2: BGR%/MGR% of ozone; Table S3: BGR%/MGR% of ozone enhanced with MpEO; Table S4: BGR%/MGR% of MpEO on ATCC strains; Figure S1: GC-MS Chromatogram.

Author Contributions: Conceptualisation: A.-D.F., O.B., D.O., A.G. and V.T.A.; methodology: O.B., D.O. and A.-D.F.; formal analysis: A.-D.F., O.B., D.O., A.G., V.T.A., I.C. and I.P.; original draft preparation: A.-D.F., O.B., D.O., A.G., V.T.A. and I.C.; writing—review and editing: A.-D.F., O.B., D.O., A.G., V.T.A., R.D., C.S., I.C., A.-G.N., I.P. and A.D.F.; visualisation: A.-D.F., O.B., D.O., A.G., V.T.A., R.D., C.S., I.C., A.-G.N., I.P. and A.-D.F.; supervision: A.G. All authors have read and agreed to the published version of the manuscript.

Funding: This research was funded by "Victor Babeş" University of Medicine and Pharmacy, Timisoara, Romania, Alin-Daniel Floare.

Institutional Review Board Statement: Not applicable.

Informed Consent Statement: Not applicable.

Data Availability Statement: Not applicable.

Acknowledgments: We were able to carry out this research with the support of the Interdisciplinary Research Platform belonging to the University of Life Sciences "King Michael I" from Timisoara, where the analyses were made.

Conflicts of Interest: The authors declare no conflict of interest.

References

1. Elsayad, I.I. Chemical analysis and surface morphology of enamel following ozone application with different concentrations and exposure times. *J. Adv. Res.* **2011**, *2*, 131–136. [CrossRef]
2. Viebahn-Haensler, R.; Fernández, O.L. Ozone in Medicine. The Low-Dose Ozone Concept and Its Basic Biochemical Mechanisms of Action in Chronic Inflammatory Diseases. *Int. J. Mol. Sci.* **2021**, *22*, 7890. [CrossRef]
3. Lacruz, R.S.; Habelitz, S.; Wright, J.T.; Paine, M.L. Dental Enamel Formation and Implications for Oral Health and Disease. *Physiol. Rev.* **2017**, *97*, 939–993. [CrossRef] [PubMed]
4. Kreulen, C.; Spijker, A.V.; Rodriguez, J.; Bronkhorst, E.; Creugers, N.; Bartlett, D. Systematic Review of the Prevalence of Tooth Wear in Children and Adolescents. *Caries Res.* **2010**, *44*, 151–159. [CrossRef]
5. Wang, X.; Liao, D.; Ji, Q.-M.; Yang, Y.-H.; Li, M.-C.; Yi, X.-Y.; Li, C.; Chen, Y.; Tao, H.-B.; Zhai, W.-H. Analysis of Bactericidal Effect of Three Medical Ozonation Dosage Forms on Multidrug-Resistant Bacteria from Burn Patients. *Infect. Drug Resist.* **2022**, *15*, 1637–1643. [CrossRef]
6. Gholamipourfard, K.; Salehi, M.; Banchio, E. *Mentha piperita* phytochemicals in agriculture, food industry and medicine: Features and applications. *S. Afr. J. Bot.* **2021**, *141*, 183–195. [CrossRef]
7. Lazăr, R.N.; Alexa, E.; Obiștioiu, D.; Cocan, I.; Pătruică, S. The Effect of the Use of Essential Oils in the Feed of Bee Families on Honey Chemical Composition and Antimicrobial Activity. *Appl. Sci.* **2022**, *12*, 1094. [CrossRef]
8. Ashrafi, B.; Rashidipour, M.; Marzban, A.; Soroush, S.; Azadpour, M.; Delfani, S.; Ramak, P. *Mentha piperita* essential oils loaded in a chitosan nanogel with inhibitory effect on biofilm formation against S. mutans on the dental surface. *Carbohydr. Polym.* **2019**, *212*, 142–149. [CrossRef] [PubMed]
9. Daferera, D.J.; Ziogas, B.N.; Polissiou, M.G. The effectiveness of plant essential oils on the growth of Botrytis cinerea, *Fusarium* sp. and *Clavibacter michiganensis* subsp. michiganensis. *Crop Prot.* **2003**, *22*, 39–44. [CrossRef]

10. Singh, R.; Shushni, M.A.; Belkheir, A. Antibacterial and antioxidant activities of *Mentha piperita* L. *Arab. J. Chem.* **2015**, *8*, 322–328. [CrossRef]
11. van Vuuren, S.; Suliman, S.; Viljoen, A. The antimicrobial activity of four commercial essential oils in combination with conventional antimicrobials. *Lett. Appl. Microbiol.* **2009**, *48*, 440–446. [CrossRef]
12. Shalayel, M.H.F.; Asaad, A.M.; Qureshi, M.A.; Elhussein, A.B. Anti-bacterial activity of peppermint (*Mentha piperita*) extracts against some emerging multi-drug resistant human bacterial pathogens. *J. Herb. Med.* **2017**, *7*, 27–30. [CrossRef]
13. Mane, M.B.; Bhandari, V.M.; Ranade, V.V. Safe water and technology initiative for water disinfection: Application of natural plant derived materials. *J. Water Process. Eng.* **2021**, *43*, 102280. [CrossRef]
14. Mane, M.B.; Bhandari, V.M.; Balapure, K.; Ranade, V.V. Destroying antimicrobial resistant bacteria (AMR) and difficult, opportunistic pathogen using cavitation and natural oils/plant extract. *Ultrason. Sonochem.* **2020**, *69*, 105272. [CrossRef] [PubMed]
15. Menéndez, S.; Falcón, L.; Maqueira, Y. Therapeutic efficacy of topical OLEOZON® in patients suffering from onychomycosis: OLEOZON® against Onychomycosis. *Mycoses* **2011**, *54*, e272–e277. [CrossRef] [PubMed]
16. Valacchi, G.; Zanardi, I.; Lim, Y.; Belmonte, G.; Miracco, C.; Sticozzi, C.; Bocci, V.; Travagli, V. Ozonated oils as functional dermatological matrices: Effects on the wound healing process using SKH1 mice. *Int. J. Pharm.* **2013**, *458*, 65–73. [CrossRef] [PubMed]
17. Travagli, V.; Zanardi, I.; Valacchi, G.; Bocci, V. Ozone and Ozonated Oils in Skin Diseases: A Review. *Mediat. Inflamm.* **2010**, *2010*, 610418. [CrossRef] [PubMed]
18. Nagayoshi, M.; Fukuizumi, T.; Kitamura, C.; Yano, J.; Terashita, M.; Nishihara, T. Efficacy of ozone on survival and permeability of oral microorganisms. *Oral Microbiol. Immunol.* **2004**, *19*, 240–246. [CrossRef]
19. Ghobashy, S.; El-Tokhey, H. In vivo study of the effectiveness of ozonized olive oil gel on inhibiting enamel demineralization during orthodontic treatment. *Egypt. Orthod. J.* **2012**, *41*, 93–113. [CrossRef]
20. El Hadary, A.A.; Yassin, H.H.; Mekhemer, S.T.; Holmes, J.C.; Grootveld, M. Evaluation of the Effect of Ozonated Plant Oils on the Quality of Osseointegration of Dental Implants Under the Influence of Cyclosporin A: An In Vivo Study. *J. Oral Implant.* **2011**, *37*, 247–257. [CrossRef] [PubMed]
21. Kumar, T.; Arora, N.; Puri, G.; Aravinda, K.; Dixit, A.; Jatti, D. Efficacy of ozonized olive oil in the management of oral lesions and conditions: A clinical trial. *Contemp. Clin. Dent.* **2016**, *7*, 51. [CrossRef]
22. Pietrocola, G.; Ceci, M.; Preda, F.; Poggio, C.; Colombo, M. Evaluation of the antibacterial activity of a new ozonized olive oil against oral and periodontal pathogens. *J. Clin. Exp. Dent.* **2018**, *10*, e1103–e1108. [CrossRef]
23. Ugazio, E.; Tullio, V.; Binello, A.; Tagliapietra, S.; Dosio, F. Ozonated Oils as Antimicrobial Systems in Topical Applications. Their Characterization, Current Applications, and Advances in Improved Delivery Techniques. *Molecules* **2020**, *25*, 334. [CrossRef]
24. Sechi, L.; Lezcano, I.; Nunez, N.; Espim, M.; Dupre, I.; Pinna, A.; Molicotti, P.; Fadda, G.; Zanetti, S. Antibacterial activity of ozonized sunflower oil (Oleozon). *J. Appl. Microbiol.* **2001**, *90*, 279–284. [CrossRef]
25. Kazi, M.; Chatzopoulou, P.; Christos, L. Effect of Ozonation on the Essential Oil Composition of Dried Aromatic Plants. In Proceedings of the 8th International Conference on Information and Communication Technologies in Agriculture, Food and Environment, Chania, Greece, 21–24 September 2017.
26. Karlsson, M.F.; Birgersson, G.; Prado, A.M.C.; Bosa, F.; Bengtsson, M.; Witzgall, P. Plant Odor Analysis of Potato: Response of Guatemalan Moth to Above- and Belowground Potato Volatiles. *J. Agric. Food Chem.* **2009**, *57*, 5903–5909. [CrossRef]
27. Sumalan, R.M.; Kuganov, R.; Obistioiu, D.; Popescu, I.; Radulov, I.; Alexa, E.; Negrea, M.; Salimzoda, A.F.; Sumalan, R.L.; Cocan, I. Assessment of Mint, Basil, and Lavender Essential Oil Vapor-Phase in Antifungal Protection and Lemon Fruit Quality. *Molecules* **2020**, *25*, 1831. [CrossRef] [PubMed]
28. Muntean, D.; Licker, M.; Alexa, E.; Popescu, I.; Jianu, C.; Buda, V.; Dehelean, C.A.; Ghiulai, R.; Horhat, F.; Horhat, D.; et al. Evaluation of essential oil obtained from *Mentha* × *piperita* L. against multidrug-resistant strains. *Infect. Drug Resist.* **2019**, *12*, 2905–2914. [CrossRef] [PubMed]
29. Yu, X.; Liang, C.; Fang, H.; Qi, X.; Li, W.; Shang, Q. Variation of trichome morphology and essential oil composition of seven Mentha species. *Biochem. Syst. Ecol.* **2018**, *79*, 30–36. [CrossRef]
30. Desam, N.R.; Al-Rajab, A.J.; Sharma, M.; Mylabathula, M.M.; Gowkanapalli, R.R.; Albratty, M. Chemical constituents, in vitro antibacterial and antifungal activity of *Mentha* × *Piperita* L. (peppermint) essential oils. *J. King Saud Univ.-Sci.* **2019**, *31*, 528–533. [CrossRef]
31. Tomescu, A.; Sumalan, R.-M.; Pop, G.; Alexa, E.; Poiana, M.-A.; Copolovici, D.-M.; Sorina, C.; Stroe, M.; Negrea, M.; Galuscan, A. Chemical Composition and Protective Antifugal Activity of *Mentha piperita* L. and *Salvia Officinalis* L. Essential Oils Against *Fusarium Graminearum* Spp. *Rev. Chim.* **2015**, *66*, 1027–1030.
32. Ebani, V.V.; Nardoni, S.; Bertelloni, F.; Tosi, G.; Massi, P.; Pistelli, L.; Mancianti, F. In Vitro Antimicrobial Activity of Essential Oils against Salmonella enterica Serotypes Enteritidis and Typhimurium Strains Isolated from Poultry. *Molecules* **2019**, *24*, 900. [CrossRef]
33. Lin, S.; Wang, Y.; Wu, K.; Yu, G.; Liu, C.; Su, C.; Yi, F. Study on the Effect of *Mentha* × *piperita* L. Essential Oil on Electroencephalography upon Stimulation with Different Visual Effects. *Molecules* **2022**, *27*, 4059. [CrossRef] [PubMed]
34. Ebani, V.V.; Najar, B.; Bertelloni, F.; Pistelli, L.; Mancianti, F.; Nardoni, S. Chemical Composition and In Vitro Antimicrobial Efficacy of Sixteen Essential Oils against *Escherichia coli* and *Aspergillus fumigatus* Isolated from Poultry. *Veter.-Sci.* **2018**, *5*, 62. [CrossRef] [PubMed]

35. Kalemba, D.; Synowiec, A. Agrobiological Interactions of Essential Oils of Two Menthol Mints: *Mentha piperita* and *Mentha arvensis*. *Molecules* **2019**, *25*, 59. [CrossRef]
36. Giuliani, G.; Ricevuti, G.; Galoforo, A.; Franzini, M. Microbiological aspects of ozone: Bactericidal activity and antibiotic/antimicrobial resistance in bacterial strains treated with ozone. *Ozone Ther.* **2018**, *3*. [CrossRef]
37. Srinivasan, S.R.; Amaechi, B.T. Ozone: A paradigm shift in dental therapy. *J. Glob. Oral Health* **2019**, *2*, 68–77. [CrossRef]
38. Prebeg, D.; Katunarić, M.; Budimir, A.; Pavelić, B.; Šegović, S.; Anić, I. Antimicrobial Effect of Ozone Made by KP Syringe of High-Frequency Ozone Generator. *Acta Stomatol. Croat.* **2016**, *50*, 134–142. [CrossRef]
39. Wilczyńska-Borawska, M.; Leszczyńska, K.; Nowosielski, C.; Stokowska, W. Ozone in Dentistry: Microbiological Effects of Gas Action Depending on the Method and the Time of Application Using the Ozonytron Device. Experimental Study. *Ann. Acad. Med. Stetin.* **2011**, *57*, 99–103.
40. Baysan, A.; Whiley, R.; Lynch, E. Antimicrobial Effect of a Novel Ozone–Generating Device on Micro–Organisms Associated with Primary Root Carious Lesions in vitro. *Caries Res.* **2000**, *34*, 498–501. [CrossRef]
41. Eick, S.; Tigan, M.; Sculean, A. Effect of ozone on periodontopathogenic species—An in vitro study. *Clin. Oral Investig.* **2012**, *16*, 537–544. [CrossRef]
42. Polydorou, O.; Halili, A.; Wittmer, A.; Pelz, K.; Hahn, P. The antibacterial effect of gas ozone after 2 months of in vitro evaluation. *Clin. Oral Investig.* **2012**, *16*, 545–550. [CrossRef]
43. Johansson, E.; Claesson, R.; van Dijken, J. Antibacterial effect of ozone on cariogenic bacterial species. *J. Dent.* **2009**, *37*, 449–453. [CrossRef]
44. Huth, K.C.; Quirling, M.; Maier, S.; Kamereck, K.; AlKhayer, M.; Paschos, E.; Welsch, U.; Miethke, T.; Brand, K.; Hickel, R. Effectiveness of ozone against endodontopathogenic microorganisms in a root canal biofilm model. *Int. Endod. J.* **2009**, *42*, 3–13. [CrossRef]
45. Hauser-Gerspach, I.; Pfäffli-Savtchenko, V.; Dähnhardt, J.E.; Meyer, J.; Lussi, A. Comparison of the immediate effects of gaseous ozone and chlorhexidine gel on bacteria in cavitated carious lesions in children in vivo. *Clin. Oral Investig.* **2009**, *13*, 287–291. [CrossRef]
46. Makeeva, I.M.; Turkina, A.Y.; Margaryan, E.G.; Paramonov, Y.O.; Polyakova, M.A. Assessment of antibacterial efficacy of ozone therapy in treatment of caries at the white spot stage. *Stomatologiya* **2017**, *96*, 7–10. [CrossRef]
47. Song, M.; Zeng, Q.; Xiang, Y.; Gao, L.; Huang, J.; Huang, J.; Wu, K.; Lu, J. The antibacterial effect of topical ozone on the treatment of MRSA skin infection. *Mol. Med. Rep.* **2017**, *17*, 2449–2455. [CrossRef] [PubMed]
48. Gupta, M. Ozone: An Emerging Prospect in Dentistry. *Indian J. Dent. Sci.* **2012**, *4*, 47–50.
49. Montevecchi, M.; Dorigo, A.; Cricca, M.; Checchi, L. Comparison of the Antibacterial Activity of an Ozonated Oil with Chlorhexidine Digluconate and Povidone-Iodine. A Disk Diffusion Test. *N. Microbiol.* **2013**, *36*, 289–302.
50. Epelle, E.I.; Emmerson, A.; Nekrasova, M.; Macfarlane, A.; Cusack, M.; Burns, A.; Mackay, W.; Yaseen, M. Microbial Inactivation: Gaseous or Aqueous Ozonation? *Ind. Eng. Chem. Res.* **2022**, *61*, 9600–9610. [CrossRef] [PubMed]
51. Schneider, H.G. Keine Resistenzbildung Der Mikroflora, Keine Allergischen Reaktionen. *Zahnarzt Woche* **2004**, *19*, 24.
52. Grandi, G.; Cavallo, R.; Zanotto, E.; Cipriani, R.; Panico, C.; Protti, R.; Scapagnini, G.; Davinelli, S.; Costagliola, C. In vitro antimicrobial activity of ozonated oil in liposome eyedrop against multidrug-resistant bacteria. *Open Med.* **2022**, *17*, 1057–1063. [CrossRef]
53. Saravani, S.; Ghaffari, M.; Valizadeh, M.; Ali-Malayeri, F.; Biabangard, A. Antimicrobial Activity of *Mentha piperita*, *Rosmarinus officinalis*, and *Withania somnifera* Prepared by Ultrasound against *Escherichia coli* Isolated from Poultry Stool. *Gene Cell Tissue* **2021**, *9*, e109104. [CrossRef]
54. Heydari, M.; Zanfardino, A.; Taleei, A.; Bushehri, A.A.S.; Hadian, J.; Maresca, V.; Sorbo, S.; Di Napoli, M.; Varcamonti, M.; Basile, A.; et al. Effect of Heat Stress on Yield, Monoterpene Content and Antibacterial Activity of Essential Oils of *Mentha* × *piperita* var. Mitcham and *Mentha arvensis* var piperascens. *Molecules* **2018**, *23*, 1903. [CrossRef]
55. Camele, I.; Gruľová, D.; Elshafie, H. Chemical Composition and Antimicrobial Properties of *Mentha* × *piperita* cv. 'Kristinka' Essential Oil. *Plants* **2021**, *10*, 1567. [CrossRef]
56. Trombetta, D.; Castelli, F.; Sarpietro, M.G.; Venuti, V.; Cristani, M.; Daniele, C.; Saija, A.; Mazzanti, G.; Bisignano, G. Mechanisms of Antibacterial Action of Three Monoterpenes. *Antimicrob. Agents Chemother.* **2005**, *49*, 2474–2478. [CrossRef]
57. Sikkema, J.; de Bont, J.A.; Poolman, B. Mechanisms of membrane toxicity of hydrocarbons. *Microbiol. Rev.* **1995**, *59*, 201–222. [CrossRef] [PubMed]
58. Hulea, A.; Obiștioiu, D.; Cocan, I.; Alexa, E.; Negrea, M.; Neacșu, A.-G.; Hulea, C.; Pascu, C.; Costinar, L.; Iancu, I.; et al. Diversity of Monofloral Honey Based on the Antimicrobial and Antioxidant Potential. *Antibiotics* **2022**, *11*, 595. [CrossRef] [PubMed]
59. Obistioiu, D.; Cocan, I.; Tîrziu, E.; Herman, V.; Negrea, M.; Cucerzan, A.; Neacsu, A.-G.; Cozma, A.; Nichita, I.; Hulea, A.; et al. Phytochemical Profile and Microbiological Activity of Some Plants Belonging to the Fabaceae Family. *Antibiotics* **2021**, *10*, 662. [CrossRef]
60. *CLSI Standard M07*; Methods for Dilution Antimicrobial Susceptibility Tests for Bacteria That Grow Aerobically, 11th ed. Clinical and Laboratory Standards Institute: Wayne, PA, USA, 2018.

61. Noites, R.; Pina-Vaz, C.; Rocha, R.; Carvalho, M.F.; Gonçalves, A.; Pina-Vaz, I. Synergistic Antimicrobial Action of Chlorhexidine and Ozone in Endodontic Treatment. *BioMed. Res. Int.* **2014**, *2014*, 592423. [CrossRef] [PubMed]
62. Baysan, A.; Lynch, E. Effect of ozone on the oral microbiota and clinical severity of primary root caries. *Am. J. Dent.* **2004**, *17*, 56–60.

Disclaimer/Publisher's Note: The statements, opinions and data contained in all publications are solely those of the individual author(s) and contributor(s) and not of MDPI and/or the editor(s). MDPI and/or the editor(s) disclaim responsibility for any injury to people or property resulting from any ideas, methods, instructions or products referred to in the content.

molecules

Article

Chemical Composition and Antibacterial Activity of Liquid and Volatile Phase of Essential Oils against Planktonic and Biofilm-Forming Cells of *Pseudomonas aeruginosa*

Malwina Brożyna [1,*], Justyna Paleczny [1], Weronika Kozłowska [2], Daria Ciecholewska-Juśko [3], Adam Parfieńczyk [4], Grzegorz Chodaczek [5] and Adam Junka [1,*]

1 Department of Pharmaceutical Microbiology and Parasitology, Medical University of Wroclaw, 50-367 Wroclaw, Poland; justyna.paleczny@student.umw.edu.pl
2 Division of Pharmaceutical Biotechnology, Department of Pharmaceutical Biology and Biotechnology, Medical University of Wroclaw, 50-367 Wroclaw, Poland; weronika.kozlowska@umw.edu.pl
3 Department of Microbiology and Biotechnology, West Pomeranian University of Technology, 70-311 Szczecin, Poland; daria.ciecholewska@zut.edu.pl
4 Faculty of Medicine, Lazarski University, 02-662 Warsaw, Poland; adam.parfienczyk@lazarski.pl
5 Bioimaging Laboratory, Łukasiewicz Research Network—PORT Polish Center for Technology Development, 54-066 Wroclaw, Poland; grzegorz.chodaczek@port.org.pl
* Correspondence: malwina.brozyna@student.umw.edu.pl (M.B.); feliks.junka@gmail.com (A.J.)

Abstract: *Pseudomonas aeruginosa* is an opportunistic pathogen causing life-threatening, hard-to-heal infections associated with the presence of a biofilm. Essential oils (EOs) are promising agents to combat pseudomonal infections because of the alleged antimicrobial activity of their volatile fractions and liquid forms. Therefore, the purpose of this paper was to evaluate the antibacterial efficacy of both volatile and liquid phases of seven EOs (thyme, tea tree, basil, rosemary, eucalyptus, menthol mint, lavender) against *P. aeruginosa* biofilm and planktonic cells with the use of a broad spectrum of analytical in vitro methods. According to the study results, the antibacterial activity of EOs in their liquid forms varied from that of the volatile fractions. Overall, liquid and volatile forms of rosemary EO and tea tree EO displayed significant antibiofilm effectiveness. The outcomes indicate that these particular EOs possess the potential to be used in the therapy of *P. aeruginosa* infections.

Keywords: biofilm; *Pseudomonas aeruginosa*; essential oil; EOs in liquid form; volatile fractions; antimicrobial activity

Citation: Brożyna, M.; Paleczny, J.; Kozłowska, W.; Ciecholewska-Juśko, D.; Parfieńczyk, A.; Chodaczek, G.; Junka, A. Chemical Composition and Antibacterial Activity of Liquid and Volatile Phase of Essential Oils against Planktonic and Biofilm-Forming Cells of *Pseudomonas aeruginosa*. *Molecules* **2022**, *27*, 4096. https://doi.org/10.3390/molecules27134096

Academic Editor: Daniela Rigano

Received: 28 April 2022
Accepted: 22 June 2022
Published: 25 June 2022

Publisher's Note: MDPI stays neutral with regard to jurisdictional claims in published maps and institutional affiliations.

Copyright: © 2022 by the authors. Licensee MDPI, Basel, Switzerland. This article is an open access article distributed under the terms and conditions of the Creative Commons Attribution (CC BY) license (https://creativecommons.org/licenses/by/4.0/).

1. Introduction

Pseudomonas aeruginosa is a Gram-negative, opportunistic bacterium responsible for a growing number of serious nosocomial infections. As an example, such different populations as patients suffering from chronic wounds, cystic fibrosis, catheter-related infections, or AIDS are at risk of developing a severe pseudomonal infection, accounting for high mortality and morbidity rates [1]. The production of surface factors, flagella, pili, lipopolysaccharide, toxin secretion and biofilm formation are considered primary virulence determinants of *P. aeruginosa* [2]. Biofilms are complex microbial communities in which cells display characteristic spatial localization and phenotypic and biochemical features differentiating them from their free-swimming (planktonic) counterparts. Cells in the biofilm are encased in extracellular polymeric substances (EPS), which serve as a scaffold for the structural integrity of the microbial community and as a barrier protecting cells from detrimental factors. The biofilm may form on living tissue and indwelling medical devices, including catheters, tracheal tubes, implants, etc. Cells in the biofilm are highly tolerant to antimicrobials (antibiotics, especially) and are resistant to host immune defense mechanisms [3]. Therefore, therapeutic options for the treatment of pseudomonal, biofilm-based infections are limited. Thus, numerous attempts have been made to devise novel strategies for combating *Pseudomonas* infections [2]. Among them, the application of various kinds of phytochemical molecules has been reported to be a promising direction with regard to *P. aeruginosa* biofilm eradication and overcoming this bacteria's resistance to antibiotics [4].

One of the types of such phytochemical molecules is referred to as essential oils (EOs). These mixtures of plants' secondary metabolites belong mainly to groups of terpenes, sesquiterpenes, and phenylpropanoids. Their broad spectrum of biological activity covers antibacterial, antifungal, anti-inflammatory, antiviral, antioxidant, and anticancer properties. Overall, EO activity depends on the content of the constituents, which is, to a major extent, impacted by such variables as plant origin, cultivation conditions, and extraction techniques [5]. EOs are characterized by high lipophilicity and volatility. Due to the lipophilic nature, they exhibit an unspecific mode of antimicrobial action, which relies on binding to the cell wall or the membrane and damaging its integrity. Concerning this broad mechanism of action and already proven effectiveness against multidrug-resistant microorganisms/biofilms, EOs are suggested to be the alternative approach to combat bacterial infections [6]. However, the topical administration of EOs may cause allergic reactions and skin irritation; thus, undiluted EOs cannot be applied directly to body parts altered by the infection process [7]. From the manufacturing perspective, EOs' lipophilic properties also restrain the development of physically stable, non-harmful skin formulations [8]. Therefore, volatile fractions of EOs are more and more considered to be suitable for the treatment of specific bacterial infections (e.g., of the respiratory tract, skin, or wounds) [9].

The already performed research demonstrated that the volatile forms of EOs may display a stronger antimicrobial effect than their liquid phases used in direct contact [10]. It is suggested that hydrophobic molecules in the aqueous phase associate and form micelles, which hinder the attachment of EOs to microorganisms. The EOs' volatile fractions are devoided of this disadvantage [11]. In addition, because EOs' volatile fractions may be administered locally (not via systemic dosage), the risk of side effects and interactions with other drugs is exiguous [8]. However, the majority of research is still mainly focused on EOs' antimicrobial activity through direct contact with the oil in its liquid form. Only a limited number of studies on antimicrobial effectiveness of the volatile forms of EOs are available [12]. Therefore, the purpose of this paper was to evaluate the antibacterial potential of both volatile and liquid phases of seven commercially available EOs against *P. aeruginosa* biofilm and planktonic cells with the use of a broad spectrum of analytical in vitro methods.

2. Results

2.1. GC–MS (Gas Chromatography–Mass Spectrometry) Analysis of the Tested EOs Composition

The percentage composition of EOs' constituents was assessed using gas chromatography–mass spectrometry. The two main ingredients of thyme oil (T-EO) are thymol and p-cymene. Terpinen-4-ol and γ-terpinene predominate in tea tree oil (TT-EO). Basil oil (B-EO) is composed of methyl chavicol and linalool only. Camphor and 1,8-cineole are the main components of rosemary oil (R-EO), while 1,8-cineole and γ-terpinene are primarily presented in eucalyptus oil (E-EO). Menthol mint oil (MM-EO) contains mostly menthol and menthone, and linalyl acetate and linalool are the main ingredients of lavender oil (L-EO). The compositions of EOs were compared to the Polish Pharmacopoeia XI standards. TT-EO was the only one whose components (and their percentage) were directly within pharmacopeial ranges. B-EO is not included in the Pharmacopoeia. Table 1 presents a list of the EOs' detected compounds. Chromatograms of the tested EOs are presented in Figure S1 in the Supplementary Materials.

Table 1. The content (% ± standard deviation) of compounds in essential oils. T-EO, thyme oil; TT-EO, tea tree oil; B-EO, basil oil; R-EO, rosemary oil; E-EO, eucalyptus oil; MM-EO, menthol mint oil; L-EO, lavender oil. Dashes (-) indicate the compounds not presented in the specific EO. The components in line with Polish Pharmacopoeia XI standards are marked in green color.

Retention Time (min)	Compound	Mean Concentration (%) ± SD						
		T-EO	TT-EO	B-EO	R-EO	E-EO	MM-EO	L-EO
6.37	α-Thujene	-	1.11 ± 0.02	-	-	-	-	-
6.56	α-Pinene	2.20 ± 0.09	2.85 ± 0.06	-	2.58 ± 0.28	2.30 ± 0.04	3.01 ± 0.17	0.94 ± 0.04
6.98	Camphene	0.73 ± 0.04	-	-	-	-	-	-
7.8	Sabinene	0.64 ± 0.03	0.75 ± 0.02	-	-	-	-	-
8.22	Cyclofenchene	-	-	-	4.99 ± 0.48	-	2.68 ± 0.17	3.66 ± 0.35
8.25	β-Pinene	0.81 ± 0.05	0.69 ± 0.01	-	-	1.09 ± 0.02	-	-
8.65	α-Phellandrene	-	0.61 ± 0.01	-	0.23 ± 0.07	2.02 ± 0.02	-	-
8.69	2-Bornene	-	-	-	3.00 ± 0.29	-	-	-
9.06	α-Terpinene	-	11.07 ± 0.17	-	0.80 ± 0.04	-	-	-
9.31	p-Cymene	26.91 ± 0.99	4.69 ± 0.07	-	-	6.89 ± 0.07	-	-
9.45	Limonene	0.77 ± 0.04	2.08 ± 0.05	-	14.26 ± 0.99	-	3.78 ± 0.27	-
9.52	1.8-Cineole	-	3.34 ± 0.06	-	30.12 ± 1.74	79.10 ± 0.61	-	2.56 ± 0.39
9.57	β -Thujene	-	-	-	2.33 ± 0.11	-	-	-
10.22	Myrcene	-	-	-	0.38 ± 0.14	-	-	-
10.22	Myrcene	-	-	-	-	-	-	0.86 ± 0.13
10.47	γ-Terpinene	8.60 ± 0.03	19.07 ± 0.27	-	8.21 ± 0.35	8.16 ± 0.07	-	-
10.83	3-Carene	-	-	-	-	-	-	1.19 ± 0.12
11.35	o-Cymene	-	-	-	3.15 ± 1.45	-	-	1.19 ± 0.12
11.44	α-Terpinolene	-	4.34 ± 0.06	-	-	-	-	-
11.85	Linalool	3.45 ± 0.15	-	10.69 ± 1.13	-	-	-	37.76 ± 1.18
13.36	Camphor	0.66 ± 0.06	-	-	21.97 ± 0.77	-	-	0.84 ± 0.10
14.58	Terpinen-4-ol	-	33.27 ± 0.79	-	-	-	-	3.03 ± 0.46
14.99	α-Terpineol	7.84 ± 0.30	3.26 ± 0.13	-	1.56 ± 0.39	-	-	-
15.98	Menthone	-	-	-	-	-	24.53 ± 0.23	-
16.08	Isoborneol	-	-	-	1.53 ± 0.06	-	-	-
16.36	Isomenthone	-	-	-	-	-	13.54 ± 1.75	-
16.40	Borneol	-	-	-	2.69 ± 0.09	-	-	2.04 ± 0.56
16.66	Menthol	-	-	-	-	-	45.57 ± 2.21	-
17.59	Methyl chavicol	-	-	89.31 ± 1.13	-	-	-	-
18.49	Thymol	44.00 ± 0.46	-	-	-	-	-	-
19.60	Linalyl acetate	-	-	-	-	-	-	41.13 ± 0.40
20.61	Bornyl acetate	-	-	-	1.17 ± 0.09	-	-	-
20.81	Lavandulyl acetate	-	-	-	-	-	-	1.80 ± 0.22
20.89	Menthyl acetate	-	-	-	-	-	5.61 ± 0.36	-
22.36	β-Caryophyllene	1.00 ± 0.05	0.53 ± 0.01	-	-	-	-	-
22.83	Aromadendrene	-	1.83 ± 0.03	-	-	-	-	-
23.32	Alloaromadendrene	-	0.82 ± 0.02	-	-	-	-	-
24.03	Viridiflorene	-	2.35 ± 0.04	-	-	-	-	-
24.55	β-Cadinene	-	2.78 ± 0.03	-	-	-	-	-
24.80	Caryophyllene	-	-	-	0.85 ± 0.21	-	1.27 ± 0.57	3.47 ± 0.66

2.2. *Biofilm Biomass and Activity Level Assessment*

In the next line of investigation, the level of *P. aeruginosa* biofilm biomass and its cells' metabolic activity were assessed. All tested strains were metabolically active and able to form biofilm under applied in vitro conditions; however, differences in both biofilm biomass and metabolic activity were observed between specific strains (Figure 1).

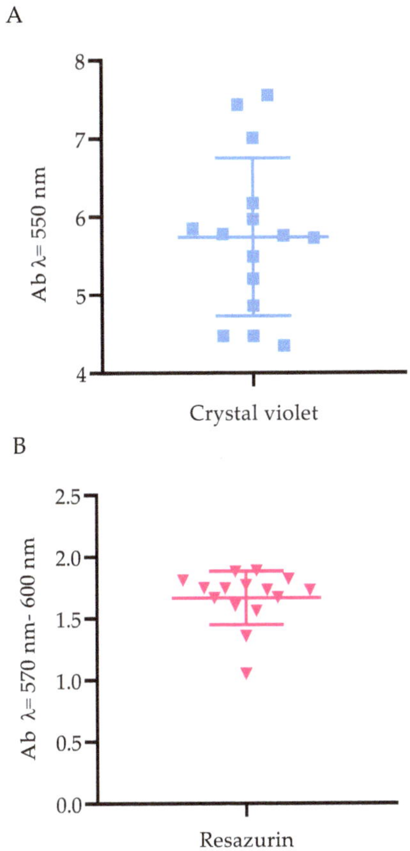

Figure 1. The ability of *Pseudomonas aeruginosa* clinical (PA 1–7, PA 13–19) and the reference (ATCC 15442) strains to form biofilm. (**A**) Biofilm biomass level assessed with the crystal violet method. (**B**) Metabolic activity of biofilm-forming cells, determined with resazurin staining. Ab, absorbance. The average and standard deviations are marked.

2.3. Antimicrobial Activity of EOs in Their Liquid Forms against P. aeruginosa Planktonic Forms

Two standard techniques were performed to evaluate the antimicrobial effectiveness of EOs in their liquid forms toward planktonic cells: disc diffusion, in which non-emulsified EOs were applied, and minimal inhibitory concentration (MIC) estimation, in which EOs emulsions were used. The results from the first methodology mentioned are presented in Table 2. R-EO exhibited the highest antimicrobial activity, and zones of growth inhibition (mm) were obtained for all strains. T-EO and E-EO displayed moderate antibacterial efficacy against most of the strains. L-EO was almost inactive against *P. aeruginosa* strains. Because Tween 20 was used as an emulsifying agent in the MIC assay, its antibacterial potential against planktonic forms of *P. aeruginosa* ATCC 15442 strain was also evaluated. The aforementioned substance did not influence *P. aeruginosa* cell growth in concentrations from 1.0% to 0.002% (*v*/*v*), as it is presented in Figure S2 in the Supplementary Materials. R-EO emulsion was the most active among the tested EOs against planktonic pseudomonal cells. MIC values of B-EO ranged from 12.5% to 1.6% (*v*/*v*). In other EOs, MIC values were detected mostly in concentrations of 12.5–25.0% (*v*/*v*). Moreover, particular strains were not sensitive to MM-EO and T-EO even if a 25.0% (*v*/*v*) concentration of these oils was applied. The detailed list of the determined MIC (%) values is presented in Table 3.

Table 2. Mean diameters of inhibition zones (mm ± standard deviation) after treatment of *P. aeruginosa* clinical (PA 1–7, PA 13–19) and the reference (ATCC 15442) strains with non-emulsified EOs in their liquid forms assessed with the disc diffusion method. T-EO, thyme oil; TT-EO, tea tree oil; B-EO, basil oil; R-EO, rosemary oil; E-EO, eucalyptus oil; MM-EO, menthol mint oil; L-EO, lavender oil. According to their susceptibility to a particular oil, the strains were divided into two groups of seven or eight samples per group. In the case of B-EO, MM-EO, and L-EO, the zone equal to 0 mm was the parameter for the low-susceptibility group and higher than 0 mm for the high-susceptibility one. The groups are marked as follows: red, low susceptibility among the tested strains; green, high susceptibility among the tested strains.

Mean Zones of Growth Inhibition (mm ± SD) after Treatment with Non-Emulsified EOs in Their Liquid Forms							
Strain	**T-EO**	**TT-EO**	**B-EO**	**R-EO**	**E-EO**	**MM-EO**	**L-EO**
PA 1	9 ± 1	8 ± 1	10 ± 2	18 ± 5	18 ± 1	8 ± 7	7 ± 1
PA 2	13 ± 3	0 ± 0	12 ± 1	17 ± 5	18 ± 2	0 ± 0	0 ± 0
PA 3	3 ± 5	13 ± 1	0 ± 0	15 ± 4	11 ± 1	0 ± 0	0 ± 0
PA 4	9 ± 1	0 ± 0	0 ± 0	29 ± 2	0 ± 0	0 ± 0	0 ± 0
PA 5	0 ± 0	0 ± 0	0 ± 0	20 ± 2	8 ± 2	6 ± 6	0 ± 0
PA 6	5 ± 5	6 ± 5	0 ± 0	20 ± 1	9 ± 1	7 ± 6	0 ± 0
PA 7	8 ± 7	0 ± 0	0 ± 0	19 ± 1	0 ± 0	11 ± 2	0 ± 0
PA 13	13 ± 4	4 ± 8	10 ± 1	15 ± 4	14 ± 1	0 ± 0	0 ± 0
PA 14	7 ± 1	0 ± 0	0 ± 0	15 ± 1	0 ± 0	0 ± 0	0 ± 0
PA 15	8 ± 0	0 ± 0	8 ± 1	14 ± 1	14 ± 1	0 ± 0	0 ± 0
PA 16	3 ± 6	9 ± 1	0 ± 0	19 ± 0	0 ± 0	0 ± 0	0 ± 0
PA 17	11 ± 1	9 ± 1	0 ± 0	19 ± 1	10 ± 8	0 ± 0	0 ± 0
PA 18	10 ± 1	14 ± 1	0 ± 0	15 ± 0	0 ± 0	4 ± 7	0 ± 0
PA 19	0 ± 0	0 ± 0	0 ± 0	13 ± 2	11 ± 0	0 ± 0	0 ± 0
ATCC 15442	7 ± 0	5 ± 4	0 ± 0	12 ± 0	0 ± 0	0 ± 0	0 ± 0

Table 3. Minimal inhibitory concentration (%) (v/v) of emulsified EOs in their liquid forms against *P. aeruginosa* clinical (PA 1–7, PA 13–19) and the reference (ATCC 15442) strains assessed with the microdilution method. T-EO, thyme oil; TT-EO, tea tree oil; B-EO, basil oil; R-EO, rosemary oil; E-EO, eucalyptus oil; MM-EO, menthol mint oil; L-EO, lavender oil. R symbols indicate EOs where the minimal inhibitory concentration values were not reached in the highest concentration (25.0% (v/v)) of applied emulsions.

Minimal Inhibitory Concentration (%) of Emulsified EOs in Their Liquid Forms							
Strain	**T-EO**	**TT-EO**	**B-EO**	**R-EO**	**E-EO**	**MM-EO**	**L-EO**
PA 1	6.3	12.5	6.3	0.4	25.0	25.0	25.0
PA 2	25.0	25.0	12.5	0.8	25.0	25.0	25.0
PA 3	25.0	12.5	3.1	0.4	25.0	25.0	25.0
PA 4	25.0	12.5	6.3	0.4	25.0	25.0	25.0
PA 5	12.5	12.5	3.1	0.4	25.0	25.0	25.0
PA 6	25.0	25.0	6.3	0.8	25.0	25.0	25.0
PA 7	0.2	12.5	1.6	6.3	25.0	R	25.0
PA 13	25.0	12.5	1.6	0.8	25.0	R	25.0
PA 14	25.0	25.0	3.1	0.4	25.0	25.0	25.0
PA 15	25.0	25.0	3.1	0.8	25.0	25.0	25.0
PA 16	R	25.0	12.5	0.8	25.0	25.0	25.0
PA 17	R	25.0	12.5	0.8	25.0	25.0	25.0
PA 18	25.0	25.0	3.1	0.8	25.0	12.5	25.0
PA 19	12.5	25.0	3.1	0.4	12.5	6.3	25.0
ATCC 15442	R	25.0	12.5	0.4	25.0	R	25.0

2.4. *Antimicrobial Activity of EOs in Their Liquid Forms against P. aeruginosa Biofilms*

The antibiofilm activity of EOs in their liquid forms was determined using the standard microdilution method and the modified A.D.A.M. (antibiofilm dressing's activity measurement) technique. To evaluate the MBEC values (minimal biofilm eradication concentration) of the emulsified EOs in their liquid forms, resazurin (7-hydroxy-3H-phenoxazin-3-one-10-oxide sodium salt) staining was applied as an indicator of the presence of metabolically active cells. Due to the observed inconsistency of the results (data not shown) obtained for strains treated with T-EO and TT-EO, and dyed with resazurin, TTC (2,3,5-triphenyl-tetrazolium chloride) solution was used instead (for these above-mentioned EOs). As the MBEC values were not found within the applied concentration spectrum, changes in the biofilm-forming cells viability (%) after exposure to selected EO concentrations are presented in the study. The significant changes in the biofilm-forming cells viability were assessed after the treatment of pseudomonal biofilm with T-EO and TT-EO emulsions in concentrations of 25.0%–6.3% (v/v). B-EO and R-EO emulsions, both in concentrations of 25.0%–12.5% (v/v) and L-EO emulsion in a concentration of 25.0% (v/v) exhibited potent antibiofilm effectiveness. The percentage changes in the pseudomonal biofilm-forming cells viability after the application of 25.0% (v/v), 12.5% (v/v), and 6.3% (v/v) oils emulsions are presented in Figure 2 and Figures S3 and S4 in the Supplementary Materials, respectively.

Finally, the antibiofilm effect exerted by volatile phases of non-emulsified EOs was determined with the AntiBioVol (antibiofilm activity of volatile compounds) methodology. As can be seen in Table 4, PA 6, PA 17–19 and the ATCC 15442 strains were the most susceptible among all tested strains to volatile EOs. Reductions of biofilm cells viability ranged from 5.40% to 52.99% were obtained for the strains after the exposure to MM-EO. Volatile fractions of 96% (v/v) ethanol reduced 100.00% of pseudomonal biofilm.

Table 4. Changes in biofilm-forming cells viability (%) of *P. aeruginosa* clinical (PA 1–7, PA 13–19) and the reference (ATCC 15442) strains after treatment with volatile non-emulsified EOs assessed with the AntiBioVol (antibiofilm activity of volatile compounds) method. T-EO, thyme oil; TT-EO, tea tree oil; B-EO, basil oil; R-EO, rosemary oil; E-EO, eucalyptus oil; MM-EO, menthol mint oil; L-EO, lavender oil. The strains were grouped by their susceptibility to the particular oil. The groups are marked as follows: red, lowest susceptibility; purple, moderate susceptibility; green, highest susceptibility among the tested strains. The negative values indicate an increase in biofilm-forming cells viability after their treatment with EOs in comparison to the growth control (untreated cells).

Changes in the Biofilm-Forming Cells Viability (%) after Treatment with Volatile Fractions of Non-Emulsified EOs							
Strain	**T-EO**	**TT-EO**	**B-EO**	**R-EO**	**E-EO**	**MM-EO**	**L-EO**
PA 1	18.13	11.08	13.23	2.65	13.21	26.86	4.49
PA 2	29.21	24.08	6.18	−0.11	22.29	40.82	2.53
PA 3	10.29	19.17	10.73	−17.52	3.75	25.85	−0.78
PA 4	12.94	14.66	−1.71	−28.46	6.75	33.56	−6.11
PA 5	1.71	50.34	15.55	45.10	18.26	9.98	34.20
PA 6	9.22	53.81	−12.83	37.88	19.24	52.99	32.67
PA 7	3.51	20.04	−83.01	12.43	−96.64	5.40	−81.17
PA 13	8.11	−1.63	22.75	−4.50	3.69	17.62	−2.68
PA 14	12.30	−3.61	−17.21	−5.96	−8.91	19.66	−4.45
PA 15	15.10	−12.49	45.13	−10.29	−1.21	37.22	−3.83
PA 16	18.57	0.17	−3.85	−3.40	0.01	9.49	1.27
PA 17	6.38	22.34	20,84	14.73	13.74	11.22	42.97
PA 18	9.12	27.69	33.45	16.28	12.29	15.00	33.00
PA 19	3.58	27.20	7.97	18.32	11.94	19.00	40.84
ATCC 15442	13.08	53.11	22.80	43.74	−7.53	46.76	−1.71

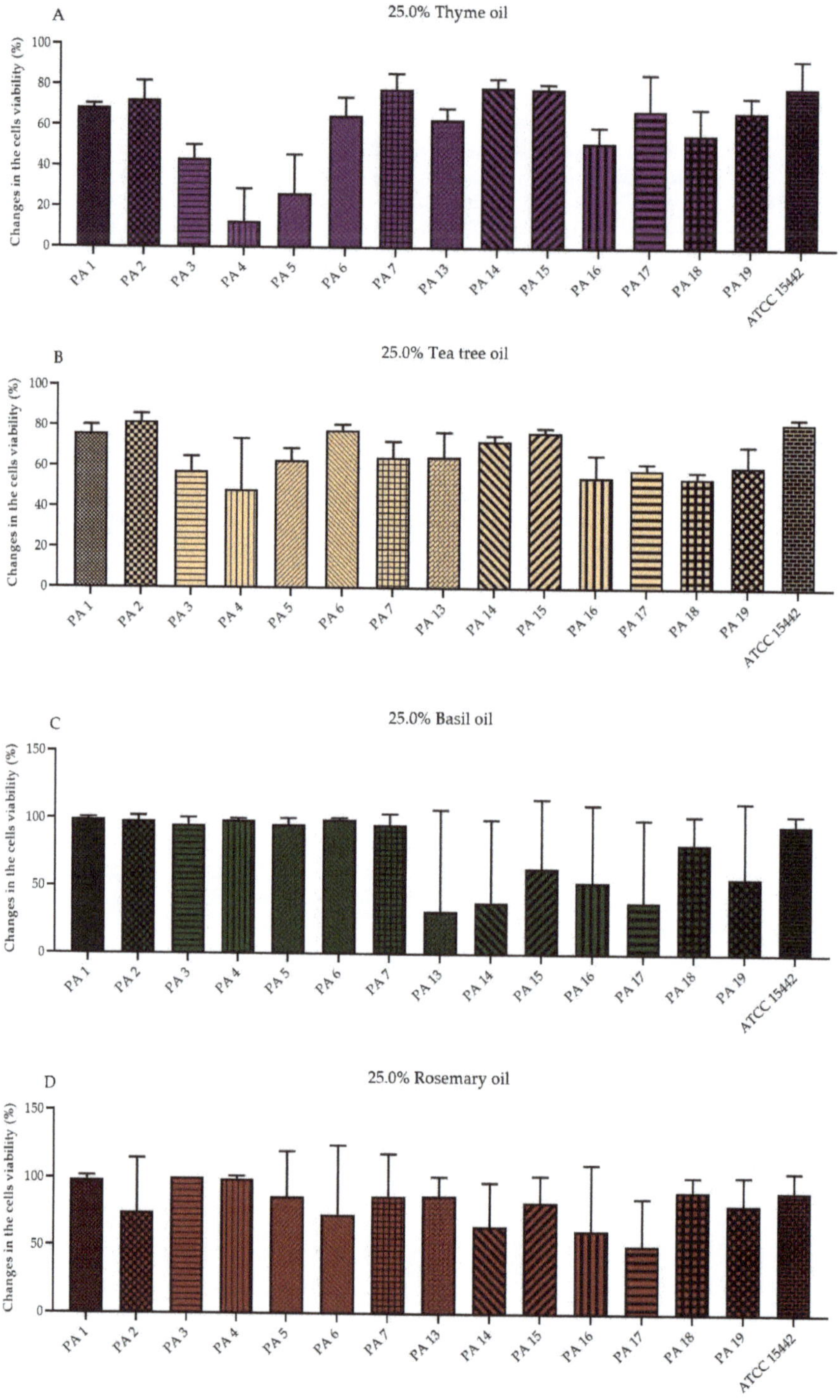

Figure 2. *Cont.*

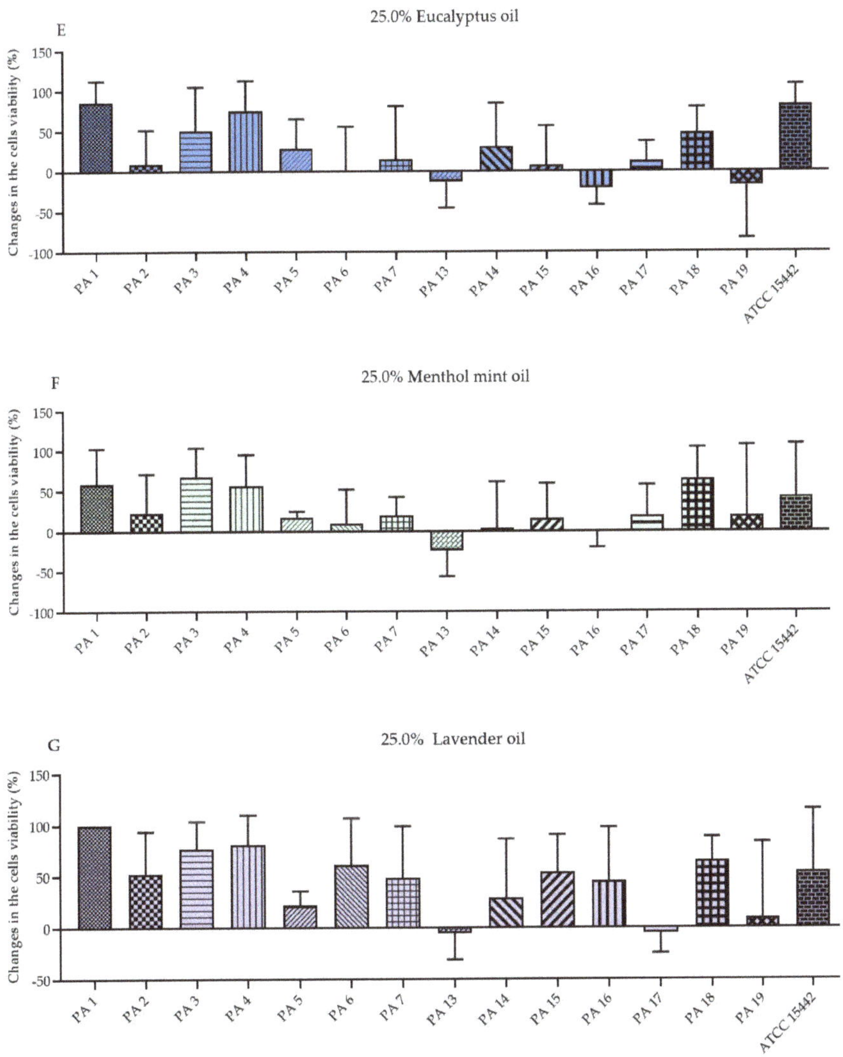

Figure 2. Changes in the biofilm-forming cells viability (%) of *P. aeruginosa* clinical (PA 1–7, PA 13–19) and the reference (ATCC 15442) strains after treatment with emulsified essential oils in their liquid forms in the concentration of 25.0% (*v*/*v*). Results of microdilution methodology with (**A**,**B**) TTC and (**C**–**G**) resazurin staining. Standard deviations are marked. The negative values indicate an increase in biofilm-forming cells viability after their treatment with EOs in comparison to the growth control (untreated cells).

To perform the modified A.D.A.M. technique, biocellulose discs were soaked with the non-emulsified EOs. The concentration of EOs released from the biocellulose discs was approximately 65.8%. The biofilms of PA 1–7 and PA 14–16 strains were the most prone to the activity of the EOs in their liquid forms (Table 5). The antibiofilm effectiveness of individual EOs was strain-dependent, although regarding all EOs except for B-EO, the reduction of cells viability equal to 60.11% or more was reported for selected strains. The mentioned B-EO displayed the lowest antibiofilm activity, and 96% (*v*/*v*) ethanol was applied against the reference strain to prove the method's usability. The mean obtained reduction of the biofilm cells viability after the treatment with ethanol was 90.33%.

Table 5. Changes in the biofilm-forming cells viability (%) of *P. aeruginosa* clinical (PA 1–7, PA 13–19) and the reference (ATCC 15442) strains after treatment with non-emulsified essential oils in their liquid forms assessed with the A.D.A.M. (antibiofilm dressing's activity measurement) method. T-EO, thyme oil; TT-EO, tea tree oil; B-EO, basil oil; R-EO, rosemary oil; E-EO, eucalyptus oil; MM-EO, menthol mint oil; L-EO, lavender oil. The strains were grouped by their susceptibility to the particular oil. The groups are marked as follows: red, lowest susceptibility; purple, moderate susceptibility; green, the highest susceptibility among the tested strains. The negative values indicate an increase in biofilm-forming cells viability after their treatment with EOs in comparison to the growth control (untreated cells).

	Changes in the Biofilm-Forming Cells Viability (%) after Treatment with Non-Emulsified EOs in Their Liquid Forms						
Strain	**T-EO**	**TT-EO**	**B-EO**	**R-EO**	**E-EO**	**MM-EO**	**L-EO**
PA 1	47.23	50.57	3.49	69.55	33.14	26.11	43.93
PA 2	52.79	54.53	32.19	52.64	38.57	35.23	38.61
PA 3	63.13	37.74	22.23	87.56	64.39	79.78	69.11
PA 4	38.37	58.28	10.05	81.75	52.14	27.31	61.08
PA 5	69.45	50.36	37.78	70.46	19.31	53.54	54.64
PA 6	60.11	60.23	25.06	37.05	29.49	72.11	60.74
PA 7	36.34	44.24	20.36	38.93	20.23	64.88	40.90
PA 13	51.79	44.36	−34.87	55.01	−16.98	−13.07	20.92
PA 14	49.02	81.27	16.06	31.39	31.71	25.78	27.83
PA 15	63.80	85.39	25.56	36.98	23.81	44.93	31.14
PA 16	77.03	80.88	37.35	53.77	21.97	−12.46	50.29
PA 17	13.07	44.30	36.54	31.32	44.17	43.41	51.57
PA 18	−49.39	−39.67	−138.15	−152.19	−168.89	−110.24	−127.17
PA 19	56.21	83.26	4.13	17.92	1.78	46.16	34.75
ATCC 15442	41.11	47.06	12.50	14.56	11.33	49.87	57.36

2.5. *Antimicrobial Activity of EOs' Volatile Fractions against P. aeruginosa*

The influence of non-emulsified EOs' volatile phases on pseudomonal planktonic cells was assessed with the inverted Petri dish method. All of the examined strains (PA 1-7, PA 13–19, ATCC 15442) were resistant to the volatile of L-EO and B-EO. Almost all of the tested strains were resistant to T-EO (PA 1-6, PA 13-19, ATCC 15442), TT-EO (PA 1-7, PA 13–17, PA 19, ATCC 15442), MM-EO (PA 1-4, PA 6-7, PA 13-19, ATCC 15442). Mean zones of growth inhibition obtained for R-EO ranged from 0 to 23 mm. Table 6 presents the mean diameters of inhibition zones (mm) measured after the treatment of planktonic cells with the volatile fractions of EOs.

Table 6. Mean diameters of inhibition zones (mm ± standard deviation) after treatment of planktonic forms of *P. aeruginosa* clinical (PA 1–7, PA 13–19) and the reference (ATCC 15442) strains with volatile fractions of non-emulsified EOs assessed with the inverted Petri dish method. T-EO, thyme oil; TT-EO, tea tree oil; B-EO, basil oil; R-EO, rosemary oil; E-EO, eucalyptus oil; MM-EO, menthol mint oil; L-EO, lavender oil. According to their susceptibility to R-EO, the strains were divided into two groups for seven or eight samples per group. For the rest of the EOs, the zone equal to 0 mm was the parameter for the low susceptibility group and higher than 0 mm for the high susceptibility one. The groups are marked as follows: red, low susceptibility among the tested strains; green, high susceptibility among the tested strains.

	Mean Zones of Growth Inhibition (mm ± SD) after Treatment with Volatile Fractions of Non-Emulsified EOs						
Strain	**T-EO**	**TT-EO**	**B-EO**	**R-EO**	**E-EO**	**MM-EO**	**L-EO**
PA 1	0 ± 0	0 ± 0	0 ± 0	18 ± 15	22 ± 2	0 ± 0	0 ± 0
PA 2	0 ± 0	0 ± 0	0 ± 0	13 ± 13	16 ± 2	0 ± 0	0 ± 0
PA 3	0 ± 0	0 ± 0	0 ± 0	13 ± 6	16 ± 2	0 ± 0	0 ± 0
PA 4	0 ± 0	0 ± 0	0 ± 0	23 ± 2	0 ± 0	0 ± 0	0 ± 0
PA 5	0 ± 0	0 ± 0	0 ± 0	19 ± 3	0 ± 0	3 ± 6	0 ± 0
PA 6	0 ± 0	0 ± 0	0 ± 0	15 ± 2	0 ± 0	0 ± 0	0 ± 0
PA 7	3 ± 5	0 ± 0	0 ± 0	15 ± 3	0 ± 0	0 ± 0	0 ± 0
PA 13	0 ± 0	0 ± 0	0 ± 0	3 ± 6	3 ± 5	0 ± 0	0 ± 0
PA 14	0 ± 0	0 ± 0	0 ± 0	0 ± 0	0 ± 0	0 ± 0	0 ± 0
PA 15	0 ± 0	0 ± 0	0 ± 0	5 ± 4	0 ± 0	0 ± 0	0 ± 0
PA 16	0 ± 0	0 ± 0	0 ± 0	12 ± 4	0 ± 0	0 ± 0	0 ± 0
PA 17	0 ± 0	0 ± 0	0 ± 0	16 ± 5	0 ± 0	0 ± 0	0 ± 0
PA 18	0 ± 0	7 ± 6	0 ± 0	14 ± 6	0 ± 0	0 ± 0	0 ± 0
PA 19	0 ± 0	0 ± 0	0 ± 0	3 ± 5	0 ± 0	0 ± 0	0 ± 0
ATCC 15442	0 ± 0	0 ± 0	0 ± 0	0 ± 0	0 ± 0	0 ± 0	0 ± 0

2.6. Microscopic Visualization of Biofilm

The R-EO's high activity against *P. aeruginosa* biofilms was additionally confirmed by fluorescence microscopy (Figure 3). While the high amount of live bacterial, biofilm-forming cells (dyed green) was observed in the untreated control setting (Figure 3A,C), the exposure of biofilm to the liquid (Figure 3B) or the volatile (Figure 3D) R-EO resulted in the high increase in dead/damaged biofilm-forming cells (dyed red/orange).

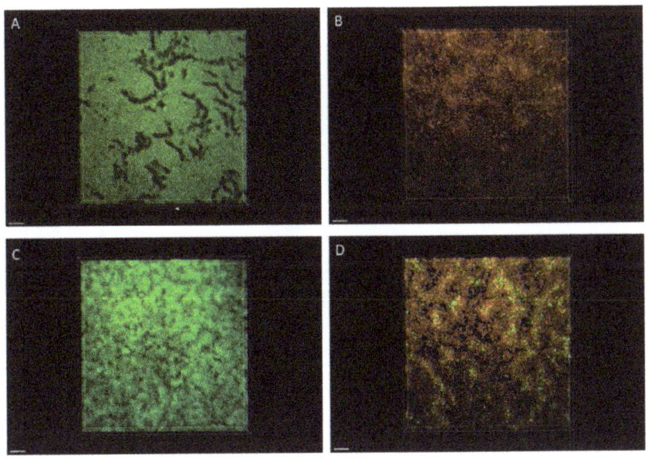

Figure 3. Impact of R-EO on *P. aeruginosa* ATCC 15442 biofilm. (**A**,**B**) Pseudomonal biofilm untreated and treated with R-EO in its liquid form, assessed with the modified A.D.A.M. (antibiofilm dressing's activity measurement) method. (**C**,**D**) Pseudomonal biofilm untreated and treated with R-EO volatiles, assessed with the AntiBioVol (antibiofilm activity of volatile compounds) assay. The red/orange color shows pseudomonal cells altered/damaged as the result of exposure to R-EO, while green-colored cells are non-altered, viable cells. Moreover, the darker (less green) picture shows that fewer live cells are captured in this particular field of vision.

2.7. Statistical Analysis

Statistical analysis was performed to evaluate statistically significant differences between EOs' antimicrobial activity. A summary of significance levels for each method is presented in Table 7.

Table 7. Significance levels of differences in changes in pseudomonal biofilm cells viability after treatment with EOs in their liquid forms and volatile fractions obtained with three methods. The differences were statistically significant for $p < 0.05$ and are referred to as $p < 0.03$ (*), $p < 0.006$ (**), $p < 0.00003$ (***); ns refers to difference being statistically insignificant. T-EO, thyme oil; TT-EO, tea tree oil; B-EO, basil oil; R-EO, rosemary oil; E-EO, eucalyptus oil; MM-EO, menthol mint oil; L-EO, lavender oil.

Comparison of the Changes in the Biofilm-Forming Cells Viability after Treatment with Emulsified EOs in Their Liquid Forms in the Concentration of 25.0% (*v*/*v*)							
	T-EO	**TT-EO**	**B-EO**	**R-EO**	**E-EO**	**MM-EO**	**L-EO**
T-EO	-	ns	***	***	ns	ns	ns
TT-EO	ns	-	**	***	*	**	ns
B-EO	***	**	-	ns	***	***	***
R-EO	***	***	ns	-	***	***	***
E-EO	ns	*	***	***	-	ns	ns
MM-EO	ns	**	***	***	ns	-	ns
L-EO	ns	ns	***	***	ns	ns	-

Table 7. *Cont.*

Comparison of the Changes in the Biofilm-Forming Cells Viability after Treatment with Emulsified EOs in Their Liquid Forms in the Concentration of 12.5 (v/v)							
	T-EO	**TT-EO**	**B-EO**	**R-EO**	**E-EO**	**MM-EO**	**L-EO**
T-EO	-	ns	ns	ns	***	***	***
TT-EO	ns	-	ns	ns	***	***	***
B-EO	ns	ns	-	ns	***	***	***
R-EO	ns	ns	ns	-	***	***	**
E-EO	***	***	***	***	-	ns	ns
MM-EO	***	***	***	***	ns	-	ns
L-EO	***	***	***	**	ns	ns	-
Comparison of the Changes in the Biofilm-Forming Cells Viability after Treatment with Emulsified EOs in Their Liquid Forms in the Concentration of 6.3 (v/v)							
	T-EO	**TT-EO**	**B-EO**	**R-EO**	**E-EO**	**MM-EO**	**L-EO**
T-EO	-	ns	***	***	***	***	***
TT-EO	ns	-	***	**	***	***	***
B-EO	***	***	-	ns	ns	ns	ns
R-EO	***	**	ns	-	***	ns	*
E-EO	***	***	ns	***	-	ns	ns
MM-EO	***	***	ns	ns	ns	-	ns
L-EO	***	***	ns	*	ns	ns	-
Comparison of the Changes in the Biofilm-Forming Cells Viability after Treatment with Non-Emulsified EOs in Their Liquid Forms							
	T-EO	**TT-EO**	**B-EO**	**R-EO**	**E-EO**	**MM-EO**	**L-EO**
T-EO	-	ns	***	ns	**	ns	ns
TT-EO	ns	-	***	ns	***	*	ns
B-EO	***	***	-	***	ns	**	***
R-EO	ns	ns	***	-	**	ns	ns
E-EO	**	***	ns	**	-	ns	*
MM-EO	ns	*	**	ns	ns	-	ns
L-EO	ns	ns	***	ns	*	ns	-
Comparison of the Changes in the Biofilm-Forming Cells Viability after Treatment with Volatile Fractions of Non-Emulsified EOs							
	T-EO	**TT-EO**	**B-EO**	**R-EO**	**E-EO**	**MM-EO**	**L-EO**
T-EO	-	ns	ns	ns	ns	**	ns
TT-EO	ns	-	ns	**	**	ns	**
B-EO	ns	ns	-	ns	ns	**	ns
R-EO	ns	**	ns	-	ns	***	ns
E-EO	ns	**	ns	ns	-	***	ns
MM-EO	**	ns	**	***	***	-	***
L-EO	ns	**	ns	ns	ns	***	-

3. Discussion

EOs are volatile plant derivatives of global medical interest due to their high antimicrobial activity. However, the number of studies focused specifically on the antimicrobial/antibiofilm potential of EOs' volatile forms is still limited. EO antimicrobial activity depends on a plethora of factors: EO state of matter (volatile or liquid one), chemical composition, hydrophilicity/hydrophobicity, species/strain of microorganism they act against, but also on the type of methodology applied to analyze the aforementioned potential [13]. Therefore, the aim of this paper was to investigate and compare the anti-pseudomonal efficacy of seven EOs in their liquid forms and their volatile fraction with the use of a diversified spectrum of techniques in order to obtain cohesive data. First, the content of the EOs' constituents was determined to confirm the presence of molecules recognized as those exhibiting antimicrobial potential (Table 1, Figure S1 in the Supplementary Materials). Next, the pseudomonal strains' ability to form biofilm was assessed (Figure 1). Having preliminary tests performed, the evaluation of the EOs in their liquid forms' antimicrobial activity against *P. aeruginosa* planktonic cells was made with the use of two different techniques. Their results indicated that R-EO was the most effective one among the tested EOs (Tables 2 and 3). In addition, it is suggested that the strong antibacterial potential of R-EO is associated with the activity of its main component, 1,8-cineole [14–16]. However, 1,8-cineole is also the major compound of E-EO (Table 1); this specific oil exhibited significantly lower anti-pseudomonal efficacy than R-EO. Other research teams indicated that the antimicrobial effectiveness of crude R-EO and E-EO was stronger than 1,8-cineole applied as a self-reliant antimicrobial agent [17,18]. Therefore, the synergistic (or at least additive) effect of remaining R-EO constituents may also account for the overall oil's activity against *P. aeruginosa*. It is hypothesized that the smaller the droplets of EO emulsions, the higher the antimicrobial effect that occurs [19]. As we have shown earlier, the droplet diameters of the E-EO emulsion were 2201 nm, while R-EO was 783 nm [20]. The minimal inhibitory concentration (MIC) values (Table 3) of L-EO were 25% (v/v). MIC values of E-EO were: 12.5% (v/v) for PA 19 strain and 25% (v/v) for the rest of the strains, whereas the zones of growth inhibition being the result of antimicrobial activity of L-EO vs. E-EO were 0-7 and 0-18 mm, respectively (Table 2). This interesting observation requires additional experiments to be elucidated because a broad spectrum of possible variables may contribute to the discrepancy between outcomes measured by two testing methods. As an example, the differences in water solubility of the main EO ingredients also have an impact on diffusion through agar. Because agar is mainly composed of water, the higher the aqueous solubility of the compounds, the better their diffusion across the agar. The aqueous solubility of 1,8-cineole, which predominates in E-EO, is 2.63 mg/mL at 293 K, while the solubility of linalool (the main component of L-EO) is 1.34 mg/mL [21]. Linalool diffuses more poorly through the agar than 1,8-cineole, thus, the growth inhibition zones of L-EO are less than E-EO.

To evaluate the antibiofilm efficacy of EOs in their liquid forms, the microdilution method and the modified A.D.A.M. (antibiofilm dressing's activity measurement) assay were conducted. In the microdilution assay, significant biofilm-forming cells viability reduction was demonstrated for the emulsions of T-EO, TT-EO, B-EO, R-EO and L-EO in their liquid forms at a concentration of 25% (v/v). In the modified A.D.A.M. methodology, levels of biofilm-forming cells viability reduction were diversified and dependent on the EO applied and on the specific strain exposed to the EO's activity. Nevertheless, by means of both methods, R-EO, T-EO and TT-EO in their liquid forms can be pointed out as the most potent against the pseudomonal biofilm, which is in line with the results of other research teams. It was reported that above 90% reduction of *P. aeruginosa* biofilm was obtained after the incubation with T-EO and that 1,8-cineole, the main component of R-EO, greatly affected *P. aeruginosa* biofilm formation and disrupted the mature biofilm [18,22–24]. The volatile fractions of R-EO and TT-EO also displayed high effectiveness against *P. aeruginosa* biofilm, which may result from their multiple mechanisms of anti-biofilm activity.

In the case of volatile fraction assays, not only does the volatility of the components determine EO activity but also the number and concentration of particular molecules adhered to the agar surface where biofilm forms. In the standard inverted Petri dish method, the EOs are applied to a small stretch (6 mm in diameter paper disc). Therefore, the obtained zones of growth inhibition also depend on the volatiles spreading. The volume of the EOs used in the inverted Petri dish method is lower, and the tightness of the setting is poorer than in the AntiBioVol (antibiofilm activity of volatile compounds) technique; thus, the loss of volatiles is higher.

As mentioned above, volatile fractions of EOs exhibited higher antimicrobial activity than those EOs in their liquid forms in some of the research reports [25]. However, opposite results are reported in the present paper (Tables 4 and 6). In turn, outcomes of other studies stay in line with the data provided in our work and indicate that EOs in their liquid forms are more active against bacteria than in volatiles, due to the direct contact of molecules with the microorganism [9]. It was also suggested that the antimicrobial activity of EOs' volatiles is related to the volatility of the EOs' compounds and their adsorption into the agar surface, which is associated with its hydrophilic character [26,27]. In the previous paper, we demonstrated that the adsorption of EO compounds to the agar changes with the time of exposure [28]. Moreover, the maximum concentration of molecules adsorbed into the agar surface was approximately 40% [28]. Therefore, the working concentration of the active substances is much lower than in the liquid tests. The obtained differences in both fractions' activity may result from various times of biofilm incubation with the applied indicators of cell metabolic activity (resazurin and tetrazolium chloride). According to the AntiBioVol results, TT-EO, R-EO, MM-EO, and L-EO were the most potent ones among the analyzed EOs. The speed of EOs evaporation is related to their vapor pressure; the higher the vapor pressure, the faster they evaporate. The vapor pressures of EOs main compounds at 25 °C are: thymol, 2.2 Pa; terpinen-4-ol, 6.4 Pa; methyl chavicol, 22.0 Pa; 1,8-cineole, 253.0 Pa; menthol, 19.0 Pa; linalyl acetate, 61.0 Pa. In our study, we found no correlation between the above-mentioned parameter and the level of antibacterial activity of EOs. Such a phenomenon may be explained by the fact that besides vapor pressure, EO's particular antimicrobial mechanism of action and affinity to agar surface should be considered to analyze the total level of antimicrobial activity. Additionally, strain-dependent tolerance to EOs was observed when AntiBioVol was applied, similar to what was observed when the modified A.D.A.M. method was used. The EOs examined in the current study were also scrutinized in our previous line of investigation toward a Gram-positive bacterium, *S. aureus* [20]. Comparing results from the recent and the present work, it has to be stated that both fractions of all EOs (except for R-EO) displayed higher antimicrobial activity against *S. aureus* than *P. aeruginosa*. It may be related to the fact that, because of the hydrophobic nature, EOs react with lipids of the cell membrane and lead to a leak of intracellular substances and to damage of the cell. Gram-negative bacteria hydrophilic cell walls hinder the penetration of lipophilic EOs, resulting in their higher tolerance to EOs than in the case of Gram-positive pathogens. Furthermore, Gram-negative *P. aeruginosa* is a ubiquitous, opportunistic microorganism, forming a robust biofilm on solid surfaces at the water–air interface [29–31]. It thrives there by developing a vast spectrum of adaptive resistance to various antimicrobial agents, probably including those secreted by plants. [32]. Contrary to other EOs, MIC values of R-EO in its liquid emulsified form were equal for *S. aureus* and *P. aeruginosa*. (Table 3) [20]. An explanation of this phenomenon may be the hypothesis that the charge of the bacterial cell surfaces and the cell shape play a role in their tolerance to EOs as well [33–35]. Hajlaouli et al. demonstrated a more evident reduction of the bacterial cells' negative charge for the Gram-negative versus Gram-positive bacteria after the treatment with EOs [34]. Rod-shaped bacterial cells were also more susceptible to EOs than the coccoid ones [35]. In the present study, EOs in their liquid forms and their volatile fractions were investigated because the antimicrobial effect of EOs may alter among different in vitro conditions. R-EO exhibited the highest antibacterial effectiveness against *P. aeruginosa* of all tested EOs. The reported data confirm the high potency of EOs against *P. aeruginosa* biofilms and planktonic cells. Therefore, we are convinced that the data presented in this paper, showing EOs as a promising alternative to the current (performed by means of antibiotics and antiseptics, mostly) anti-pseudomonal therapies, will finally pave the way for novel solutions and approaches aiming to significantly reduce the risk associated with pseudomonal biofilm-related infections.

4. Materials and Methods

4.1. Bacterial Strains and Culture Conditions

For research purposes, one reference strain of *Pseudomonas aeruginosa* 15442 from the American Type and Culture Collection (ATCC) and fourteen clinical strains (later referred to as PA 1–7 and PA 13–19) of this bacterial species were applied. The strains were part of the Strain and Line Collection of Pharmaceutical Microbiology and Parasitology Department of the Medical University of Wroclaw. The clinical strains were obtained in the year 2016 during the internal Wroclaw Medical University SUB. D198.16.001 project: "The insight into biofilm-related properties of clinical microorganisms and possibilities of their eradication". All patients provided written consent to participate in the trial and allowed the material obtained during the study (exudate, biopsy specimen, microorganisms) to be used for scientific purposes. The study was approved by the Bioethical Committee of Wroclaw Medical University, protocol # 8/2016.

In each of the performed experiments, the microorganisms' overnight cultures in the TSB medium (Tryptic Soy Broth, Biomaxima, Lublin, Poland) were prepared, and 0.5 McFarland suspensions were established afterward in a 0.9% (*w/v*) solution of sodium chloride (NaCl, Stanlab, Lublin, Poland) using a densitometer (Densilameter II Erba Lachema, Brno, the Czech Republic).

4.2. Essential Oils

Seven commercially available essential oils in their liquid forms and their volatile fractions were tested in the research. The applied EOs are listed in Table 8.

Table 8. List of the essential oils analyzed in the paper.

Common Name of EO	Plant Origin	Part of the Plant	Abbreviation	Manufacturer, City, Country
thyme oil	*Thymus vulgaris* L.	herb	T-EO	Etja, Elblag, Poland
tea tree oil	*Melaleuca alternifolia* Cheel.	leaves	TT-EO	Pharmatech, Zukowo, Poland
basil oil	*Ocimum basilicum* L.	flowers	B-EO	Nanga, Zlotow, Poland
rosemary oil	*Rosmarinus officinalis* L.	flowering shoots	R-EO	Nanga, Zlotow, Poland
eucalyptus oil	*Eucalyptus globulus* Labill.	leaves and twigs	E-EO	Pharmatech, Zukowo, Poland
lavender oil	*Lavandula angustifolia* Mill.	flowering herb	L-EO	Kej, Cirkowice, Poland
menthol mint oil	*Mentha arvensis* L.	leaves	MM-EO	Optima Natura, Grodki, Poland

Due to the volatility of EOs, individual essential oils and control settings were examined on separate plates. Moreover, to prevent EO evaporation, in the experiments where their volatile phases were investigated, plates were sealed with parafilm.

4.3. GC–MS (Gas Chromatography–Mass Spectrometry) Analysis of the Tested EOs Composition

Essential oils (EOs) were diluted with hexane (JTB, GB), vortexed, and immediately analyzed. All analyses were performed in triplicate.Analysis was carried out using the system Agilent 7890B GC coupled with 7000GC/TQ system connected to PAL RSI85 autosampler (Agilent Technologies, Palo Alto, CA, USA).

The applied column was HP-5 MS; 30 m × 0.25 mm × 0.25 µm (J&W, Agilent Technologies, Palo Alto, CA, USA). Helium was used as a carrier gas at a total flow of 1 mL/min. Chromatographic conditions were applied as follows: split injection in a ratio 100:1, the injector was set on 250 °C, oven temperature program was: 50 °C held for 1 min, then 4 °C/min up to 130 °C, 10 °C/min to 280 °C, and then isothermal for 2 min. The MS detector operated in the electronic impact ionization mode at 70 eV; transfer line, source, and quadrupole temperatures were set at 320, 230, and 150 °C, respectively. Masses were registered in a range from 30 to 400 m/z. Peaks identification was performed using MassHunter Workstation Software Version B.08.00 coupled with the NIST17 mass spectra library and accomplished by comparison with linear retention indexes. The relative abundance of each EO constituent was expressed as percentage content based on the peak area normalization. Regarding the obtained outcomes, only TT-EO met the requirement of pharmacopeial standards. Although, the analysis was not performed in accordance with the normalization procedure from Polish Pharmacopea XI (different column and different temperature program).

4.4. Biofilm Biomass Level Assay

The crystal violet assay was performed in 96-well plates (Wuxi Nest Biotechnology, Wuxi, China) to evaluate the total biofilm mass. Briefly, 0.5 McFarland bacterial suspensions, prepared as described in Materials and Methods Section 4.1., were diluted 1000 times in TSB (Tryptic Soy Broth, Biomaxima, Lublin, Poland) medium, and 100 µL was poured into the wells of the plates. The plates were then incubated under static conditions for 24 h at 37 °C. Subsequently, the medium above the biofilm was gently pipetted out, and the plates were dried for 10 min at 37 °C. Then, 100 µL of 20% (*v*/*v*) aqueous crystal violet solution (Aqua-med, Lodz, Poland) was added to each well for 10 min at room temperature. Biofilm cells were rinsed twice with 100 µL of 0.9% (*w/v*) solution of sodium chloride (NaCl, Stanlab, Lublin, Poland) to remove unbounded cells and excess stain. The plates were transferred to the incubator (37 °C) for 10 min. Next, 100 µL of 30% (*v*/*v*) water solution of acetic acid (Chempur, Piekary Slaskie, Poland) was introduced to the wells, and the plates were shaken for 30 min at room temperature at 450 rpm (Mini-shaker PSU-2T, Biosan SIA, Riga, Latvia). The solution was transferred to fresh 96-well plates, and the absorbance was measured at 550 nm using a spectrophotometer (Multiskan Go, Thermo Fisher Scientific, Vantaa, Finland). One independent experiment was performed with six technical replicates.

4.5. Biofilm Activity Level Assessment

The presented protocol of biofilm culturing in Materials and Methods Section 4.4. was applied. After the biofilm formation, 10 µL of 0.1% (*w/v*) resazurin sodium salt (7-Hydroxy-3H-phenoxazin-3-one-10-oxide sodium salt, Acros Organics, Geel, Belgium) solution in TSB (Tryptic Soy Broth, Biomaxima, Lublin, Poland) was added to the wells, and the plates were incubated at 37 °C for two hours. The solution was transferred to fresh 96-well plates (Wuxi Nest Biotechnology, Wuxi, China), and its absorbance was measured at 570 and 600 nm using a spectrophotometer (Multiskan Go, Thermo Fisher Scientific, Vantaa, Finland). To assess the cells viability, the absorbance value at 600 nm was subtracted from the value at 570 nm. One independent experiment was carried out with six technical repetitions.

4.6. Methods for the Assessment of the Activity of EOs in Their Liquid Forms

4.6.1. The Disc Diffusion Method

In the experiment, 90 mm diameter, 14.2 mm height Petri dishes (Noex, Komorniki, Poland) with 5 mm thick Mueller–Hinton agar layers (Biomaxima, Lublin, Poland) were used. Standard paper discs (diameter of 6 mm, 0.5 mm thickness) were introduced to the wells of 48-well plates (Thermo Fisher Scientific, Waltham, MA, USA), and 0.2 mL of each non-emulsified EOs or saline (control of bacterial growth) (NaCl, Stanlab, Lublin, Poland) was added. The plates were wrapped with tape and kept at 4 °C for 30 min to soak the discs. In the second step of the tests, the Petri dish plates were inoculated with the bacterial suspensions at density 0.5 McFarland prepared as described in Materials and Methods Section 4.1. Next, the soaked paper discs were placed onto the agar layer for assessing the antimicrobial activity of pure EOs in their liquid forms. The dishes were incubated for 24 h at 37 °C, and bacterial growth inhibition zones were measured in mm with a ruler. If no total inhibition was obtained, zones of partial growth inhibition were assessed (in mm). When unequal zones were observed, a shorter diameter was included. One independent experiment with three repetitions was performed, and the mean diameters were calculated.

4.6.2. Assessment of the Minimal Inhibitory Concentrations of EOs Emulsions

To determine the minimal inhibitory concentration (MIC) values of the EOs in their liquid forms, their emulsions in Tween 20 (Zielony Klub, Kielce, Poland) were prepared. In the first step, each EO was combined with the solution of 1.0% (*v*/*v*) Tween 20 in TSB medium (Tryptic Soy Broth, Biomaxima, Lublin, Poland) in ratio 1:1 and mixed with a vortex (Micro-shaker type 326 m, Premed, Marki, Poland) for 30 min at room temperature. Following, geometric dilutions were prepared in TSB and shaken for 30 s. Subsequently, 0.5 MacFarland bacterial suspensions were made according to the description in Materials and Methods Section 4.1. and diluted 1000 times in the TSB medium. Next, 100 μL of the suspensions was added to the wells of 96-well plates (Jet Bio-Filtration Co. Ltd., Guangzhou, China), and the same volume of diluted emulsions was poured. Therefore, the final concentration of each EO applied in the test ranged from 25.0% (*v*/*v*) to 0.01% (*v*/*v*). The following additional samples were included: control of bacterial growth (bacteria with medium), control of medium sterility (medium only), control of 1.0–0.002% (*v*/*v*) Tween 20 antimicrobial activity and samples of emulsions only. The absorbance of the solution was measured at 580 nm using a spectrophotometer (Multiskan Go, Thermo Fisher Scientific, Vantaa, Finland). During the 24 h incubation at 37 °C, the plates were shaken at 350 rpm (Mini-shaker PSU-2T, Biosan SIA, Riga, Latvia). The absorbance was measured immediately after the incubation at the same wavelength. The MIC value was assessed in the concentration (%) (*v*/*v*) in which the difference between the absorbance after and before the sample's incubation was equal to or lower than zero. Two independent experiments were performed, each with three technical replicates.

4.6.3. Assessment of the Minimal Biofilm Eradication Concentrations of EOs Emulsions

The EOs emulsions and bacterial suspensions applied to assess the Minimal Biofilm Eradication Concentration (MBEC) values were prepared as was elaborated in Materials and Methods Section 4.6.2.. On the first day of the experiment, 100 μL of the aforementioned suspensions was added to the wells of 96-well plates (Jet Bio-Filtration Co. Ltd., Guangzhou, China) containing 100 μL of TSB medium (Tryptic Soy Broth, Biomaxima, Lublin, Poland). The plates were incubated at 37 °C for 24 h under static conditions. Subsequently, the medium was aspirated, and 200 μL of EOs emulsions geometric dilutions were added to the wells. The concentration of each EO applied in the test ranged from 25.0% (*v*/*v*) to 0.01% (*v*/*v*). Control of bacterial growth (bacteria with medium) and medium sterility (medium only) were also prepared. The plates were re-incubated for the next 24 h. Based on the results of the preliminary study (data not shown), the MBEC values of T-EO and TT-EO were determined using TTC (2,3,5-triphenyl-tetrazolium chloride, AppliChem Gmbh, Darmstadt, Germany) as the indicator, while resazurin sodium salt (7-Hydroxy-3H-phenoxazin-3-one-10-oxide sodium salt, Acros Organics, Geel, Belgium) was applied to other EOs. The following steps were executed for the resazurin assay: 40 μL of 0.05% (*w/v*) resazurin solution in TSB medium was added to the biofilm wells treated with EOs emulsions and to the control wells. The plates were incubated for 2 h at 37 °C with continuous shaking at 400 rpm (Mini-shaker PSU-2T, Biosan SIA, Riga, Latvia). The absorbance was measured at 570 and 600 nm using a spectrophotometer (Multiskan Go, Thermo Fisher Scientific, Vantaa, Finland). To calculate the final absorbance of the tested substance, the absorbance value at 600 nm was subtracted from the value at 570 nm. The MBEC values were determined as the lowest concentration of the emulsions where the obtained difference was equal to or lower than zero. For the TTC methodology, 0.2% (*w/v*) TTC solution in TSB was added in the volume of 150 μL to the biofilm after its incubation with EOs emulsions. The plates were incubated for 2 h at 37 °C. Next, the MBEC value was visually determined in the lowest concentration where no red color was observed. Due to the fact that in no case was the MBEC assessed, the three highest concentrations of each EO emulsion for each strain were chosen for further analysis. The contents of the mentioned wells were transferred to 1.5 mL Eppendorf tubes, and 350 μL of 0.1% (*w/v*) solution of saponin (VWR Chemicals, Radnor, PA, USA) was added, and the Eppendorf tubes were vortexed for 1 min. Then, 700 μL of methanol (Stanlab, Lublin, Poland) was added, and the Eppendorf tubes were vortexed for 30 min. Finally, the Eppendorf tubes were centrifuged for 1 min at 3000 rpm, and 100 μL of the supernatant was transferred in three replicates to the wells of the 96-well plates. The same procedure was performed for the control growth wells. The absorbance was measured at 490 nm.

The level of the reduction of biofilm-forming cells viability was calculated according to the formula.

$$\text{Cells viability reduction } (\%) = 100\% - \left(\frac{\text{AbS}}{\text{AbC}} * 100\%\right) \quad (1)$$

AbS, absorbance of the tested substance;
AbC, mean absorbance of growth control.
Two independent experiments were performed, each with three technical replicates.

4.6.4. Evaluation of Antibiofilm Activity of Non-Emulsified EOs Using the Modified A.D.A.M. (Antibiofilm Dressing's Activity Measurement) Assay

The methodology was a modification of the procedure displayed in our previous research [36]. The following steps of the experiment were performed:

Preparation before the Experiment

To treat the biofilm with the EOs, biocellulose discs (BC) were produced. A *Komagataeibacter xylinus* ATCC 53524 (American Type and Culture Collection) strain was used for this purpose. A Herstin–Schramm (H-S) medium composed of 2% (*w/v*) glucose (Chempur, Piekary Slaskie, Poland), 0.5% (*w/v*) yeast extract (VWR, Radnor, PA, USA), 0.5% (*w/v*) bactopeptone (VWR, Radnor, PA, USA), 0.115% (*w/v*) citric acid monohydric (POCH, Gliwice, Poland), 0.27% (*w/v*) Na2HPO4 (POCH, Gliwice, Poland), 0.05% (*w/v*) MgSO4·7H2O (POCH, Gliwice, Poland), and 1% (*v/v*) ethanol (Chempur, Piekary Slaskie, Poland) was used for bacterial culturing. Then, 1 mL of H-S medium was added to the wells of a 24-well plate (Wuxi Nest Biotechnology, Wuxi, China) and inoculated with *K.xylinus*. The plate was incubated for 7 days at 28 °C to obtain 14 mm BC discs. Then, the discs were removed and washed with 0.1 M NaOH (Chempur, Piekary Slaskie, Poland) at 80 °C. Next, BC discs were washed with double-distilled water to neutralize their pH and autoclaved. To evaluate the concentration of substances adsorbed into the BC discs, six of them were weighed, dried at 37 °C and weighed again. The average concentration (%) of the adsorbed liquid was calculated with the formula:

$$\text{Compound concentration } (\%) = [\text{EV}/((\text{WBC} - \text{DBC}) + \text{EV})] * 100\% \quad (2)$$

EV, a volume of essential oil (mL);
WBC, the weight of wet BC disc (g);
DBC, the weight of dry BC disc (g).

First Day of the Experiment

In the first line of the investigation, 1.5 mL of Brain Heart Infusion Broth (BHI, Biomaxima, Lublin, Poland) with 2% (*w/v*) of Bacteriological Lab Agar (Biomaxima, Lublin, Poland) was poured into the wells of a 24-well plate (further referred to as Agar Plate). After the agar solidifies, 8 mm in diameter plugs were cut out of each well using a cork-borer. The plugs were removed and discarded to make 8 mm in diameter tunnels in agar, and the Agar Plate was kept refrigerated for the next day. Simultaneously, the same agar formulation was used to prepare a 6 mm in height agar layer on a Petri dish (Noex, Komorniki, Poland). Then, agar plugs 8 mm in diameter were cut out from the agar Petri dish and placed in a fresh 24-well plate (further referred to as Plugs Plate). The microorganisms' suspensions, prepared according to the description in Materials and Methods Section 4.1., were diluted 1000 times in TSB medium (Tryptic Soy Broth, Biomaxima, Lublin, Poland), introduced to the wells of the Plugs Plate in the volume of 2 mL and transferred to the incubator (37 °C) for biofilm formation. The BC discs, prepared as described above, were placed in a fresh 24-well plate (further referred to as Discs Plate), and 1 mL of undiluted, non-emulsified essential oils or saline (control of bacterial growth), or 96% (*v/v*) ethanol

(Chempur, Piekary Slaskie, Poland) was added, and the plate was sealed with parafilm and placed at 4 °C. Both plates (Plugs Plate 2 and Discs Plate) were incubated for 24 h.

Second Day of the Experiment

Subsequently, biofilm plugs were taken out from the Plugs Plate and placed in the bottom of agar tunnels in the Agar Plate so that the biofilm was on the top of the plugs. Then, 50 µL of TSB medium was added to fill up the space in the tunnels. The BC discs soaked with non-emulsified EOs/saline/ethanol were transferred from the Disc Plate and placed on the top of the agar wells in the Agar Plate. The biofilm was adhered to the plug and had no direct contact with the BC. Substances were gradually released from the BC to the medium. The experimental setting was incubated for 24 h/ 37 °C under static conditions.

Third Day of the Experiment

After the biofilm treatment with the tested substances, the BC discs were removed. The content of each agar tunnel (medium and agar plugs) was transferred to the wells of 48-well plates (Wuxi Nest Biotechnology, Wuxi, China) and 1 mL of 0.002% (*w/v*) resazurin sodium salt (7-Hydroxy-3H-phenoxazin-3-one-10-oxide sodium salt, Acros Organics, Geel, Belgium) in TSB was added. The plates were shaken at 350 rpm (Mini-shaker PSU-2T, Biosan SIA, Riga, Latvia) for 4 h and 15 min at 37 °C. Then, 100 µL of the color solution was transferred to 96-well plates (Jet Bio-Filtration Co. Ltd., Guangzhou, China) in three replications from each 48-well. The absorbance was measured at 570 and 600 nm using a spectrophotometer (Multiskan Go, Thermo Fisher Scientific, Vantaa, Finland). To calculate the final absorbance of the tested substance, the absorbance at 600 nm was subtracted from the value at 570 nm. The formula calculated the reduction of cells viability:

$$\text{Cells viability reduction }(\%) = 100\% - \left(\frac{\text{AbS}}{\text{AbC}} * 100\%\right) \tag{3}$$

AbS, absorbance of the tested substance;

AbC, mean absorbance of growth control.

One independent experiment was performed with six technical replicates. Antibiofilm activity of ethanol was examined only against the *P. aeruginosa* 15442 strain.

4.7. *Methods for the Assessment of EOs' Volatile Fractions Activity*

4.7.1. The Inverted Petri Dish Methodology

The assay was performed similarly to the method presented in Materials and Methods Section 4.6.1. The paper disc soaked with the non-emulsified tested substances was solely placed onto the lid of the Petri dish.

4.7.2. Evaluation of Antibiofilm Activity of Non-Emulsified EOs Using the AntiBioVol (Antibiofilm Activity of Volatile Compounds) Method

The assay was performed based on the protocol presented in our previous study [37].

First Day of the Experiment

In this part of the investigation, the Agar Plate, the Petri dish with the plugs and the Plugs Plate were prepared as mentioned in Materials and Methods Section 4.6.4. The following modifications were made. First, wells of the Agar Plate were filled with BHI (Brain Heart Infusion Broth, Biomaxima, Lublin, Poland) and agar to full. Second, twice as many agar plugs were cut out from the Petri dish, and part of them remained sterile. They were placed on the bottom of the agar tunnels (Agar Plate) and kept refrigerated until the next day.

Second Day of the Experiment

Biofilm plugs were taken out from the Plugs Plate and placed in the agar tunnels of the Agar Plate on the top of the sterile ones. Then, 0.5 mL of undiluted, non-emulsified EOs or saline (control of growth) or ethanol (Chempur, Piekary Slaskie, Poland) was added to the wells of a fresh 24-well plate (later referred to as Substance Plate) (Wuxi Nest Biotechnology, Wuxi, China). The Agar Plate was placed upside down on the Substance Plate, and the agar wells were set directly above the wells with the tested substances. The plugs' diameters were the same as the tunnels in which they were placed, and plates were gently transferred to protect the plugs from dropping into the wells of the Substance Plate. The plates were sealed and incubated for 24 h at 37 °C under static conditions.

Third Day of the Experiment

As the incubation was completed, the plates were separated, and the upper plugs (containing biofilms) were transferred to fresh 48-well plates (Wuxi Nest Biotechnology, Wuxi, China), and 1 mL of 0.002% (*w/v*) resazurin sodium salt (7-Hydroxy-3H-phenoxazin-3-one-10-oxide sodium salt, Acros Organics, Geel, Belgium) in TSB was poured. The incubation was continued for 2 h/ 37 °C with shaking at 350 rpm. Then, 100 µL of the color solution was transferred to 96-well plates (Jet Bio-Filtration Co. Ltd Guangzhou, China)) in three replications from each 48-well. The absorbance was measured at 570 and 600 nm using a spectrophotometer (Multiskan Go, Thermo Fisher Scientific, Vantaa, Finland). The absorbance at 600 nm was subtracted from the value at 570 nm. The reduction of cells viability was calculated by the formula:

$$\text{Cells viability reduction } (\%) = 100\% - (\frac{\text{AbS}}{\text{AbC}} * 100\%) \quad (4)$$

AbS, absorbance of the tested substance;

AbC, mean absorbance of growth control.

One independent experiment was performed with six technical replicates. Antibiofilm activity of ethanol was examined only against the *P. aeruginosa* 15442 strain.

4.8. Microscopic Visualization of Biofilm

The pseudomonal biofilm (ATCC 15442 strain) treated with R-EO or saline by means of the A.D.A.M. (antibiofilm dressing's activity measurement) or the AntiBioVol (antibiofilm activity of volatile compounds) technique (as presented in Materials and Methods Section 4.6.4. or Section 4.7.2, respectively) was immersed in 1 mL of Filmtracer™ LIVE/DEAD™ Biofilm Viability Kit (Invitrogen, Thermo Fisher Scientific, USA) solution and incubated at room temperature for 15 min. After incubation, the solution was removed, and the biofilms were gently rinsed once with sterile water. The biofilms were analyzed using a confocal microscope (Leica, SP8, Wetzlar, Germany) with a 25× water dipping objective, using sequential mode for 488 nm laser line and 500–530 nm emission to detect SYTO-9 and 552 nm laser line and 575–627 nm emission to detect propidium iodide (PI) within microbial cells.

4.9. Statistical Analysis

Statistical analysis was performed using Statistica (Version 13; TIBCO Software Inc., Palo Alto, CA, USA). Normality distribution was assessed with the Shapiro–Wilk test. To compare EOs' efficacy, a non-parametric ANOVA Kruskal–Wallis test and post hoc Dunn's analysis were performed. Results with a significance level $p < 0.05$ were considered significant. The statistical analysis is presented in Table 7.

5. Conclusions

The antimicrobial activity of EOs' volatile fractions varies from that of EOs in their liquid forms. The antimicrobial effectiveness of EOs may depend on the type of methodology applied to analyze the antimicrobial activity. Rosemary and tea tree EOs in their liquid forms and their volatile fractions displayed significant antibiofilm effectiveness against *P. aeruginosa*. Volatile fractions of menthol mint EO exhibited the most potent antibiofilm effect among all tested EOs.

Supplementary Materials: The following supporting information can be downloaded at: https://www.mdpi.com/article/10.3390/molecules27134096/s1, Figure S1: Chromatograms of tested EOs measured with GC–MS (Gas Chromatography–Mass Spectrometry); Figure S2: The antimicrobial activity of different concentrations (%) (*v*/*v*) of Tween 20 against planktonic forms of *P. aeruginosa* ATCC 15442 strains. Ab, absorbance; C+, untreated cells. The absorbance of untreated cells is marked with a red line; Figure S3: Changes in the biofilm-forming cells viability (%) of *P. aeruginosa* clinical (PA 1-7, PA 13-19) and the reference (ATCC 15442) strains after treatment with emulsified essential oils in their liquid forms in the concentration of 12.5% (*v*/*v*). Results of microdilution methodology with (A,B) TTC and (C–G) resazurin staining. Standard deviations are marked. The negative values indicate an increase in biofilm-forming cells viability after their treatment with EOs in comparison to the growth control (untreated cells); Figure S4: Changes in the biofilm-forming cells viability (%) of *P. aeruginosa* clinical (PA 1–7, PA 13–19) and the reference (ATCC 15442) strains after treatment with emulsified essential oils in their liquid forms in the concentration of 6.3% (*v*/*v*). Results of microdilution methodology with (A,B) TTC and (C–G) resazurin staining. Standard deviations are marked. The negative values indicate an increase in biofilm-forming cells viability after their treatment with EOs in comparison to the growth control (untreated cells).

Author Contributions: Conceptualization, M.B. and A.J.; methodology, M.B. and A.J.; software, M.B.; validation, M.B. and A.J.; formal analysis, M.B. and A.J.; investigation, M.B., W.K., J.P., G.C., D.C.-J., A.P., and A.J.; resources, M.B. and A.J.; data curation, M.B.; writing—original draft preparation, M.B.; writing—review and editing, M.B. and A.J.; visualization, M.B.; supervision, A.J.; project administration, M.B. and A.J.; funding acquisition, M.B. and A.J. All authors have read and agreed to the published version of the manuscript.

Funding: This research was funded by National Science Centre in Poland, grant number PREL.D230.22.003.

Institutional Review Board Statement: Not applicable.

Informed Consent Statement: Not applicable.

Data Availability Statement: All necessary data are presented in the manuscript and in the supplementary data and can be provided from the authors upon reasonable request.

Conflicts of Interest: The authors declare no conflict of interest.

Sample Availability: Samples of the essential oils analyzed in this research are available from the authors upon reasonable request.

References

1. Bassetti, M.; Vena, A.; Croxatto, A.; Righi, E.; Guery, B. How to Manage Pseudomonas Aeruginosa Infections. *Drugs Context* **2018**, *7*, 212527. [CrossRef] [PubMed]
2. Harmsen, M.; Yang, L.; Pamp, S.J.; Tolker-Nielsen, T. An Update on Pseudomonas Aeruginosa Biofilm Formation, Tolerance, and Dispersal. *FEMS Immunol. Med. Microbiol.* **2010**, *59*, 253–268. [CrossRef] [PubMed]
3. Vestby, L.K.; Grønseth, T.; Simm, R.; Nesse, L.L. Bacterial Biofilm and Its Role in the Pathogenesis of Disease. *Antibiotics* **2020**, *9*, 59. [CrossRef] [PubMed]
4. Bharadwaj, K.K.; Rabha, B.; Choudhury, B.K.; Rosalin, R.; Sarkar, T.; Baishya, D.; Chanu, N.B.; Singh, Y.D.; Panda, M.K.; Pati, S. Current Strategies in Inhibiting Biofilm Formation for Combating Urinary Tract Infections: Special Focus on Peptides, Nano-Particles and Phytochemicals. *Biocatal. Agric. Biotechnol.* **2021**, *38*, 102209. [CrossRef]
5. Baptista-Silva, S.; Borges, S.; Ramos, O.L.; Pintado, M.; Sarmento, B. The Progress of Essential Oils as Potential Therapeutic Agents: A Review. *J. Essent. Oil Res.* **2020**, *32*, 279–295. [CrossRef]
6. Yu, Z.; Tang, J.; Khare, T.; Kumar, V. The Alarming Antimicrobial Resistance in ESKAPEE Pathogens: Can Essential Oils Come to the Rescue? *Fitoterapia* **2020**, *140*, 104433. [CrossRef]
7. Sarkic, A.; Stappen, I. Essential Oils and Their Single Compounds in Cosmetics-a Critical Review. *Cosmetics* **2018**, *5*, 11. [CrossRef]
8. Das, S.; Horváth, B.; Šafranko, S.; Jokić, S.; Széchenyi, A.; Koszegi, T. Antimicrobial Activity of Chamomile Essential Oil: Effect of Different Formulations. *Molecules* **2019**, *24*, 4321. [CrossRef]
9. Ács, K.; Balázs, V.L.; Kocsis, B.; Bencsik, T.; Böszörményi, A.; Horváth, G. Antibacterial Activity Evaluation of Selected Essential Oils in Liquid and Vapor Phase on Respiratory Tract Pathogens. *BMC Complementary Altern. Med.* **2018**, *18*, 227. [CrossRef]

10. Amat, S.; Baines, D.; Alexander, T.W. A vapour phase assay for evaluating the antimicrobial activities of essential oils against bovine respiratory bacterial pathogens. *Lett. Appl. Microbiol.* **2017**, *65*, 489–495. [CrossRef]
11. Sakkas, H.; Papadopoulou, C. Antimicrobial Activity of Basil, Oregano, and Thyme Essential Oils. *J. Microbiol. Biotechnol.* **2017**, *27*, 429–438. [CrossRef] [PubMed]
12. Abers, M.; Schroeder, S.; Goelz, L.; Sulser, A.; St. Rose, T.; Puchalski, K.; Langland, J. Antimicrobial Activity of the Volatile Substances from Essential Oils. *BMC Complementary Med. Ther.* **2021**, *21*, 124. [CrossRef] [PubMed]
13. Reyes-Jurado, F.; Franco-Vega, A.; Ramírez-Corona, N.; Palou, E.; López-Malo, A. Essential Oils: Antimicrobial Activities, Extraction Methods, and Their Modeling. *Food Eng. Rev.* **2015**, *7*, 275–297. [CrossRef]
14. de Sousa, J.P.; de Azerêdo, G.A.; de Araújo Torres, R.; da Silva Vasconcelos, M.A.; da Conceição, M.L.; de Souza, E.L. Synergies of Carvacrol and 1,8-Cineole to Inhibit Bacteria Associated with Minimally Processed Vegetables. *Int. J. Food Microbiol.* **2012**, *154*, 145–151. [CrossRef]
15. de Azeredo, G.A.; Stamford, T.L.M.; Nunes, P.C.; Gomes Neto, N.J.; de Oliveira, M.E.G.; de Souza, E.L. Combined Application of Essential Oils from *Origanum vulgare* L. and *Rosmarinus officinalis* L. to Inhibit Bacteria and Autochthonous Microflora Associated with Minimally Processed Vegetables. *Food Res. Int.* **2011**, *44*, 1541–1548. [CrossRef]
16. de Sousa, J.P.; de Araújo Torres, R.; de Azerêdo, G.A.; Figueiredo, R.C.B.Q.; da Silva Vasconcelos, M.A.; de Souza, E.L. Carvacrol and 1,8-Cineole Alone or in Combination at Sublethal Concentrations Induce Changes in the Cell Morphology and Membrane Permeability of Pseudomonas Fluorescens in a Vegetable-Based Broth. *Int. J. Food Microbiol.* **2012**, *158*, 9–13. [CrossRef] [PubMed]
17. Jiang, Y.; Wu, N.; Fu, Y.J.; Wang, W.; Luo, M.; Zhao, C.J.; Zu, Y.G.; Liu, X.L. Chemical Composition and Antimicrobial Activity of the Essential Oil of Rosemary. *Environ. Toxicol. Pharmacol.* **2011**, *32*, 63–68. [CrossRef] [PubMed]
18. Hendry, E.R.; Worthington, T.; Conway, B.R.; Lambert, P.A. Antimicrobial Efficacy of Eucalyptus Oil and 1,8-Cineole Alone and in Combination with Chlorhexidine Digluconate against Microorganisms Grown in Planktonic and Biofilm Cultures. *J. Antimicrob. Chemother.* **2009**, *64*, 1219–1225. [CrossRef]
19. Mutlu-Ingok, A.; Firtin, B.; Karbancioglu-Guler, F.; Altay, F. A Study on Correlations between Antimicrobial Effects and Diffusion Coefficient, Zeta Potential and Droplet Size of Essential Oils. *Int. J. Food Eng.* **2020**, *16*, 1–9. [CrossRef]
20. Brożyna, M.; Paleczny, J.; Kozłowska, W.; Chodaczek, G.; Dudek-Wicher, R.; Felińczak, A.; Gołębiewska, J.; Górniak, A.; Junka, A. The Antimicrobial and Antibiofilm in Vitro Activity of Liquid and Vapour Phases of Selected Essential Oils against Staphylococcus Aureus. *Pathogens* **2021**, *10*, 1207. [CrossRef]
21. Cal, K. Aqueous Solubility of Liquid Monoterpenes at 293 K and Relationship with Calculated Log P Value. *Yakugaku Zasshi* **2006**, *126*, 307–309. [CrossRef] [PubMed]
22. Alibi, S.; Ben Selma, W.; Ramos-Vivas, J.; Smach, M.A.; Touati, R.; Boukadida, J.; Navas, J.; Ben Mansour, H. Anti-Oxidant, Antibacterial, Anti-Biofilm, and Anti-Quorum Sensing Activities of Four Essential Oils against Multidrug-Resistant Bacterial Clinical Isolates. *Curr. Res. Transl. Med.* **2020**, *68*, 59–66. [CrossRef] [PubMed]
23. Ceylan, O.; Uğur, A.; Saraç, N.; Ozcan, F.; Baygar, T. The in Vitro Antibiofilm Activity of *Rosmarinus officinalis* L. Essential Oil against Multiple Antibiotic Resistant *Pseudomonas* sp. and *Staphylococcus* sp. *J. Food Agric. Environ.* **2014**, *12*, 82–86.
24. Karuppiah, V.; Thirunanasambandham, R.; Thangaraj, G. Anti-Quorum Sensing and Antibiofilm Potential of 1,8-Cineole Derived from Musa Paradisiaca against Pseudomonas Aeruginosa Strain PAO1. *World J. Microbiol. Biotechnol.* **2021**, *37*, 66. [CrossRef]
25. Janssen, A.M.; Chin, N.L.J.; Scheffer, J.J.C.; Svendsen, A.B. Screening for Antimicrobial Activity of Some Essential Oils by the Agar Overlay Technique—Statistics and Correlations. *Pharm. Weekbl. Sci. Ed.* **1986**, *8*, 289–292. [CrossRef]
26. Cavanagh, H.M.A. Antifungal Activity of the Volatile Phase of Essential Oils: A Brief Review. *Nat. Prod. Commun.* **2007**, *2*, 1297–1302. [CrossRef]
27. Goñi, P.; López, P.; Sánchez, C.; Gómez-Lus, R.; Becerril, R.; Nerín, C. Antimicrobial Activity in the Vapour Phase of a Combination of Cinnamon and Clove Essential Oils. *Food Chem.* **2009**, *116*, 982–989. [CrossRef]
28. Dudek-Wicher, R.; Paleczny, J.; Kowalska-Krochmal, B.; Szymczyk-Ziółkowska, P.; Pachura, N.; Szumny, A.; Brożyna, M. Activity of Liquid and Volatile Fractions of Essential Oils against Biofilm Formed by Selected Reference Strains on Polystyrene and Hydroxyapatite Surfaces. *Pathogens* **2021**, *10*, 515. [CrossRef]
29. Reyes-Jurado, F.; Navarro-Cruz, A.R.; Ochoa-Velasco, C.E.; Palou, E.; López-Malo, A.; Ávila-Sosa, R. Essential Oils in Vapor Phase as Alternative Antimicrobials: A Review. *Crit. Rev. Food Sci. Nutr.* **2020**, *60*, 1641–1650. [CrossRef]
30. Gagné-Thivierge, C.; Barbeau, J.; Levesque, R.C.; Charette, S.J. A New Approach to Study Attached Biofilms and Floating Communities from Pseudomonas Aeruginosa Strains of Various Origins Reveals Diverse Effects of Divalent Ions. *FEMS Microbiol. Lett.* **2018**, *365*, fny155. [CrossRef]
31. Azam, M.W.; Khan, A.U. Updates on the Pathogenicity Status of Pseudomonas Aeruginosa. *Drug Discov. Today* **2019**, *24*, 350–359. [CrossRef] [PubMed]
32. Poole, K. Pseudomonas Aeruginosa: Resistance to the Max. *Front. Microbiol.* **2011**, *2*, 65. [CrossRef] [PubMed]
33. Kim, J.; Marshall, M.R.; Wei, C.-i. Antibacterial Activity of Some Essential Oil Components against Five Foodborne Pathogens. *J. Agric. Food Chem.* **1995**, *43*, 2839–2845. [CrossRef]
34. Hajlaoui, H.; Trabelsi, N.; Noumi, E.; Snoussi, M.; Fallah, H.; Ksouri, R.; Bakhrouf, A. Biological Activities of the Essential Oils and Methanol Extract of Tow Cultivated Mint Species (Mentha Longifolia and Mentha Pulegium) Used in the Tunisian Folkloric Medicine. *World J. Microbiol. Biotechnol.* **2009**, *25*, 2227–2238. [CrossRef]

35. Cristani, M.; D'Arrigo, M.; Mandalari, G.; Castelli, F.; Sarpietro, M.G.; Micieli, D.; Venuti, V.; Bisignano, G.; Saija, A.; Trombetta, D. Interaction of Four Monoterpenes Contained in Essential Oils with Model Membranes: Implications for Their Antibacterial Activity. *J. Agric. Food Chem.* **2007**, *55*, 6300–6308. [CrossRef] [PubMed]
36. Junka, A.F.; Żywicka, A.; Szymczyk, P.; Dziadas, M.; Bartoszewicz, M.; Fijałkowski, K.A.D.A.M. Test (Antibiofilm Dressing's Activity Measurement)—Simple Method for Evaluating Anti-Biofilm Activity of Drug-Saturated Dressings against Wound Pathogens. *J. Microbiol. Methods* **2017**, *143*, 6–12. [CrossRef]
37. Brożyna, M.; Żywicka, A.; Fijałkowski, K.; Gorczyca, D.; Oleksy-Wawrzyniak, M.; Dydak, K.; Migdał, P.; Dudek, B.; Bartoszewicz, M.; Junka, A. The Novel Quantitative Assay for Measuring the Antibiofilm Activity of Volatile Compounds (Antibiovol). *Appl. Sci.* **2020**, *10*, 7343. [CrossRef]

molecules

MDPI

Article

Antifungal Volatilomes Mediated Defense Mechanism against *Fusarium oxysporum* f. sp. *lycopersici*, the Incitant of Tomato Wilt

Praveen Thangaraj [1,*], Akshaya Subbaih Balamurali [1], Krishnamoorthy Akkanna Subbiah [1,*], Nakkeeran Sevugapperumal [1], Thiribhuvanamala Gurudevan [1], Sivakumar Uthandi [2] and Haripriya Shanmugam [3]

[1] Department of Plant Pathology, Tamil Nadu Agricultural University, Coimbatore 641 003, India; akshayaagri14@gmail.com (A.S.B.); nakkeeranayy@tnau.ac.in (N.S.); ragumala2000@gmail.com (T.G.)
[2] Department of Agricultural Microbiology, Tamil Nadu Agricultural University, Coimbatore 641 003, India; usivakumartnau@gmail.com
[3] Department of Nanoscience and Technology, Tamil Nadu Agricultural University, Coimbatore 641 003, India; haripriya.s@tnau.ac.in
* Correspondence: tpraveen1993@gmail.com (P.T.); milkmushapk2@gmail.com (K.A.S.)

Abstract: In this study, the volatilomes of naturally growing plant leaves were immobilized in a suitable substrate to enhance vapors' diffusion in the soil to eradicate the *Fusarium* wilt pathogens in Tomato. Volatilomes produced by *Mentha spicata* leaves immobilized in vermiculite ball was found to be effective and exhibit 92.35 percent inhibition on the mycelial growth of *Fusarium oxysporum* f. sp. *lycopersici* (FOL). Moreover, the volatilomes of *M. spicata* immobilized vermiculite balls were tested based on the distance traveled by the diffused volatilomes from the ball and revealed that the volatilomes of *M. spicata* traveled up to 20 cm distance from the center of PVC (Polyvinly chloride) chamber showed maximum reduction in colony growth of FOL at 12th day after inoculation. Tomato plants inoculated with FOL revealed increased expressions of defense gene, pathogenesis related protein (PR1) with 2.63-fold after 72 h and the gene, transcription factor (WRKY) increased with 2.5-fold after 48 h on exposure to the volatilomes of *M. spicata* vermiculite balls. To the best of our knowledge, this is the first report on development of volatilomes based vermiculite ball formulations. This result indicated that the volatilomes of *M. spicata* are promising phyto-fumigants for management of Tomato *Fusarial* wilt.

Keywords: volatilomes; phyto-fumigants; antifungal action; *Fusarium oxysporum*

Citation: Thangaraj, P.; Balamurali, A.S.; Subbiah, K.A.; Sevugapperumal, N.; Gurudevan, T.; Uthandi, S.; Shanmugam, H. Antifungal Volatilomes Mediated Defense Mechanism against *Fusarium oxysporum* f. sp. *lycopersici*, the Incitant of Tomato Wilt. *Molecules* **2022**, *27*, 3631. https://doi.org/10.3390/molecules27113631

Academic Editor: Ladislav Kokoska

Received: 11 April 2022
Accepted: 6 May 2022
Published: 6 June 2022

Publisher's Note: MDPI stays neutral with regard to jurisdictional claims in published maps and institutional affiliations.

Copyright: © 2022 by the authors. Licensee MDPI, Basel, Switzerland. This article is an open access article distributed under the terms and conditions of the Creative Commons Attribution (CC BY) license (https://creativecommons.org/licenses/by/4.0/).

1. Introduction

Wilt caused by *Fusarium oxysporum* f. sp. *lycopersici* (Sacc.) (FOL) is a devastating disease infecting tomatoes at all stages from seedling to fruiting around the world. FOL being hemi-biotrophic xylem colonizing pathogen, exhibits a wider host range inducing characteristic symptoms such as chlorosis of leaves, browning of vascular tissue and death of plants. In India, a huge loss of up to 45 percent in tomato crops grown under greenhouse conditions was recorded due to FOL incidence [1–3]. A considerable yield loss occurs due to the prolonged survival of the pathogen propagules in soil, their rapid germination and infection of plants under conducive conditions.

Cultural practices and regular applications of chemical fungicides are partially effective; however, application of fungicides repeatedly may cause damage to the natural ecosystem, human health and aquatic life, and also be detrimental to the beneficial microflora in soil [4]. Soil disinfestation (SD) by fumigation with highly toxic synthetic chemicals may lead to the development of fungicidal resistance in plant pathogens. However, carbon disulfide, methyl bromide (MB), methyl iodide, formaldehyde and hydrogen cyanide were also banned because of safety concerns and due to their effect on depletion of the ozone layer [5].

Large-scale application of beneficial microorganisms with antagonistic properties against plant pathogens could maintain the ecosystem more productive; keep the crop healthy and develop sustainable crop protection and production [6,7]. Thus, bio-control agents are considered to be the best alternatives for the management of soil-borne fungal pathogens by replacing synthetic chemical fungicides. Although these bio-inoculants enforce several modes of action, the novel concept of plant disease management through Volatile Organic Compounds (VOCs) becomes an attractive topic for research and debate. VOCs produced by the natural plant and microbial origins will have a greater stakeholder in the management of soil-borne plant pathogens. VOCs could have different biological and ecological functions in solemnizing defense mechanisms against plant pathogens and even plant growth promotion [8–10].

VOCs are low molecular weight, carbon structured organic compounds with high vapor pressure and lipophilic character [11]. VOCs are chemically diverse and belong to the large group of terpenoids and their derivatives, fatty acid-derived volatiles, phenyl propanoid aromatic compounds, alkanes, alkenes, alcohols, benzenoids, pyrazines, acids, esters, aldehydes, ketones and sulfur-containing compounds [12]. Some VOCs act as signaling mediators for cell-to-cell communications. These compounds are naturally produced through diverse biosynthetic pathways such as methylerythritol phosphate pathway (MEP), fatty acid metabolism, shikimate pathway, mevalonic acid pathway (MVA), lipo-oxygenase pathway (LOX), cinnamate pathway and amino acid oxidation [13,14].

VOCs produced by odorous plants and microbes have received more attention in the management of plant disease, plant growth promotion and induction of defense genes in plants [15]. VOCs extracted from the leaves of *Achyranthes aspera, Lawsonia inermis,* and *Mimosa pudica* [16,17] have been explored for the management of fungal pathogens. VOCs of essential oil such as *Foeniculum vulgare, Laurus nobilis, Lavandula stoechas* subsp., *stoechas, Origanum syriacum* var. *bevanii, Rosmarinus officinalis,* and *Thymbra spicata* subsp., *spicata* showed antifungal activities against *Phytopthora infestans* [18]. Similarly, VOCs produced by some microbial biocontrol agents and endophytic fungi act as elicitors that prime defense in plants against pathogens and promote growth [19]. VOCs produced by *Bacillus pumilus* exhibited antimicrobial activities against *F. oxysporum* f. sp. *lycopersici* and also elicited defense mechanisms by up-regulation of defense genes [20]. Despite the efficacy of VOCs in enhancing plant defense under in vitro laboratory, their effectiveness to enhance defenses mechanism against plant diseases has not been explored in the field. To date, a VOCs-mediated push–pull method of management strategy has been developed for attraction and repulsion of herbivores from the plants [21]. Recently, new techniques on micro encapsulation of VOCs have been developed, which allow controlled release of synthetic blends of VOCs, but their efficacy was not known, since it is not yet clear how VOCs inhibit plant diseases in nature.

With this background, the present study was carried out to develop suitable plant and microbes associated volatilome formulations for the management of *Fusarium* wilt pathogen infecting tomato crops under protected cultivation.

2. Results

2.1. Volatilomes Formulation

In earlier studies, the natural plant and microbe-associated volatilomes were screened individually against *FOL* using a sealed plate assay. The volatilomes produced by the leaves of *Mentha spicata, Cymbopogon citratus* and mycelia of *Trichoderma asperellum* showed the maximum higher inhibitory potential on mycelial growth of *FOL*. The leaves of *M. spicata* and *C. citratus* and the mycelial culture of *T. asperellum* were individually immobilized in a vermiculite ball in the form of Volatilomes Immobilized Vermiculite Ball formulation and further tested against the mycelial growth of pathogen under in-vitro. The vermiculite was served as volatile absorbent material.

Moreover, the volatilomes produced by vermiculite and castor oil used in the volatilome ball formulation did not show any relative abundance peak area, apart from the column contaminants as displayed in the heatmap (Figure 1). The prepared vermiculite ball-based volatilome formulations were well packed in an airtight zip lock polybag to prevent the diffusion of volatilomes.

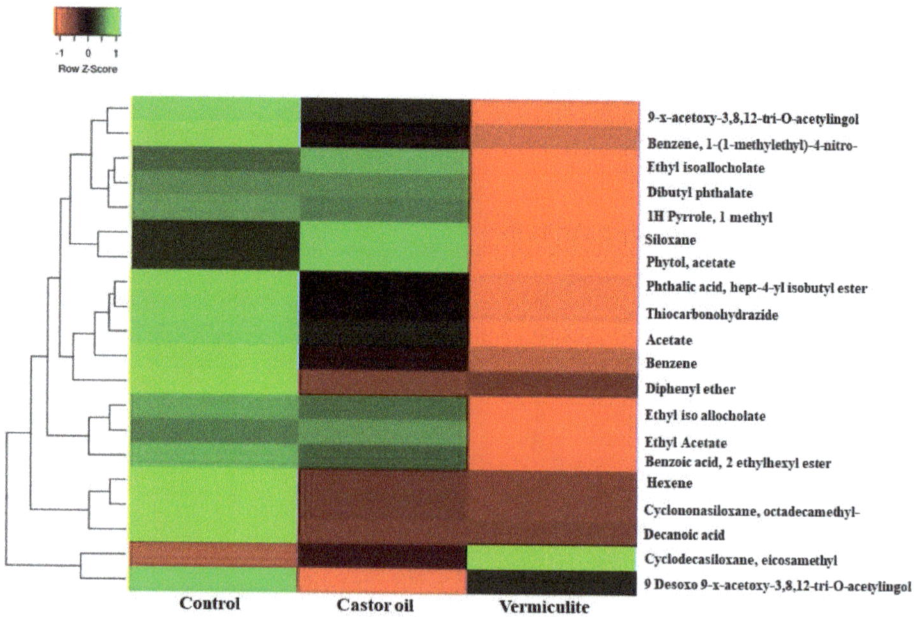

Figure 1. Comparative analysis of VOCs produced by vermiculite and castor oil.

2.2. Efficacy of Volatilomes Immobilized Vermiculite Balls against Pathogens

The efficacy of volatilomes immobilized vermiculite balls were tested against *FOL* using an olfactory chamber. The vermiculite balls of *M. spicata* volatilomes inhibited up to 92.35 percent on the mycelial growth of *FOL* (Figure 2). Vaporous inhibitory action was conspicuous at all six outlets, indicating the diffusion of volatiles all around the olfactory chamber. Subsequently, the vermiculite ball of *C. citratus* leaves showed 75.29 percent inhibition on the mycelial growth of *FOL*. The mycelia of *FOL* were completely altered with distorted growth on experimentation with the olfactory chamber (Figure 3), as compared to normal linear thick, pink-colored mycelial growth in control plates (Figure S1). Conclusively, the mycelial growth and sporulation behavior of the *FOL* was inhibited on exposure to the volatilomes of *M. spicata* leaves immobilized in the vermiculite balls.

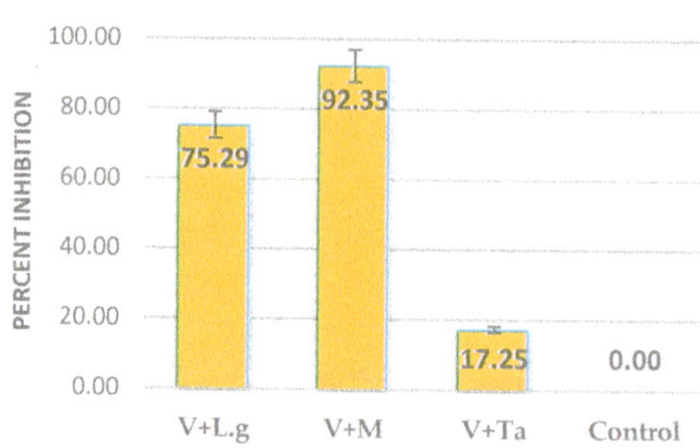

Figure 2. In vitro efficacy of VOCs immobilized vermiculite ball using olfactory chamber, V—Vermiculite, L.g—*C. citratus*, M—*M. spicata*, Ta—*Trichoderma asperellum*.

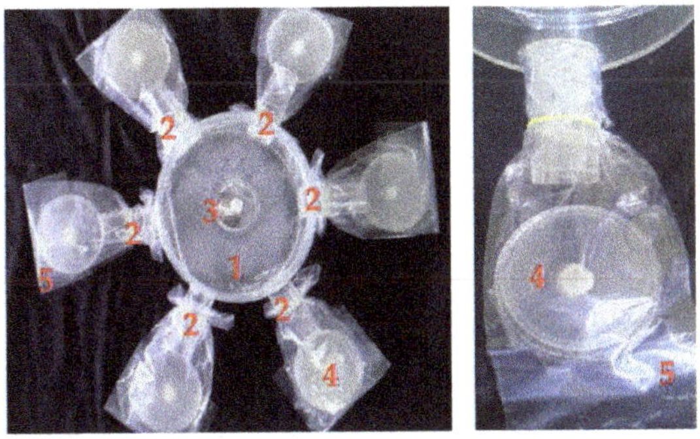

Figure 3. Effect of volatilomes immobilized vermiculite ball using olfactory chamber. 1—Olfactory chamber, 2—Outlet flow of chamber (6 no's), 3—Vermiculite ball formulation, 4—Petri dish containing pathogen and 5—Sterilized ploy cover.

2.3. GC MS Analysis of M. spicata Volatilomes Immobilized Vermiculite Ball

Based on the significant antagonistic potential on the inhibition of mycelial growth as well as sporulation behavior of *FOL*, the volatilomes immobilized ball of *M. spicata* and *C. citratus* were subjected to GC-MS analysis to find out the presence of major antifungal compounds.

In *M. spicata* immobilized ball, carvone was frequently represented with a peak area of 8–8.26% at 10.23 RT (Figure S2) up to 96 h of trapping (Figure 4, Table S1), which was comparatively less as compared to the peak area of 56% at 9.89 RT when standard carvone was used. The volatilomes of *C. citratus* showed the least diffusion of citronellol volatiles with less peak area of 0.98% at 9.79 RT, which was comparatively very less in comparison to that of the peak area of 31.82% at 9.96 RT, when standard citronellol was used. Therefore, the volatilomes of *M. spicata* ball recorded the maximum diffusion of carvone volatiles with a high peak area percent. However, the volatilomes produced by *C. citratus* (Table S2) did not diffuse citronellol volatiles with similar peak area percent as represented by the standard citronellol. Based on their higher antifungal activity against the pathogen, the volatilomes of *M. spicata* leaves immobilized vermiculite ball was taken to test their efficacy under pot culture conditions.

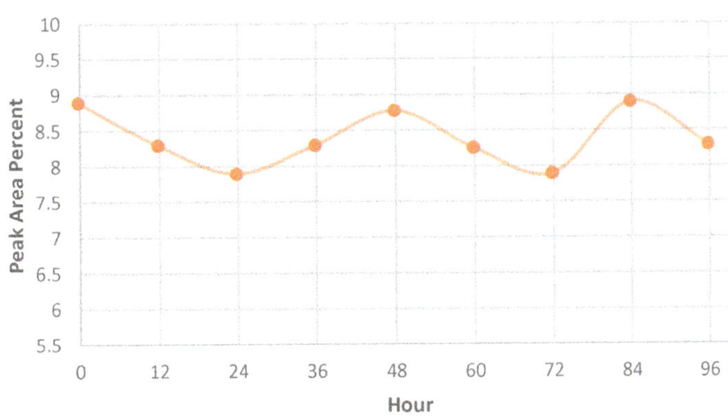

Figure 4. Peak area percent of carvone volatiles produced by the leaves of *M. spicata* immobilized vermiculite ball.

2.4. *SEM Imaging of M. spicata Volatilomes Immobilized Vermiculite Ball*

The structural changes of vermiculite loaded with the leaves of *M. spicata* were recorded concerning vertical and horizontal sheet expansion of vermiculite content. The vermiculite sheet ultra-structure was found to be raised by 2.7 folds. The volatilomes immobilized formulation consisting of vermiculite sample for its superior volatile absorbing capacity, to which finely grounded leaves of *M. spicata* was added and blended with castor oil. The volatile permeability and diffusion coefficients of the *M. spicata* immobilized vermiculite balls were altered due to the high loading of vermiculites compared to vermiculite balls without *M. spicata* volatilomes (Figure 5A,C). The volatilomes permeability was increased by 2–3 fold by the vermiculite (Figure 5E). The expansion coefficients of vermiculite were computed from *M. spicata* leaves immobilized ball, and the crystal structure of vermiculite was increased (Figure 5B,D). The expansion of crystal structure from the volatilomes of *M. spicata* ball significantly increased the absorption efficiency with vermiculite content. This expansion is mainly due to a change in the structure of the octahedral and tetrahedral layer, when volatilomes were immobilized in a vermiculite sample in the form of a ball. The excess sorption appears to be the effect of absorption of *M. spicata* volatilomes by the vermiculite and further helps to find out their efficacy against *FOL*.

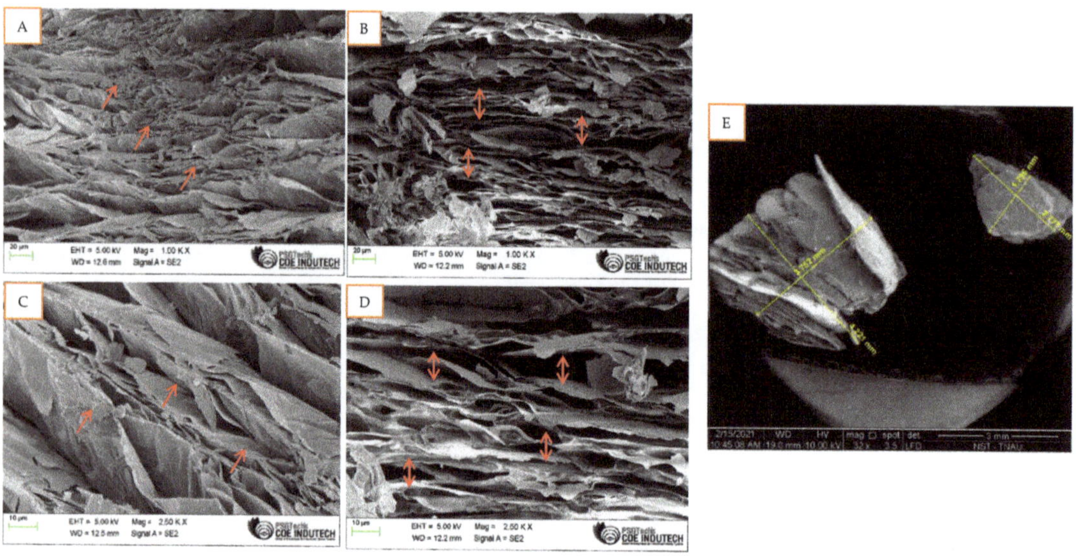

Figure 5. SEM imaging of *M. spicata* vermiculite ball. (**A**,**C**)—Control (Without immobilized) at 1.00 Kx magnification; (**B**,**D**)—*M. spicata* immobilized vermiculite ball at 2.50 Kx magnification. The double arrow indicates expanding, and the single arrow indicates non-expanding. (**E**)—Fold change in control (**Right**) and *M. spicata* (**Left**) immobilized vermiculite ball.

2.5. *Testing the Distance Traveled by the Volatilomes Immobilized Vermiculite Ball*

In this study, the volatilomes of *M. spicata* immobilized vermiculite ball were kept at the center of the PVC chamber to find out the exact distance traveled by the volatile that diffused from the ball.

The volatilomes of *M. spicata* balls recorded the maximum inhibition of colony growth of *FOL* at 10 and 20 cm distances away from the center of the PVC chamber. This indicates that *M. spicata* leaves immobilized vermiculite ball traveled in the PVC chamber up to a distance of 20 cm and suppressed the colony growth of the pathogen (0×10^{-3} CFU at 12 DAI (Day After Inoculation)) than in control (228×10^{-3} CFU) (Figure S3). However, the volatilomes did not travel at a distance above 30 to 60 cm from the center of the PVC chamber and exhibited the highest number of colony regrowth of *FOL* (213 × 10^{-3} CFU on 2 DAI; 87 × 10^{-3} CFU on 12 DAI) (Figure 6). However, the volatilomes of *M. spicata* leaves immobilized vermiculite balls proved to have fungicidal action in reducing the spore count of *FOL* up to a distance of 20 cm from the center of the PVC chamber. From these results, it is inferred that the volatilomes of *M. spicata* balls need to be placed at a distance of 20 cm from the root zone of tomato plants to reduce the infections and pathogen populations present in the soil.

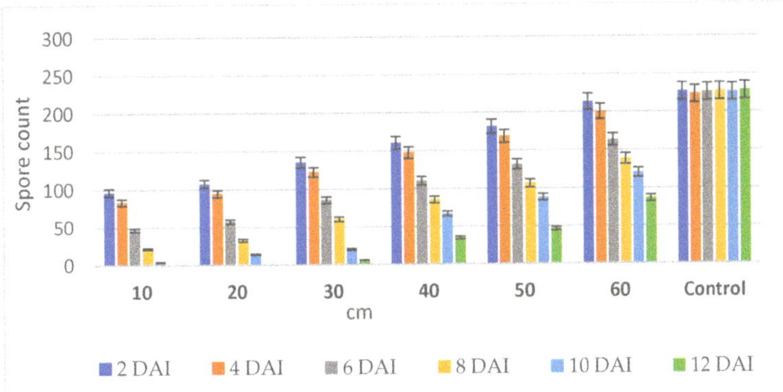

Figure 6. Performance of *M. spicata* volatilomes diffusion from the vermiculite ball over a distance against *F. oxysporum* f. sp. *lycopersici*.

2.6. *Effect of Volatilomes Immobilized Vermiculite Balls under Glasshouse Condition*

In this investigation, Phytoformulation of *M. spicata* volatilomes immobilized vermiculite ball was used to validate their efficacy against *FOL* in tomato under glass conditions using a controlled volatile growth chamber. Among the treatments, *M. spicata* immobilized balls recorded a lesser wilt incidence with 8.33 percent than in control (100 percent wilt incidence) (Table 1). The pathogen-inoculated pot expressed and observed with the symptom of chlorosis, drying of the vascular system followed by wilting of the plants after 45 days of inoculation (Figure S4). The growth parameters were highly promoted on exposure to the volatilomes of *M. spicata* immobilized vermiculite balls with 9.33 no of branches, shoot (31.83 cm) and root length (22.67 cm) as compared to pathogen-un-inoculated and pathogen-inoculated tomato plant.

Table 1. Efficacy of VOCs immobilized vermiculite ball against *F. oxysporum* f. sp. *lycopersici*.

Treatments	Shoot Length(cm)	Root Length (cm)	No of Branches	Percent Wilt Incidence	Percent Reduction over Control
Vermiculite + *M. spicata* in pathogen inoculated	31.83	22.67	9.33	8.33 [c] (16.78)	91.67 [a] (73.22)
Uninoculated control	25.83	15.87	7.33	83.33 [b] (65.91)	16.67 [b] (24.09)
Inoculated control	21.67	3.10	3.33	100.00 [a] (90.00)	0.00 [c] (0.02)

Values are the mean of five replications. Data in parentheses are arc sine transformed values. Means in a column followed by the same superscript letters are not significantly different according to the DMRT test at $p = 0.05$. Superscript letters in Percent Wilt Incidence a, b—Higher Wilt Incidence, c—Lower Wilt Incidence. Superscript letters (a, b, c) in Percent Reduction over Control represent higher to lower inhibition of pathogenic propagules in the soil.

2.7. *Defense Genes Expression*

The tomato plants on exposure to *M. spicata* leaves immobilized vermiculite balls and challenge inoculated with pathogens (*FOL*) were analyzed for their expression with further quantification of defense genes by using reverse transcriptase quantitative PCR (q-RT PCR). During the quantitative analysis of the PCR product, the melt curve analysis of each gene was performed to eradicate non-specific amplification of the sample. To understand the plant sample-based volatilomes mediated defense mechanisms in tomato plants, four defense responsive genes *viz.*, WRKY transcription factor (WRKY), thaumatin-like protein (TLP), pathogenesis-related protein (PR1) and lipoxygenase (LOX) were studied for their expression patterns. All the genes that were focused on were found to be up-regulated in tomato plants after 48 and 72 h on exposure to *M. spicata* volatilomes immobilized vermiculite balls. Plants challenge inoculated with pathogens and those which are exposed to volatilomes of *M. spicata* recorded higher levels of expression patterns of all the four defense responsive genes studied.

Tomato seedlings exposed to the volatilomes of *M. spicata* immobilized vermiculite ball formulation and challenges inoculated with *FOL* were assessed for the expression of *SlWRKY*, *SlTLP*, *SlPR 1* and *SlLOX* gene in roots due to the tripartite interactions. The relative quantification of the defense gene was normalized to *SlACTIN*. Among the different HPT, the volatilomes of *M. spicata* exposed tomato plants and challenge inoculated with pathogen highly induced the expression of *SlWRKY* gene at 48 HPT (2.5-fold) followed by pathogen-un-inoculated with plants (0.67-fold). The pathogen-un-inoculated and inoculated tomato plants showed a lower level of *SlWRKY* gene expression during all hours of post treatment. However, during the tri-trophic interaction, i.e., volatilomes of *M. spicata* exposed plants infected with the pathogen showed an up-regulation of the expression of the *SlWRKY* gene (Figure 7A), which conclusively indicated that the volatilomes of *M. spicata* could trigger and activate defense mechanism in tomato plants. The defense gene *SlTLP* was expressed at 72 HPT (2.00 fold) and 48 HPT (1.80 fold). A lower level of *SlTLP* gene expression was displayed in all the treatments at 0 and 24 HPT. This result indicated that the volatilomes of *M. spicata* up-regulated the expression of *SlTLP* gene in pathogen-inoculated tomato plants (Figure 7B). In the case of the defense gene *SlPR1*, higher level of *SlPR1* expression at 72 HPT (Figure 7C) was followed by 48 HPT (2.63-fold and 2-fold, respectively). Hence, the volatilome of *M. spicata* exposed plants significantly up-regulated the *Sl*PR1 gene, showing the potential of priming defense against pathogen infection. Similarly, the higher level of expressions of *SlLOX* gene was observed at 0, 24, 48 and 72 HPT on the volatilomes of *M. spicata* exposed plant. Meanwhile, the pathogen-un-inoculated, pathogen-inoculated and the volatilomes of *M. spicata* exposed un-inoculated plants showed a lower level of *SlLOX* gene expression (Figure 7D). Nevertheless, the volatilomes of *M. spicata* exposed plants strongly expressed the activation of immune response in tomato plants. Conclusively, the results indicated that the tomato plants exposed to volatilomes of *M. spicata* triggered a resistance mechanism against *F. oxysporum* f. sp. *lycopersici*.

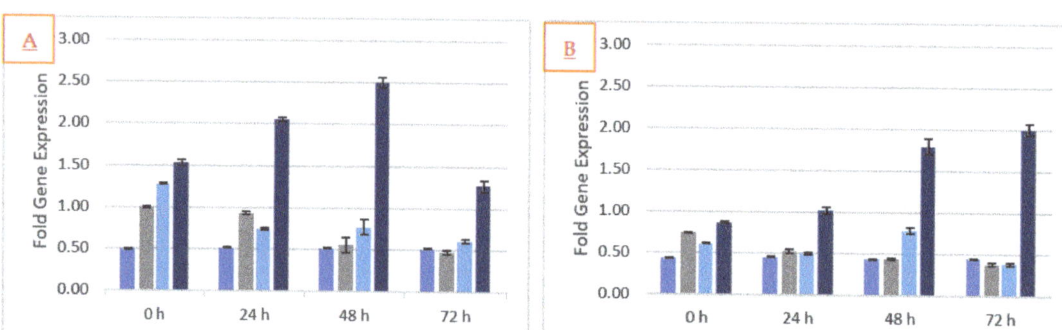

Figure 7. *Cont.*

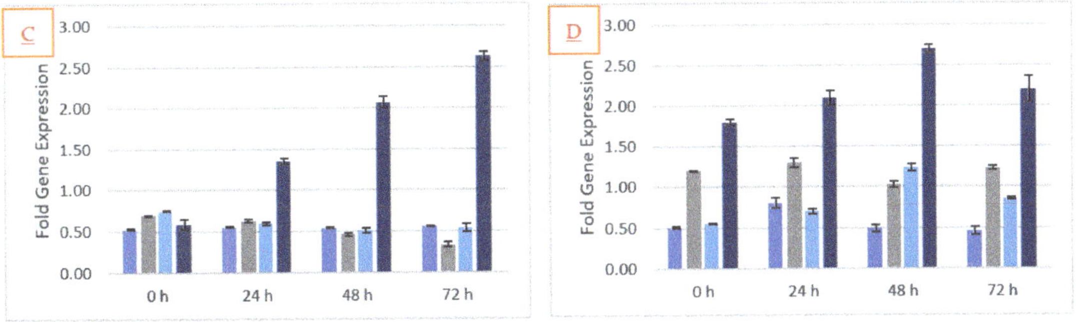

Figure 7. Defense gene expression of (**A**) WRKY, (**B**) TLP, (**C**) PR1 and (**D**) LOX during interaction of volatilomes of *M. spicata* immobilized vermiculite ball with *F. oxysporum* f. sp. *lycopercisi* inoculated Tomato Plant.

3. Discussion

As reported by several authors, the natural plant and microbial samples are well known to produce a wide variety of antimicrobial activity, as they are naturally active against a large spectrum of phytopathogenic microorganisms [22,23]. In the present study, the antifungal volatile of *M. spicata*, *C. citratus* and mycelia of *T. asperellum* was found to be effective on the mycelial growth of *FOL*. Based on this significant inhibition of the pathogen, the volatilomes produced by the leaves of *M.* spicata and *C. citratus* and mycelial cultures of *T. asperellum* were immobilized in a vermiculite ball-based formulation for further testing under pot culture conditions.

Suitable immobilization substrates have been developed to induce high vaporous diffusion of volatile compounds. [24] reported that vermiculite coated copper nanoparticle showed the maximum antifungal activity against the pathogens. Earlier studies reported that vermiculite could act as a superabsorbent composite and it swelled rapidly on increasing the concentration of volatilomes [25]. Vermiculite based nanoparticles, coated with CuO and ZnO could act as antibacterial activities against *Staphylococcus aureus* with long-acting up to 24 h [26,27]. In the existing study, vermiculite was used as a volatilomes immobilizing substrates in the form of organic-based formulation against the fungal pathogens. The SEM study also proved that the vermiculite sheet was expanded two fold when immobilized with *M. spicata* leaves compared to the vermiculite ball without volatilomes.

Furthermore, the volatilomes of *M. spicata* leaves immobilized vermiculite balls effectively inhibited the mycelial growth of *FOL* (92.35 percent) using an olfactory chamber. Leaves of *C. citratus* immobilized vermiculite balls showed 75 percent inhibition of *FOL*. However, this was the first report on mycelial growth inhibition of the pathogens using an olfactory chamber assay. In contrast to the present investigation, an olfactory chamber mediated assay was reported to attract and repel the insect population using novel volatile compounds [28]. VOCs, 1-octen-3-ol and benzaldehyde were produced by lemongrass oil, which attracts and repels *Aedes aegypti* mosquitos [29]. Additionally, the volatilomes of *M. spicata* and *C. citratus* were quantified based on the production of the highest area percent of major VOCs. In support to the present findings, [30] also recorded that VOCs of carvone was highly produced by diverse plant samples, but greater in *M. spicata*, and also validated that less peak area percent of citronellol was produced by *C. citratus* than earlier reported *C. nardus* [31]. Hence, our study proved that the vermiculite balls immobilized with *M. spicata* produced carvone volatiles with constant production of relative peak area abundance (8.00%) even up to 96 h, which could effectively suppress *FOL* in tomato plants.

In addition, the defense responsive gene used in the present study could explore a significant role in the activation of defense mechanisms in tomato plants. Gene responsible for the induction of salicylic acid (SAR) and pathogenesis-related protein (PR 1) are widely used as molecular markers and greatly up-regulated when exposed to VOCs as reported by [32]. It has been reported that volatiles of 3-pentanol exposure strongly induced the up-regulation of the pathogenesis-related protein PR 1, PR 2 and proteinase inhibitor PIN 2 during the pathogen attack in the pepper plants [33]. In the present study, *M. spicata* immobilized vermiculite balls when exposed to pathogen-inoculated tomato plants exhibited the highest level of LOX gene expressions. The expression of *Sl*LOX gene steadily increased after inoculation with *FOL* (2.7-fold). It has been reported that the expression of LOX gene led to the activation of the jasmonic acid pathway on interaction with volatilomes of *M. spicata*. The higher RNA transcript level was observed at 48 HPI and increases after 72 HPI in tomato plants. The defense gene expression pattern of *Sl*LOX gene varied significantly between the treatments as reported by [32]. The authors of [34] reported that broad beans inoculated with *Puccinia striiformis* f. sp. *tritici* induced the expression of PR 1 genes. This was evident from the present findings that up-regulation of *Sl*PR1 genes was increased in tomato on interaction with volatilomes of *M. spicata* in the presence of pathogens. This result supports the hypothesis that the induction of the SAR pathway may be responsible for the activation of the resistance mechanism in tomato plants to pathogen attacks. In tomato, defense expression gene LOX and PR-1 were up-regulated after exposure with volatilomes of *M. spicata* and challenge inoculated with pathogen during 48 and 72 HPT, as observed by [32]. Thus, VOC-treated plants against *FOL* expressed a higher level of PR-1 gene through an SA-dependent signaling pathway. The present findings are further supported by [35], who reported that VOCs produced by *Ampelomyces* and *Cladosporium* prime the defense mechanism by activating the salicylic acid and jasmonic acid pathway in *Arabidopsis*. The volatiles of methyl benzoate (MeBA) and m-cresol treated in the *Arabidopsis* plant activates the defense-related genes PR1. Similarly, VOCs of methyl benzoate produced by *T. asperellum* induced the expression of the PR1 gene in *Arabidopsis* plants [36]. As for *Sl*PR 1 gene response, the defense responsive gene *Sl*WRKY was expressed in the present study on treatment with the volatilomes of *M. spicata* after 48 and 72 HPT in pathogen-inoculated tomato plants. This indicates the activation of the defense gene (*Sl*WRKY) in plants against pathogens as observed by [32]. The present findings hypothesize that expression of the *Sl*WRKY gene in tomato inoculated with *FOL* might induce the activation of the gene containing W box in the promotors as reported by [37]. Thaumatin-like proteins (TLP) are a highly complex protein family, which plays a significant role in amino acid composition and induces defense mechanisms in plants. The expressions of the TLP gene could strongly reduce pathogen infections in plants [38]. In the current study, the defense gene *Sl*TLP differentially expressed after 24 HPT and considerably increased on 48 HPT on interaction with volatilomes of *M. spicata* and it might be acted as an elicitor. Conclusively, the defense-related gene such as *Sl* WRKY, *Sl* PR 1, *Sl* LOX, and *Sl* TLP were differentially expressed on interaction with vermiculite ball of *M. spicata* with tomato plants after 48 HPT. Among the treatments, a low level of defense gene expression was noticed in volatilomes exposed tomato plants without pathogens.

Hence, the volatilomes of *M. spicata* vermiculite balls induced all the defense genes in tomato plants, but the maximum level of defense gene expressions was more in *FOL* inoculated plants than un-inoculated tomato plants. It might be attributed due to the activation of systemic acquired resistance in the tomato plants when exposed to the volatilomes of *M. spicata* leaves immobilized vermiculite balls.

4. Materials and Methods

4.1. Plant and Microbial Volatilomes

The volatilomes produced by the leaves of *Mentha spicata, Cymbopogon citratus, Vitex negundo, Coleus amboinicus, Vetiveria zizanioides, Ocimum tenuiflorum, Azadirachta indica*, mycelia of *Auricularia auriculata, Coprinus cinereus, Ganoderma lucidum, Lentinus edodus, Trichoderma asperellum* and cell cultures of *Bacillus subtilis, Streptomyces rochei* have been collected, selected and bio-efficacy explored [39]. Among the plant and microbial samples, *M. spicata, C. citratus* and *T. asperellum* were taken further to develop a suitable formulation against the mycelial growth of *F. oxysporum* f. sp. *lycopersici*.

4.2. Immobilization of Eluted and Trapped Volatile Compounds

Volatile absorbing materials such as expanded vermiculite were used in this study. As a major portion of the void volume of vermiculite, could absorb natural volatiles with high retention properties at optimum temperature, it was further used to develop a novel formulation in the form of balls for the immobilization of eluted and trapped volatilomes. Seven parts of Vermiculite were mixed well with three parts of finely grounded volatile producing leaf sample (7:3 ratios). The well-mixed sample was made into a ball (3 mm diameter) by using one mL of castor oil. The vermiculite and castor oil-associated volatiles were also estimated by following the procedure of volatile trapping as mentioned by [39] and optimized for positive release of volatilomes produced by the plant samples used. The vermiculite based volatilomes immobilized ball formulation was further used to study the interactions with the test pathogens. Further, the volatilomes produced by the leaves of plant samples were trapped and analyzed using air-entrainment technique [39] and immediately subjected to HS-GCMS analysis using Thermo GC injector coupled with a Mass Spectrophotometer. The important VOCs, carvone produced were noticed based on the percent area and probability of compounds at different hours to validate on hours of carvone production. Further, the leaves of *M. spicata* immobilized vermiculite ball were examined under Scanning Electron Microscope (Sigma with gemini column, Carl zeiss (Oberkochen, Germany); Resolution 1.5 nm) to observe the structural changes during volatilomes absorption and diffusion.

4.3. Testing the Volatilomes Immobilized Vermiculite Balls against Pathogens

The vermiculite balls immobilized with VOCs were tested for their inhibitory effect against test pathogens by the olfactory chamber method. The olfactory chamber was designed with six outlets on all sides, each measuring 3.5 cm in diameter and an inlet at the center of the chamber with 5.5 cm diameter as shown in Figure 8. Each of the outlets was interconnected to the 20 cm diameter circular chamber, so as to facilitate the free flow of volatiles from the central inlet to all the six outlets. The volatilomes immobilized vermiculite ball was placed at the center of the circular chamber. A 5 mm mycelial disc from 7 days culture of the test pathogens separately, was placed on sterile plastic Petri plates poured with PDA medium. Each of the Petri plates after removing the top lid was placed inside a sterilized poly bag and the mouth of the bag was kept intact with the outlet mouth and sealed airtight. The entire setup with Petri plates was incubated at 28 ± 2 °C until the mycelia in control covered the plate fully. The percent reduction in mycelial growth was recorded by using the formula:

$$\text{Percent inhibition (PI)} = \frac{C-T}{C} \times 100$$

where, C is the mycelial coverage of pathogen in control, T is the mycelial growth of the pathogen in treatment. The experiment was performed thrice with three replications to confirm the efficacy of volatilomes produced by vermiculite ball.

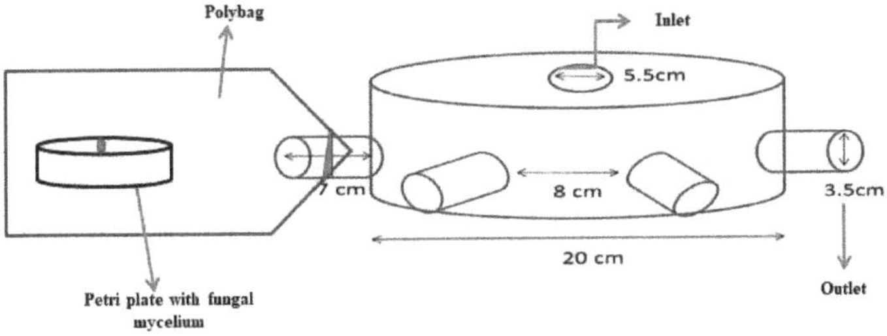

Figure 8. Model of olfactory chamber.

4.4. Testing the Distance on Volatilomes Travel

The distance traveled by the volatilomes mixture immobilized in vermiculite balls used against the test pathogens and the efficacy was tested in a PVC chamber. A PVC pipe having 20 cm diameter was cut into different lengths *viz.*, 10, 20, 40 and 60 cm (Figure 9). The PVC pipes were given with a flap cut to 10 cm width and length vary depending upon the size of PVC pipe used. In the flap cut PVC pipe, autoclaved coir compost was filled and inoculated with the spore suspension of the pathogens, separately and incubated for 7 days to induce the mycelial growth of pathogens. Later, one of the vermiculite balls with immobilized VOCs was buried in the pathogen-inoculated coir compost taken in the PVC pipes. The flap-cut portion of the PVC pipe was carefully covered with Parafilm. The experimental setup was incubated at 28 ± 2 °C and replicated thrice for each of the vermiculite ball-based formulations. A similar experimental setup was followed for control by placing vermiculite balls without the volatilomes fraction. After 4 days of incubation, the compost was collected from the PVC pipe at different distances surrounding the vermiculite ball, serially diluted with sterilized water and 10^{-3} dilution was plated on a PDA medium. The CFU count was recorded in both treatment and control at varied distances per interval and tabulated. Based on the results, a separate experiment was conducted on tomato crops grown under the poly house.

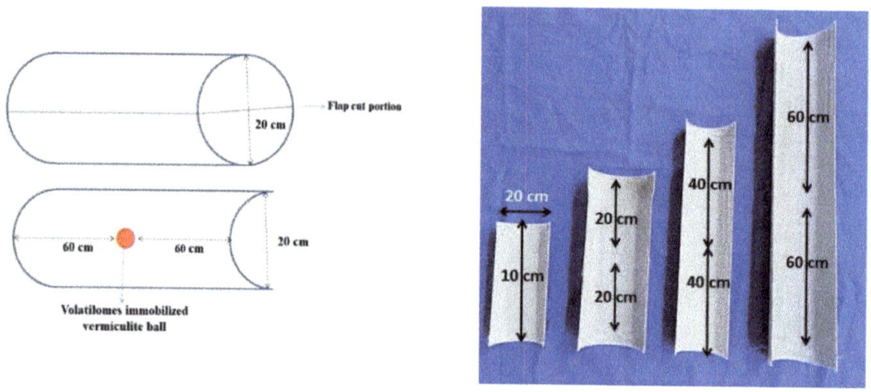

Figure 9. Model of PVC Pipe Chamber at different lengths.

4.5. Effect of Volatilomes Immobilized Vermiculite Balls Tested in Growth Chamber

The interaction of volatilomes immobilized vermiculite ball was evaluated under a pot culture experiment in tomato plants. The tomato seedling was transplanted in a pot filled with sterilized coir compost. The pots planted were interconnected with transparent polythene tubes and placed inside the volatilomes diffusing chamber (Figure 10). Arrangements were made prior to placing the volatilomes immobilized vermiculite ball inside the transparent polythene tube facilitating the free flow of volatilomes emitted by the ball between the interconnected pots. The *M. spicata* leaves immobilized vermiculite ball were imposed with four replications. Control pots interconnected with polythene tube but without VOCs immobilized vermiculite ball was maintained inside the chamber as control treatments. The pathogen was inoculated to 20 days old tomato seedlings in treatment (*M. spicata* exposing chamber) and control pots. The pathogen was not inoculated in the Pathogen un-inoculation pot, but the volatilomes of *M. spicata* ball were exposed to the pathogen-un-inoculated pot to compare the difference in volatilomes treated tomato seedling inoculated with the pathogen. Then, the volatile chamber was closed completely until the symptoms are noticed in control pots inside the chamber. The spore suspension of *FOL* was inoculated in a sterilized coir compost and the volatilomes formulation was exposed after a day of pathogen inoculation. The result of percent disease incidence and percent reduction over control was calculated after symptom expression in the pathogen-inoculated pots with treatment. In addition, the growth parameters such as root length, shoot length and number of branches produced by tomato plants were observed.

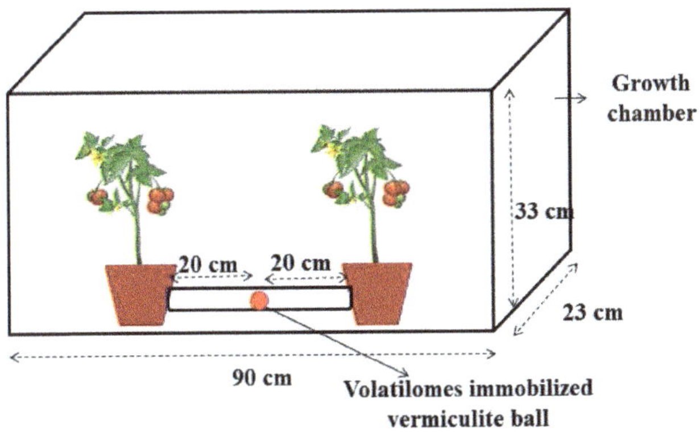

Figure 10. Volatilomes diffusing growth chamber model.

4.6. *Defense Gene Expression Studies*

A pot culture experiment was conducted with tomato plants (PKM 1) to evaluate the defense genes expression as triggered by the volatilomes immobilized in vermiculite balls containing *M. spicata* leaf volatilomes. The root samples from the plants inoculated with pathogen were taken for determining defense genes expression due to the tripartite interaction between plant, pathogen and immobilized volatilomes. The control plants were maintained with and without pathogen inoculation.

4.6.1. RNA Extraction, cDNA Conversion

Total RNA was extracted separately, from tomato plants under different treatments by following the procedures described by [32]. The root samples were collected on the first day after pathogen inoculation, rinsed in sterile distilled water and dried on filter paper. The dried root samples were weighed (200 g) and ground to a fine powder in a sterilized pestle and mortar using liquid nitrogen. The ground sample was transferred to a microfuge tube (1.5 mL) and added with one mL of cell lysis solution, followed by gentle shaking and centrifugation at 11,000 rpm at 4 °C for 15 min. After centrifugation, the supernatant was transferred to a new microfuge tube and added with 250 µL of phase separation solution. The contents were vortexed well and centrifuged again at 11,000 rpm at 4 °C for 15 min. The top aqueous portion was pipetted out to a new microfuge tube and added with an equal volume of precipitation solution and sodium hydroxide before incubating at −20 °C for 45 min. The precipitated RNA was centrifuged at 11,000 rpm at 4 °C for 15 min and the supernatant was discarded. The pellet was washed by adding one mL of 70 percent ethanol and centrifuged at 11,000 rpm at 4 °C for 15 min. The supernatant was decanted and the pellet air-dried for 10 min and resuspended with 30 µL of sterile distilled water. The quantity and quality of RNA were confirmed using Nanodrop™ 2000 spectrophotometer (Thermo-Fisher, Waltham, MA, USA). The extracted RNA was stored at −80 °C for further studies.

The DNA admixed in the extracted RNA sample was removed by following the procedure given in DNAse I removal Kit (Sigma Aldrich, Saint Louis, MO, USA). The extracted RNA (8 µL) was treated with one µL each of 10× DNAse reaction buffer and DNAse I. The DNAse treated RNA sample was incubated at 28 ± 2 °C for 15 min. After incubation, the DNAse reaction was stopped by adding 1 µL of EDTA as a stop solution to inactivate the DNAse I and incubated at 70 °C for 10 min. Then, the DNA-free RNA sample was used for cDNA synthesis. A quantity of 12 µL of DNA-free RNA sample was taken for activation of cDNA synthesis. The RNA sample was added with one µL of random primer and incubated at 65 °C for one min in a PCR thermocycler. The incubated contents were placed in ice and 4 µL of 5× reverse transcriptase (RT) reaction buffer; 2 µL of 10 mM dNTP mixture; 1 µL of revert aid were added and the contents were completely mixed and incubated in PCR thermocycler at three different step process: 42 °C for 60 min; 50 °C for 15 min; 70 °C for 15 min. The cDNA synthesized samples were stored at −80 °C.

4.6.2. Quantification of Defense Related Genes Expression

The defense gene expression in tomato plants, both in pathogen-infected and treatment imposed with volatilomes was quantified by quantitative PCR (Bio-rad, California, United States of America) using SYBR Green (KAPA SYBR FAST LC 480). The housekeeping gene ACTIN was used as internal control and the diluted cDNA samples were used for qPCR analysis. Each of the 10 µL reaction mix contained 5 µL SYBR green master mix (KAPA SYBR FAST LC 480, Sigma Aldrich, Saint Louis, MO, USA); 2 µL of primers (Forward and Reverse defense primers); one µL of sterile distilled water and 2 µL of diluted cDNA. The quantification of defense genes was performed in Biorad RT PCR with the following settings: initial denaturation (95 °C for 10 min), 39 cycles of PCR (includes denaturation 95 °C for 30 s and amplification at 58 °C for 30 s), melting curve (includes denaturation 95 °C for 5 s, amplification at 65 °C for 5 s). The relative expression of defense genes was compared with internal control (ACTIN gene) and further assessed. The changes in expression levels were analyzed using the relative quantification ($^{\Delta\Delta}$Ct) method. The analysis was carried out with three biological replicates and three technical replicates for each of the defense primers. The relative expression of each defense gene was compared among the treatments. The details of defense primers of tomato plants are listed in Table 2.

Table 2. Defense gene primers used for quantitative RT-PCR.

Defense Gene	Forward and Reverse Primer Sequence	Reference
*Sl*PR 1	Forward: 5′-ACGTCTTGGTTGTGCTAGGG-3′ Reverse: 5′-TCAAAAGCCGGTTGATTTC-3′	[32]
*Sl*LOX	Forward: 5′-TGGGATTAAACTGCCAGACC-3′ Reverse: 5′-GGCATCGGAAATTTGAGAAA-3′	
*Sl*WRKY	Forward:5′-TCTCGATCTGACCAGGTTCC-3′ Reverse: 5′-TTGCCGTCTTCGTTCTCTTT-3′	
*Sl*TLP	Forward:5′-CCATCTTTGCTTCCCACATT-3′ Reverse: 5′-ATCGGTTTACCTGCACTTGG-3′	
*Sl*Actin	Forward:5′-AGGCACACAGGTGTTATGGT-3′ Reverse: 5′-AGCAACTCGAAGCTCATTGT-3′	

4.7. *Statistical Analysis*

In-vitro experiments were performed in triplicate and analyzed using a single ANOVA with the data obtained from the radial growth of pathogenic inhibition. The result was presented as mean ± standard deviation. The GC-MS data were statistically analyzed using MetaboAnalyst V5.0 (Genome Canada, Ottawa, ON, Canada). The results of the GC-MS data were presented as a heat map using heatmapper.ca.in.

5. Conclusions

In conclusion, volatilomes produced by the leaves of *M. spicata* could induce diverse antifungal VOCs for the management of plant pathogens. Diverse volatile biomolecules have been characterized from the leaves of *M. spicata*, but some VOCs such as carvone play a major role in the suppression of pathogenic microorganisms. VOC of carvone has been proved as antimicrobial activity for the management of plant disease in horticultural crops. Thus, volatilomes produced by *M. spicata* need to be explored.

In the present study, we assessed the various immobilization substrate for volatilomes based product development to eradicate the presence of pathogenic propagules present in the soil. Further, we conclude that volatilomes of *M. spicata* leaves immobilized in a vermiculite ball were found to be effective in managing the plant disease under glasshouse conditions. The present findings could be useful for the precise analysis of *M. spicata* leaves immobilized vermiculite ball against soil-borne plant pathogens.

Supplementary Materials: The following supporting information can be downloaded at: https://www.mdpi.com/article/10.3390/molecules27113631/s1, Figure S1: Morphological Characterization of mycelial growth of *Fusarium* Pathogen using olfactory chamber; Figure S2: GCMS chromatogram of headspace volatile compounds produced by the leaves of *Mentha spicata*; Figure S3: Colony growth of *Fusarium oxysporum* (CFU) using PVC chamber; Figure S4: Efficacy of Volatilomes Immobilized Vermiculite Balls under Glass house Condition in Tomato Plant; Table S1: Area Percent of carvone produced at different hours by the leaves of *M. spicata* immobilized vermiculite ball; Table S2: GC-MS profiling of VOCs produced by leaves of *Cymbopogon citratus* immobilized vermiculite ball.

Author Contributions: K.A.S. Planned the concept of antifungal activity of volatilomes to assess against the Phyto-pathogens. N.S., T.G., S.U. and H.S.; contributed to the experimental analysis. P.T. and A.S.B. experimented. All authors have read and agreed to the published version of the manuscript.

Funding: This research received no external funding.

Institutional Review Board Statement: Not applicable.

Informed Consent Statement: Not applicable.

Data Availability Statement: All the data is available in the manuscript.

Acknowledgments: The authors thank for the support of UGC-SAP-DRS 1, DST-FIST and ICAR-AICRP on Mushroom laboratory facilities at the Department of Plant Pathology, TNAU, Coimbatore. The authors extend their sincere thanks to Vice Chancellor, Registrar, Director of Research, Dean (SPGS) of TNAU and the authors for their support. Facilities extended by Department of Agricultural Microbiology, Nano science and Technology and Agricultural Entomology are also greatly acknowledged.

Conflicts of Interest: All the authors declare that they have no conflict of interests.

References

1. Vignesh, K.; Rajamohan, K.; Balabaskar, P.; Anandan, R.; Udhayakumar, R. Survey on the incidence of fusarium wilt of tomato incited by *Fusarium oxysporum* f. sp. *lycopersici* (*FOL*) in major tomato growing areas of krishnagiri district. *Plant Arch.* **2021**, *21*, 2369–2376.
2. Srinivas, C.; Devi, D.N.; Murthy, K.N.; Mohan, C.D.; Lakshmeesha, T.; Singh, B.; Kalagatur, N.K.; Niranjana, S.; Hashem, A.; Alqarawi, A.A. *Fusarium oxysporum* f. sp. *lycopersici* causal agent of vascular wilt disease of tomato: Biology to diversity—A Review. *Saudi J. Biol. Sci.* **2019**, *26*, 1315–1324. [CrossRef] [PubMed]
3. Yucel, S.; Ozarslandan, A.; Colak, A.; Ay, T.; Can, C. Effect of solarization and fumigant applications on soilborne pathogens and root-knot nematodes in greenhouse-grown tomato in turkey. *Phytoparasitica* **2007**, *35*, 450–456. [CrossRef]
4. Panth, M.; Hassler, S.C.; Baysal-Gurel, F. Methods for management of soilborne diseases in crop production. *Agriculture* **2020**, *10*, 16. [CrossRef]
5. Katan, J. Diseases caused by soilborne pathogens: Biology, management and challenges. *J. Plant Pathol.* **2017**, *99*, 305–315.
6. Larkin, R.P.; Brewer, M.T. Effects of crop rotation and biocontrol amendments on rhizoctonia disease of potato and soil microbial communities. *Agriculture* **2020**, *10*, 128. [CrossRef]
7. Mokhtari, W.; Achouri, M.; Chtaina, N. *Rhizoctonia solani*-green beans pathosystem uncover bio control efficacy of *Trichoderma* spp. *Adv. Plants Agric. Res.* **2019**, *9*, 96–99.
8. Ameye, M.; Audenaert, K.; De Zutter, N.; Steppe, K.; Van Meulebroek, L.; Vanhaecke, L.; De Vleesschauwer, D.; Haesaert, G.; Smagghe, G. Priming of wheat with the green leaf volatile z-3-hexenyl acetate enhances defense against *Fusarium graminearum* but boosts deoxynivalenol production. *Plant Physiol.* **2015**, *167*, 1671–1684. [CrossRef]
9. Brilli, F.; Loreto, F.; Baccelli, I. Exploiting plant volatile organic compounds (VOCs) in agriculture to improve sustainable defense strategies and productivity of crops. *Front. Plant Sci.* **2019**, *10*, 264. [CrossRef]
10. Jamiołkowska, A. Natural compounds as elicitors of plant resistance against diseases and new biocontrol strategies. *Agronomy* **2020**, *10*, 173. [CrossRef]

11. Vivaldo, G.; Masi, E.; Taiti, C.; Caldarelli, G.; Mancuso, S. The network of plants volatile organic compounds. *Sci. Rep.* **2017**, *7*, 11050. [CrossRef] [PubMed]
12. Adamova, T.; Hradecky, J.; Panek, M. Volatile organic compounds (VOCs) from wood and wood-based panels: Methods for evaluation, potential health risks, and mitigation. *Polymers* **2020**, *12*, 2289. [CrossRef] [PubMed]
13. Bouwmeester, H.; Schuurink, R.C.; Bleeker, P.M.; Schiestl, F. The role of volatiles in plant communication. *Plant J.* **2019**, *100*, 892–907. [CrossRef] [PubMed]
14. Schmidt, R.; De Jager, V.; Zühlke, D.; Wolff, C.; Bernhardt, J.; Cankar, K.; Beekwilder, J.; Van Ijcken, W.; Sleutels, F.; De Boer, W. Fungal volatile compounds induce production of the secondary metabolite sodorifen in *Serratia plymuthica* pri-2c. *Sci. Rep.* **2017**, *7*, 862. [CrossRef] [PubMed]
15. Tholl, D.; Hossain, O.; Weinhold, A.; Rose, U.S.; Wei, Q. Trends and applications in plant volatile sampling and analysis. *Plant J.* **2021**, *106*, 314–325. [CrossRef]
16. Akter, A.; Neela, F.; Khan, M.; Islam, M.; Alam, M. Screening of ethanol, petroleum ether and chloroform extracts of medicinal plants, *Lawsonia inermis* L. and *Mimosa pudica* L. For antibacterial activity. *Indian J. Pharm. Sci.* **2010**, *72*, 388.
17. Khan, M.; Neela, F.; Aktar, A.; Rahman, M.; Alam, M. Antibacterial activity of *Achyranthes aspera* L.: An in vitro study. *J. Environ. Sci. Nat. Resour.* **2009**, *2*, 45–48.
18. Soylu, E.M.; Soylu, S.; Kurt, S. Antimicrobial activities of the essential oils of various plants against tomato late blight disease agent *Phytophthora infestans*. *Mycopathologia* **2006**, *161*, 119–128. [CrossRef]
19. Prsic, J.; Ongena, M. Elicitors of plant immunity triggered by beneficial bacteria. *Front. Plant Sci.* **2020**, *11*, 1675. [CrossRef]
20. Jangir, M.; Pathak, R.; Sharma, S.; Sharma, S. Biocontrol mechanisms of *Bacillus* sp., isolated from tomato rhizosphere, against *Fusarium oxysporum* f. sp. *lycopersici*. *Biol. Control* **2018**, *123*, 60–70. [CrossRef]
21. Pickett, J.A.; Khan, Z.R. Plant volatile-mediated signalling and its application in agriculture: Successes and challenges. *New Phytol.* **2016**, *212*, 856–870. [CrossRef] [PubMed]
22. Raza, W.; Yuan, J.; Ling, N.; Huang, Q.; Shen, Q. Production of volatile organic compounds by an antagonistic strain *Paenibacillus polymyxa* wr-2 in the presence of root exudates and organic fertilizer and their antifungal activity against *Fusarium oxysporum* f. sp. *niveum*. *Biol. Control* **2015**, *80*, 89–95. [CrossRef]
23. Schalchli, H.; Tortella, G.; Rubilar, O.; Parra, L.; Hormazabal, E.; Quiroz, A. Fungal volatiles: An environmentally friendly tool to control pathogenic microorganisms in plants. *Crit. Rev. Biotechnol.* **2016**, *36*, 144–152. [CrossRef] [PubMed]
24. Li, B. Antibacterial vermiculite nano-material. *J. Miner. Mater. Charact. Eng.* **2002**, *1*, 61–68.
25. Xie, Y.; Wang, A. Effects of modified vermiculite on water absorbency and swelling behavior of chitosan-g-poly (acrylic acid)/vermiculite superabsorbent composite. *J. Compos. Mater.* **2009**, *43*, 2401–2417. [CrossRef]
26. Barabaszova, K.C.; Holesova, S.; Bily, M.; Hundakova, M. Cuo and cuo/vermiculite based nanoparticles in antibacterial PVAC nanocomposites. *J. Inorg. Organomet. Polym. Mater.* **2020**, *30*, 4218–4227. [CrossRef]
27. Cech Barabaszova, K.; Holesova, S.; Sulcova, K.; Hundakova, M.; Thomasova, B. Effects of ultrasound on zinc oxide/vermiculite/chlorhexidine nanocomposite preparation and their antibacterial activity. *Nanomaterials* **2019**, *9*, 1309. [CrossRef]
28. Conchou, L.; Lucas, P.; Meslin, C.; Proffit, M.; Staudt, M.; Renou, M. Insect odorscapes: From plant volatiles to natural olfactory scenes. *Front. Physiol.* **2019**, *10*, 972. [CrossRef]
29. Afify, A.; Potter, C.J. Insect repellents mediate species-specific olfactory behaviours in mosquitoes. *Malar. J.* **2020**, *19*, 127. [CrossRef]
30. Cirlini, M.; Mena, P.; Tassotti, M.; Herrlinger, K.A.; Nieman, K.M.; Dall'Asta, C.; Del Rio, D. Phenolic and volatile composition of a dry spearmint (*Mentha spicata* L.) extract. *Molecules* **2016**, *21*, 1007. [CrossRef]
31. Muttalib, S.A.; Edros, R.; Azah, N.; Kutty, R.V. A review: The extraction of active compound from *Cymbopogon* sp. and its potential for medicinal applications. *Int. J. Eng. Technol. Sci.* **2018**, *5*, 82–98. [CrossRef]
32. Kalleli, F.; Ghassen, A.; Salem, I.B.; Boughalleb-M'hamdi, N.; M'Hamdi, M. Essential oil from fennel seeds (*Foeniculum vulgare*) reduces fusarium wilt of tomato (*Solanum lycopersicon*). *Phytopathol. Mediterr.* **2020**, *59*, 63–76.
33. Choi, H.K.; Song, G.C.; Yi, H.S.; Ryu, C.M. Field evaluation of the bacterial volatile derivative 3-pentanol in priming for induced resistance in pepper. *J. Chem. Ecol.* **2014**, *40*, 882–892. [CrossRef] [PubMed]
34. Cheng, Y.; Zhang, H.; Yao, J.; Wang, X.; Xu, J.; Han, Q.; Wei, G.; Huang, L.; Kang, Z. Characterization of non-host resistance in broad bean to the wheat stripe rust pathogen. *BMC Plant Biol.* **2012**, *12*, 96. [CrossRef] [PubMed]
35. Naznin, H.A.; Kiyohara, D.; Kimura, M.; Miyazawa, M.; Shimizu, M.; Hyakumachi, M. Systemic resistance induced by volatile organic compounds emitted by plant growth-promoting fungi in *Arabidopsis thaliana*. *PLoS ONE* **2014**, *9*, e86882. [CrossRef] [PubMed]
36. Kottb, M.; Gigolashvili, T.; Grobkinsky, D.K.; Piechulla, B. *Trichoderma* volatiles effecting *Arabidopsis*: From inhibition to protection against phytopathogenic fungi. *Front. Microbiol.* **2015**, *6*, 995. [CrossRef]
37. Aamir, M.; Singh, V.K.; Dubey, M.K.; Kashyap, S.P.; Zehra, A.; Upadhyay, R.S.; Singh, S. Structural and functional dissection of differentially expressed tomato WRKY transcripts in host defense response against the vascular wilt pathogen (*Fusarium oxysporum* f. sp. *lycopersici*). *PLoS ONE* **2018**, *13*, e0193922. [CrossRef]

38. Ojola, P.O.; Nyaboga, E.N.; Njiru, P.N.; Orinda, G. Overexpression of rice thaumatin-like protein (OSTLP) gene in transgenic cassava results in enhanced tolerance to *Colletotrichum gloeosporioides* f. sp. *manihotis*. *J. Genet. Eng. Biotechnol.* **2018**, *16*, 125–131. [CrossRef]
39. Praveen, T.; Krishnamoorthy, A.S.; Nakkeeran, S.; Sivakumar, U.; Amirtham, D. Antifungal volatiles from medicinal herbs suppress *Fusarium oxysporum* f. sp. *lycopersici*. *J. Entomol. Zool. Stud.* **2021**, *9*, 1083.

molecules

Article

Assessment of Chemical Composition and Anti-Penicillium Activity of Vapours of Essential Oils from Abies Alba and Two Melaleuca Species in Food Model Systems

Veronika Valková [1], Hana Ďúranová [2], Nenad L. Vukovic [3], Milena Vukic [3], Maciej Kluz [4] and Miroslava Kačániová [1,4,*]

[1] Institute of Horticulture, Faculty of Horticulture and Landscape Engineering, Slovak University of Agriculture, Tr. A. Hlinku 2, 94976 Nitra, Slovakia; veronika.valkova@uniag.sk
[2] AgroBioTech Research Centre, Slovak University of Agriculture, Tr. A. Hlinku 2, 94976 Nitra, Slovakia; hana.duranova@uniag.sk
[3] Department of Chemistry, Faculty of Science, University of Kragujevac, 34000 Kragujevac, Serbia; nvchem@yahoo.com (N.L.V.); milena.vukic@pmf.kg.ac.rs (M.V.)
[4] Department of Bioenergy, Food Technology and Microbiology, Institute of Food Technology and Nutrition, University of Rzeszow, 4 Zelwerowicza Str., 35-601 Rzeszow, Poland; kluczyk82@op.pl
* Correspondence: miroslava.kacaniova@gmail.com

Abstract: The possibilities of the practical utilization of essential oils (EOs) from various plant species in the food industry have attracted the attention of the scientific community. Following our previous studies, the antifungal activities of three further commercial EOs, *Melaleuca armillaris* subsp. *armillaris* (rosalina; REO), *Melaleuca quinquenervia* (niaouli; NEO), and *Abies alba* (fir; FEO), were evaluated in the present research in respect to their chemical profiles, over four different concentrations, 62.5 μL/L, 125 μL/L, 250 μL/L, and 500 μL/L. The findings revealed that the major compounds of REO, NEO, and FEO were linalool (47.5%), 1,8-cineole (40.8%), and α-pinene (25.2%), respectively. In vitro antifungal determinations showed that the inhibition zones of a *Penicillium* spp. mycelial growth ranged from no inhibitory effectiveness (00.00 ± 00.00 mm) to 16.00 ± 1.00 mm, indicating a very strong antifungal activity which was detected against *P. citrinum* after the highest REO concentration exposure. Furthermore, the in situ antifungal efficacy of all EOs investigated was shown to be dose-dependent. In this sense, we have found that the highest concentration (500 μL/L) of REO, NEO, and FEO significantly reduced ($p < 0.05$) the growth of all *Penicillium* strains inoculated on the bread, carrot, and potato models. These results indicate that the investigated EOs may be promising innovative agents in order to extend the shelf life of different types of food products, such as bread, carrot and potato.

Keywords: volatile compounds; in vitro antifungal activities; in situ efficacy; food model systems; disc diffusion method

Citation: Valková, V.; Ďúranová, H.; Vukovic, N.L.; Vukic, M.; Kluz, M.; Kačániová, M. Assessment of Chemical Composition and Anti-Penicillium Activity of Vapours of Essential Oils from Abies Alba and Two Melaleuca Species in Food Model Systems. *Molecules* **2022**, *27*, 3101. https://doi.org/10.3390/molecules27103101

Academic Editor: Ladislav Kokoska

Received: 25 March 2022
Accepted: 10 May 2022
Published: 12 May 2022

Publisher's Note: MDPI stays neutral with regard to jurisdictional claims in published maps and institutional affiliations.

Copyright: © 2022 by the authors. Licensee MDPI, Basel, Switzerland. This article is an open access article distributed under the terms and conditions of the Creative Commons Attribution (CC BY) license (https://creativecommons.org/licenses/by/4.0/).

1. Introduction

In recent decades, there have been dramatic changes in demands for food quality and safety of consumers who are increasingly aware of the impact of food on their health [1]. Bakery products and various types of vegetables, as important constituents of the human diet, provide a substantial amount of essential nutrients [2,3]. However, these goods are mainly sensitive to microbial contamination reducing their shelf life [4]. In particular, the presence of microscopic filamentous fungi (including *Penicillium* spp.) can compromise human health due to the production of mycotoxins [5]. Moreover, it is well known that microorganisms have become increasingly resistant to common antifungals, as reported in numerous studies. Taking into account this trend, some ways of stabilization and preservation techniques, such as modified atmosphere, smart packaging, and bioactive or antifungal agent coatings, have been innovated [6]. For such purposes, plant essential oils

(EOs) can serve as one of the possibilities of effective natural substance applications with antifungal activities [7]. In effect, their prediction of effective antimicrobials against yeasts and fungi have been reported in many studies [8–10].

Essential oils are mainly uncoloured fluids composed of volatile and aromatic substances which naturally occur in different parts of the plants (such as the stem, flowers, seeds, and peels) [11]. Owing to the complex chemical profile, their application appears to be a viable way in the prevention and elimination of antifungal food spoilage [12]. Indeed, the chemical composition of EOs is very complex, consisting of a mixture of more than 50 volatiles at very discrepant levels. Of these constituents, the most relevant for antifungal activities are terpenoid compounds and their derivatives, also designated as isoprenoids since the classification of terpenoids is based on the number of isoprene units [13]. There are several methods for the extraction of EOs, each exhibiting certain advantages and determining physicochemical and biological properties of the extracted oils [14]. The most widely used EO isolation techniques include traditional hydrodistillation, steam distillation extraction, organic solvent extraction, and microwave-assisted hydrodistillation [15].

Currently, more than 3000 EOs have been described, and only around 300 of them are of relevance for use in various industries. However, considering the enormous global diversity of medical plant species, as well as the industrial and commercial concern of EOs, this count is expected to increase radically [16].

Myrtaceae family, including more than 5500 species and approximately 150 genera, is considered the eighth largest flowering plant family with ecological and economic importance related to its production of EOs [17]. Among them, EOs obtained from *Melaleuca (M.) armillaris* subsp. *armillaris* (rosalina) [18] and *M. quinquenervia* (niaouli) [19] have been shown to possess significant antifungal potential. Additionally, EOs of fir, *Abies (A.) alba*, belonging to the Pinaceae family has attracted an increasing interest for its distinctive and refreshing pine–forest fragrance. Having an easing and soothing effect on muscles, it is beneficial for the respiratory system [20]. Moreover, its strong antioxidant and antimicrobial activities also indicate its significant phytomedicine potencial [21].

In this context, as well as following our previous experiments to find new interesting naturally food antifungal agents, the aim of the present study is to characterize three commercially available EOs by evaluating their chemical composition and antifungal efficacies against selected microscopic filamentous fungi of genus *Penicillium* (*P. expansum*, *P. citrinum*, and *P. crustosum*). Finally, their use by the food industry to extend the shelf life of food products will be evaluated on food-based models (bread, carrot, and potato).

2. Results

2.1. Volatile Substances of EOs

All the EOs were analyzed by gas chromatography/mass spectrometry (GC/MS), and their detailed components are summarized in Table 1; in Table 2, the amounts of volatiles in percentage for each class of compounds are presented. From them, it is evident that the identified compounds represented 99.3%, 99.4%, and 99.7% of the oils from *M. armillaris* subsp. *armillaris*, *M. quinquenervia*, and *A. alba*, respectively. The most abundant compounds were shown to be linalool (47.5%), 1,8-cineole (16.9%), and α-terpineol (5.0%) in REO; 1,8-cineole (40.8%), α-terpineol (14.6%), and viridiflorol (12.0%) in NEO; α-pinene (25.2%), β-pinene (18.3%), and α-limonene (18.1%) in FEO.

Table 1. Chemical composition of analyzed EOs.

No	Compound [a]	REO (%)	NEO (%)	FEO (%)	RI (lit.)	RI (calc.) [b]
1	santene	/	/	1.6	888	889
2	2-bornene	/	/	0.2	907	909
3	tricyclene	/	/	3.1	926	924
4	α-thujene	Tr [c]	tr	/	930	926
5	α-pinene	1	4.8	25.2	939	938
6	β-fenchene	/	/	tr	940	942
7	α-fenchene	/	/	2.6	952	947
8	camphene	0.1	0.5	13.4	954	948
9	benzaldehyde	/	tr	/	960	958
10	sabinene	tr	0.4	tr	975	977
11	β-pinene	0.9	2.2	18.3	979	980
12	β-myrcene	0.1	0.8	0.6	990	992
13	α-phellandrene	/	tr	tr	1002	1004
14	pseudolimonene	/	/	tr	1003	1003
15	δ-3-carene	/	tr	0.7	1011	1009
16	α-terpinene	tr	tr	0.1	1017	1016
17	*p*-cimene	/	3.2	/	1024	1023
18	*o*-cymene	1.3	/	1.5	1026	1026
19	α-limonene	1.3	6.4	18.1	1029	1028
20	β-phellandrene	/	/	1.1	1029	1030
21	1,8-cineole	16.9	40.8	/	1031	1033
22	(*E*)-β-ocimene	/	tr	/	1050	1047
23	γ-terpinene	1.5	1.8	tr	1059	1060
24	cis-linalool oxide	1.2	/	/	1072	1074
25	α-terpinolene	0.2	1.2	0.4	1088	1088
26	trans-linalool oxide	1.4	/	/	1086	1089
27	linalool	47.5	/	/	1096	1098
28	α-thujone	0.4	/	/	1102	1101
29	β-thujone	tr	/	/	1114	1114
30	cis-limonene oxide	/	/	0.2	1136	1136
31	trans-pinocarveol	/	/	0.2	1139	1140
32	trans-verbenol	/	/	tr	1144	1145
33	camphor	1.3	tr	/	1146	1148
34	menthone	/	tr	/	1152	1151
35	iso-menthone	/	tr	/	1162	1162
36	pinocarvone	/	/	tr	1164	1163
37	borneol	0.3	/	0.1	1169	1170
38	menthol	/	tr	/	1171	1173
39	4-terpinenol	2.9	1.5	tr	1171	1178
40	*p*-cymen-8-ol	/	/	tr	1182	1183
41	α-terpineol	5	14.6	0.6	1188	1189
42	verbenone	/	/	tr	1208	1208
43	endo-fenchyl acetate	/	/	0.2	1220	1221
44	nerol	0.2	/	/	1229	1227
45	carvone	/	/	tr	1243	1241
46	linalool acetate	0.4	/	/	1257	1255
47	2-phenyl ethyl acetate	tr	/	/	1258	1258
48	geranial	tr	/	/	1267	1263
49	(*E*)-cinnamaldehyde	tr	/	/	1270	1269
50	bornyl acetate	/	/	2.8	1285	1286
51	isobornyl acetate	tr	/	/	1285	1287
52	methyl geranate	0.1	/	/	1324	1321
53	α-terpinyl acetate	/	2.4	/	1349	1341
54	isoledene	0.4	/	/	1376	1371
55	α-copaene	/	tr	tr	1376	1379

Table 1. *Cont.*

No	Compound [a]	REO (%)	NEO (%)	FEO (%)	RI (lit.)	RI (calc.) [b]
57	longifolene	/	/	0.5	1407	1408
56	α-gurjunene	0.3	tr	/	1409	1408
58	(*E*)-caryophyllene	0.8	2.4	7	1419	1422
59	β-gurjunene	0.6	/	/	1433	1427
60	β-humulene	0.2	/	/	1438	1436
61	2-phenyl propyl isobutanoate	0.5	/	/	1440	1438
62	aromadendrene	5.3	0.6	/	1441	1443
63	α-humulene	/	tr	0.2	1454	1456
64	α-amorphene	/	tr	tr	1484	1485
65	β-selinene	/	tr	/	1490	1490
66	α-selinene	0.3	tr	/	1492	1492
67	valencene	0.4	/	/	1496	1497
68	ledene	1.5	1.6	/	1496	1498
69	bicyclogermacrene	0.8	/	/	1500	1503
70	α-muurolene	0.1	/	tr	1500	1504
71	δ-amorphene	0.1	/	/	1512	1513
72	δ-cadinene	0.5	0.5	/	1523	1525
73	cis-calamenene	0.1	/	/	1529	1530
74	α-cadinene	/	/	tr	1538	1542
75	palustrol	0.9	tr	/	1568	1570
76	spathulenol	0.5	/	/	1578	1577
77	caryophyllene oxide	1.2	tr	1	1583	1583
78	viridiflorol	0.3	12	/	1592	1593
79	widdrol	0.3	/	/	1599	1601
80	rosifoliol	0.2	/	/	1600	1604
81	ledol	/	1.7	/	1602	1605
82	α-cadinol	/	tr	/	1654	1656
	total	99.3	99.4	99.7		

[a] Identified compounds; [b] values for retention indices on HP-5MS column; [c] tr—compounds identified in amounts less than 0.1%.

Table 2. Total amount of volatiles presented in percentage for each class of compounds.

Class of Compounds	REO (%)	NEO (%)	FEO (%)
non-terpenic compounds	0.5	tr	1.8
hydrocarbons	/	/	1.8
aromatic compounds	0.5	tr	/
monoterpenes	84	80.6	89.2
monoterpene hydrocarbons	6.4	21.3	85.1
oxygenated monoterpenes	77.6	59.3	4.1
monoterpene epoxide	19.5	40.8	0.2
monoterpene alcohols	55.9	16.1	0.9
monoterpene ketones	1.7	tr	tr
monoterpene esters	0.5	2.4	3
sesquiterpenes	14.8	18.8	8.7
sesquiterpene hydrocarbons	11.4	5.1	7.7
oxygenated sesquiterpenes	3.4	13.7	1
sesquiterpene alcohols	2.2	13.7	/
sesquiterpene epoxides	1.2	tr	1
total	99.3	99.4	99.7

2.2. *In Vitro Antifungal Potential of EOs*

In the current research, a disc diffusion method was applied to evaluate the antifungal activities of the selected EOs (REO, NEO, FEO) against *P. expansum*, *P. citrinum*, and

P. crustosum. As shown in Table 3, the growth inhibition of the *Penicillium* strains depended on the type and concentration of the EO analyzed ($p < 0.05$); with increasing concentrations, the antifungal activities increased. A very strong antifungal effectiveness was observed for REO, where the highest concentration (500 μL/L) inhibited the growth of *P. citrinum* with an inhibition zone of 16.00 ± 1.00 mm. Moderate values for antifungal potential were detected for the highest concentration (500 μL/L) of REO and NEO against *P. expansum* (10.33 ± 0.58 mm) and *P. crustosum* (10.67 ± 0.58 mm), and *P. expansum* (11.00 ± 1.00 mm) and *P. citrinum* (10.00 ± 1.00 mm), respectively. The EOs exhibited weak zones of inhibition as follows: REO against *P. citrinum* (5.33 ± 1.53 mm in 125 μL/L; 9.33 ± 0.58 mm in 250 μL/L), and *P. crustosum* (5.33 ± 0.58 mm in 125 μL/L; 6.57 ± 0.58 mm in 250 μL/L); NEO against *P. expansum* (8.00 ± 1.00 mm in 250 μL/L), *P. citrinum* (5.33 ± 0.58 mm in 250 μL/L), and *P. crustosum* (7.33 ± 0.58 mm in 500 μL/L); FEO against *P. citrinum* (from 5.33 ± 0.58 mm in 125 μL/L to 7.67 ± 0.58 mm in 500 μL/L), *P. crustosum* (5.67 ± 1.15 mm in 250 μL/L and 8.33 ± 0.58 mm in 500 μL/L), and *P. expansum* (5.67 ± 1.15 mm in 500 μL/L). Remaining values for fungal growth inhibition indicated weak or very weak antifungal actions of the EOs.

Table 3. Antifungal activity of EO samples in analyzed concentrations (inhibition zone in mm).

	P. expansum				*P. citrinum*				*P. crustosum*			
Con. (μL/L)	62.5	125	250	500	62.5	125	250	500	62.5	125	250	500
REO	0.00 ± 0.00 ^{aA}	0.00 ± 0.00 ^{aA}	3.67 ± 0.58 ^{aB}	10.33 ± 0.58 ^{aC}	2.68 ± 0.58 ^{aA}	5.33 ± 1.53 ^{aB}	9.33 ± 0.58 ^{aC}	16.00 ± 1.00 ^{aD}	4.67 ± 0.58 ^{aA}	5.33 ± 0.58 ^{aA}	6.57 ± 0.58 ^{aB}	10.67 ± 0.58 ^{aC}
NEO	0.00 ± 0.00 ^{aA}	0.00 ± 0.00 ^{aA}	8.00 ± 1.00 ^{bB}	11.00 ± 1.00 ^{aC}	3.33 ± 0.58 ^{aA}	4.33 ± 0.58 ^{aAB}	5.33 ± 0.58 ^{bB}	10.00 ± 1.00 ^{bC}	0.00 ± 0.00 ^{bA}	0.00 ± 0.00 ^{bA}	4.67 ± 0.58 ^{bB}	7.33 ± 0.58 ^{bC}
FEO	0.00 ± 0.00 ^{aA}	2.33 ± 0.58 ^{bB}	3.33 ± 0.58 ^{aB}	5.67 ± 1.15 ^{bC}	4.00 ± 1.00 ^{aA}	5.33 ± 0.58 ^{aA}	7.00 ± 1.00 ^{cB}	7.67 ± 0.58 ^{cB}	0.00 ± 0.00 ^{bA}	0.00 ± 0.00 ^{bA}	5.67 ± 1.15 ^{abB}	8.33 ± 0.58 ^{bC}

Note: Mean ± standard deviation. REO: rosalina essential oil; NEO: niaouli essential oil; FEO: fir essential oil. Values in the same column with different small letters, and those in the same row (for the same type of fungi strains) with different upper-case letters, are significantly different ($p < 0.05$). Con.—concentration; 0.00—total growth; N—without growth.

2.3. Moisture Content and Water Activity of Food Models

Generally, the food quality depends on water activity (a_w) and moisture content (MC) directly affecting the microbial growth connected with the shelf life of the foods. Both parameters were significantly different ($p < 0.05$) depending on the type of substrate analyzed (Table 4). In this line, values for MC ranged from 43.12 ± 0.35% (bread) to 86.83 ± 0.42% (carrot), and aw values varied from 0.942 ± 0.001 (bread) to 0.946 ± 0.002 (potato).

Table 4. Moisture content and water activity in food models analyzed.

Parameters	Bread	Carrot	Potato
MC (%)	43.12 ± 0.35 ^{a}	86.83 ± 0.42 ^{b}	81.55 ± 1.65 ^{c}
a_w	0.942 ± 0.001 ^{a}	0.945 ± 0.002 ^{b}	0.946 ± 0.002 ^{c}

Note: Mean ± standard deviation. MC—moisture content; a_w—water activity. Values in the same row with different small letters are significantly different ($p < 0.05$).

2.4. In Situ Antifungal Potential of EOs

In situ antifungal activity of all EOs investigated against filamentous fungi strains growing on three selected food models (bread, carrot, potato) was shown to be dose-dependent. In effect, we have found that the highest concentration (500 μL/L) of REO, NEO, and FEO significantly reduced ($p < 0.05$) the growth of *P. expansum*, *P. citrinum*, and *P. crustosum* inoculated on the bread model (Table 5). In *P. expansum*, the actions were additionally identified to be an EO type-dependent. In this sense, REO appeared to be the most effective ($p < 0.05$) in the fungus growth inhibition, and the EOs effectiveness decreased in the following order: REO > FEO > NEO. Against *P. citrinum*, the highest concentration of FEO exhibited the same antifungal activity as NEO and REO; however, both EOs (NEO, REO) had conversely stronger impacts ($p < 0.05$) on *P. crustosum* growth inhibition as compared to FEO.

Table 5. In situ antifungal activity of the EO samples in analyzed concentrations against the growth of selected *Penicillium* spp. inoculated on bread.

Fungi Strain	*P. expansum*				*P. citrinum*				*P. crustosum*			
Con. (μL/L)	62.5	125	250	500	62.5	125	250	500	62.5	125	250	500
REO	8.14 ± 0.95 ^{aA}	1.20 ± 0.80 ^{aB}	29.27 ± 1.56 ^{aC}	98.50 ± 3.82 ^{aD}	−2.81 ± 1.40 ^{aA}	24.36 ± 3.57 ^{aB}	95.31 ± 4.03 ^{aC}	89.50 ± 3.73 ^{aC}	−30.43 ± 4.11 ^{aA}	−47.37 ± 5.98 ^{aB}	90.02 ± 3.42 ^{aC}	97.32 ± 4.51 ^{aC}
NEO	18.21 ± 1.65 ^{bA}	5.11 ± 2.33 ^{bB}	13.07 ± 3.92 ^{bA}	46.53 ± 4.85 ^{bC}	4.71 ± 1.19 ^{bA}	26.55 ± 1.37^{aB}	23.89 ± 2.64 ^{bB}	79.84 ± 3.66 ^{bC}	13.87 ± 3.79 ^{bA}	11.90 ± 1.61 ^{bA}	84.04 ±6.97^{aB}	95.60 ± 4.12 ^{aC}
FEO	19.92 ± 6.51 ^{bA}	19.22 ± 6.76 ^{cA}	18.62 ± 4.87 ^{bA}	81.72 ± 4.17 ^{cB}	50.87 ± 1.47 ^{cA}	49.95 ± 1.73 ^{bA}	41.23 ± 1.18 ^{cB}	87.94 ± 10.3 ^{abC}	54.25 ± 4.61 ^{cA}	25.64 ± 1.15 ^{cB}	37.62 ± 1.65 ^{bC}	45.13 ± 1.81 ^{bD}

Note: Mean ± standard deviation. REO: rosalina essential oil; NEO: niaouli essential oil; FEO: fir essential oil. Values in the same column with different small letters, and those in the same row (for the same type of fungi strains) with different upper-case letters, are significantly different ($p < 0.05$). Con.—concentration.

The growth of *P. expansum*, *P. citrinum*, and *P. crustosum* on a carrot as a food model was completely inhibited by the highest concentration of REO (Table 6). Additionally, NEO was found to be able to inhibit the growth of *P. expansum* and *P. citrinum* but to a lesser extent ($p < 0.05$). On the other hand, it showed only a slight effectiveness against the growth of *P. crustosum*. FEO had the strongest antifungal action against *P. expansum*, which was similar to that of REO. By contrast, the inhibitory effect of FEO on the mycelial growth of *P. citrinum* and *P. crustosum* was lower than those of REO and NEO, and higher than that of NEO, respectively.

Table 6. In situ antifungal activity of the EO samples in analyzed concentrations against the growth of selected *Penicillium* spp. inoculated on carrot.

Fungi Strain	*P. expansum*				*P. citrinum*				*P. crustosum*			
Con. (μL/L)	62.5	125	250	500	62.5	125	250	500	62.5	125	250	500
REO	52.17 ± 1.27 ^{aA}	60.00 ± 1.55 ^{aB}	100.00 ± 0.00 ^{aC}	100.00 ± 0.00 ^{aC}	54.05 ± 1.35 ^{aA}	72.73 ± 6.10 ^{aB}	87.84 ± 7.71 ^{aC}	100.00 ± 0.00 ^{aD}	54.88 ± 2.24 ^{aA}	59.74 ± 3.47 ^{aA}	96.51 ± 4.41 ^{aB}	100.00 ± 0.00 ^{aB}
NEO	40.35 ± 7.56 ^{bA}	0.00 ± 0.00 ^{bB}	39.39 ± 8.89 ^{bA}	72.22 ± 5.4 5^{bC}	14.75 ± 2.94 ^{bA}	26.32 ± 7.57 ^{bB}	6.90 ± 1.11 ^{bC}	62.00 ± 1.98 ^{bD}	0.00 ± 0.00 ^{bA}	0.00 ± 0.00 ^{bA}	2.78 ± 0.48 ^{bB}	14.87 ± 1.65 ^{bC}
FEO	48.44 ± 6.36 ^{abA}	8.64 ± 3.69 ^{cB}	9.86 ± 5.82 ^{cB}	97.67 ± 6.22 ^{aC}	7.69 ± 2.76 ^{cA}	0.00 ± 0.00 ^{cB}	9.80 ± 3.58 ^{bA}	41.00 ± 4.97 ^{cC}	0.00 ± 0.00 ^{bA}	20.00 ± 8.94 ^{cB}	39.68 ± 4.19 ^{cC}	64.47 ± 6.37 ^{cD}

Note: Mean ± standard deviation. REO: rosalina essential oil; NEO: niaouli essential oil; FEO: fir essential oil. Values in the same column with different small letters, and those in the same row (for the same type of fungi strains) with different upper-case letters, are significantly different ($p < 0.05$). Con.—concentration.

On a potato model (Table 7), the growth of *Penicillium* spp. was markedly inhibited by the highest concentrations of all EOs investigated (except for FEO in *P. crustosum*). Regarding this, the results showed the strongest antifungal activity of REO and NEO against the growth of *P. expansum*, *P. citrinum*, and *P. crustosum*. However, the inhibitory action of the highest concentration of FEO was species-dependent. Indeed, the EO displayed a very strong effectiveness against *P. expansum*, whilst its efficacy against the growth of *P. citrinum* and *P. crustosum* was only weak.

Table 7. In situ antifungal activity of the EO samples in analyzed concentrations against the growth of selected *Penicillium* spp. inoculated on potato.

Fungi Strain	*P. expansum*				*P. citrinum*				*P. crustosum*			
Con. (μL/L)	62.5	125	250	500	62.5	125	250	500	62.5	125	250	500
REO	23.21 ± 5.41 ^{aA}	4.00 ± 1.17 ^{aB}	53.18 ± 9.83 ^{aC}	83.02 ± 2.74 ^{aD}	0.00 ± 0.00 ^{aA}	0.00 ± 0.00 ^{aA}	86.32 ± 1.83 ^{aB}	76.67 ± 6.44 ^{aC}	0.00 ± 0.00 ^{aA}	29.17 ± 2.86 ^{aB}	40.82 ± 4.89 ^{aC}	70.97 ± 6.37 ^{aD}
NEO	0.00 ± 0.00 ^{bA}	0.00 ± 0.00 ^{bA}	0.00 ± 0.00 ^{bA}	71.43 ± 9.59 ^{aB}	0.00 ± 0.00 ^{aA}	0.00 ± 0.00 ^{aA}	0.00 ± 0.00 ^{bA}	86.67 ± 8.96 ^{aB}	0.00 ± 0.00 ^{aA}	0.00 ± 0.00 ^{bA}	15.79 ± 1.32 ^{bB}	92.00 ± 6.83 ^{bC}
FEO	7.81 ± 1.47 ^{cA}	9.62 ± 1.45 ^{cA}	82.76 ± 3.12 ^{cB}	98.31 ± 8.92 ^{bC}	10.91 ± 1.36 ^{bA}	1.79± 0.62 ^{bB}	7.32 ± 1.33 ^{cC}	18.75 ± 2.26 ^{bD}	0.00 ± 0.00 ^{aA}	0.00 ± 0.00 ^{bA}	2.00 ± 1.17 ^{cB}	9.59 ± 1.52 ^{cC}

Note: Mean ± standard deviation. REO: rosalina essential oil; NEO: niaouli essential oil; FEO: fir essential oil. Values in the same column with different small letters, and those in the same row (for the same type of fungi strains) with different upper-case letters, are significantly different ($p < 0.05$). Con.—concentration.

3. Discussion

Generally, it is well known that the antifungal activities of diverse EOs depend on their chemical profile and concentration of the individual constituents [22]. Regarding the *M. armillaris* subsp. *armillaris* EO, the literature reports the occurrence of three chemotypes of this species based on the proportions of 1,8-cineole, linalol, and methyleugenol [23]. According to Brophy and Doran [24], *M. armillaris* subsp. *armillaris* presents EO that is mainly monoterpenoid in character. The authors analyzed the EO samples obtained from two latitude areas. From their findings, it can be seen that the chemical composition of the samples was qualitatively very similar throughout the species range; however, quantitatively, the EO differed considerably in an apparent association with the latitude of collection. In line with our results, REO samples from the north of the species range contained linalool (56.2%) as the major component, and also significant amounts of 1,8-cineole (13.3%). On the other hand, REO from the south comprised 1,8-cineole (43.6%) as the principal component along with significant amounts of α-pinene (13.1%). Supporting our findings, linalool (41.6%), and 1,8-cineole (25.2%) as primary compounds of REO were also detected in the study by Zhao et al. [25].

The research of Ireland et al. [26], which examined the needle EO of *M. quinquenervia* over its geographical range in Australia and Papua New Guinea, has displayed a wide variation in chemical composition, and only two major EO chemotypes. Indeed, chemotype 1 was composed of E-nerolidol (74–95%) and linalool (14–30%), and chemotype 2 contained 1,8-cineole (10–75%), viridiflorol (13–66%), and α-terpineol (0.5–14%) in varying proportions and dominance order in the EO samples, which is in accordance with our findings. Similarly, Zhao et al. [25] found that 1,8-cineole (59.2%) and α-terpineol (9.6%) were primary compounds of NEO obtained from Australia.

FEO commonly contains various volatile substances, with α-pinene, β-pinene, camphene, limonene, and bornyl acetate being the most important [20,27,28]. In this context, the study by Zeneli et al. [29] showed the chemical variability of FEO from an Albania area, in which four samples of the genus *A. alba* were evaluated. Their results demonstrated that the samples were dominated on average by β-pinene (22.73%) and α-pinene (10.55%), which agrees with our study. Likewise, the studies by Roussis et al. [30] and Tsasi et al. [31] revealed β-pinene (19.8%) and α-pinene (10.9%), and β-pinene (26.9%), bornyl acetate (12.5%) and α-pinene (10.1%) to be the main components in FEOs obtained from the South Balkans and Greece areas, respectively.

Many studies indicated antifungal efficacy of different EOs types against a wide range of fungal strains such as *Penicillium* spp. (*P. polonicum*, *P. expansum*, *P. citrinum*, *P. crustosum*, *P. funiculosum*, *P. brevicompactum*, *P. glabrum*, *P. oxalicum*, *P. chrysogenum*) [32,33]. Very strong antifungal action of the highest REO concentration against *P. citrinum* growth observed in our study is reinforced by the work of Farag et al. [18], whose REO (in the concentration of 5 μL on the filter paper discs) exhibited the highest inhibitory effects (inhibition zone 18.3 mm) on the growth of *Aspergillus (A.) niger* among the EOs obtained from other

Melaleuca species. We hypothesize that the very high antifungal activity of our REO may be due to the high proportion of linalool (47.5%). Indeed, results of the study by Dias et al. [34] suggest the antifungal efficacy of linalool itself against *Candida* (*C.*) *albicans*, *C. tropicalis*, and *C. krusei*. However, values for the minimal inhibitory concentration (MIC) differed among the strains tested. In this line, linalool displayed the lowest inhibitory concentration against *C. tropicalis* (500 μg/mL) followed by *C. albicans* (1000 μg/mL) and *C. krusei* (2000 μg/mL), and the growth of *C. tropicalis* was completely inhibited by linalool. The underlying mechanism of the linalool action is associated with its ability to damage the cell wall of microorganisms accompained by a reduction in membrane potential, leakage of alkaline phosphatase and the release of macromolecules (including DNA, RNA, and protein), how it was found in *Pseudomonas fluorescens* [35] and *Shewanella putrefaciens* [36]. Additionally, bacterial metabolic and oxidative respiratory perturbations interfering in cellular functions and even causing cell death have been demonstrated in these studies. The very strong antifungal activity of REO can also be attributed to the high proportion of 1,8-cineole (16.9%) in its composition, which is also known for its fungicidal properties. This fact was also confirmed by Vilela et al. [37] who recorded a partial inhibitory action (5.5%; independent of organism) of 1.35 μL of 1,8-cineole (isolated from *Eucalyptus globulus* leaves) on the mycelial growth of *A. flavus* and *A. parasiticus*, thereby recognizing the component to be one of those responsible for *E. globulus* antifungal activity. In this view, it can be concluded that the presence of this substance in our NEO (40.8%) may greatly contribute to its antifungal efficacy. In this regard, Tančínová et al. [38] analyzed three EOs from the genus *Melaleuca* (tea tree, cajeput, and niaouli). They have found that *M. alternifolia* Cheel (tea tree) EO showed a strong inhibitory effect (from 84.8% to 100% inhibition) on the strains of *P. commune*. A weaker inhibitory action was reported for *M. leucadendra* (cajeput) EO and the weakest one for NEO. As in our study, plant species- and pathogen species-dependent variation in in vitro antifungal potential of 11 Myrtaceae EOs has also been reported by Lee et al. [19]. Regarding 1,8-cineole, it has been found that *Staphylococcus aureus* treated with the chemical substance exhibited prominent outer membrane disintegration with a concentrated/reduced/agglomerated nucleoplasm [39]. Additionally, the study performed by Yu et al. [40] showed a destructive effect of 1,8-cineole on organelles along with the appearance of many unidentifiable vesicular structures in *Botrytis cinerea*. Using transmission electron microscopy, the authors found that 1,8-cineole may penetrate the cell membrane and damage cellular organelles without causing lesions on the membrane. Moreover, Nikolova et al. [41] observed very weak inhibitory activity of acetone extract from *A. alba* needles against the growth of *Alternaria alternata*, corresponding with our findings dealing with weak antifungal effectiveness of FEO against the selected *Penicillium* strains investigated. A remarkable selective antimycotic effect of FEO against a clinical strain of *C. albicans* has also been demonstrated in the study of Salamon et al. [42]. Although the underlying mechanism of antifungal action of the EOs has not been yet fully elucidated, we propose that in our EOs, it may be linked to the inhibition of microscopic filamentous fungi respiration and disruption of the permeability barriers of their cell membrane structures [43]. On the basis of the promising data from in vitro antifungal activity, the samples of our EOs were also applied to determine their vapour-phase inhibitory effects on the *Penicillium* spp. growth inoculated on the selected food models.

Moisture content and aw are the pivotal parameters in predicting the quality and stability of food products [44]. Moreover, they are very valuable indicators because their values commonly correlate well with the potential for fungi growth and metabolic activity [45]. Concretely, MC defines the amount of water in a product, thus, providing information about its yield, quantity, and texture; however, it does not provide credible data regarding microbial safety. On the other hand, aw expresses the volume of water that is available not only for reaction with other molecules but also for food spoilage processes including enzymatic browning and microbial growth. Taking into account these aspects, aw is an indicator of food stability with respect to microbial growth, biochemical reaction rates, and physical properties [46]. Generally, it is known that the growth of microorganisms occurs in product samples with aw values higher than 0.60 [47], which is consistent with our detected data. In line with these findings, the approximate value for aw in white bread is within the range of 0.94 to 0.97 [48], indicating its susceptibility to microbial spoilage with the main effect coming from the growth of various molds. With respect to our result obtained from the bread analysis, similar values for aw (0.944; 0.948) were observed in two bread samples from registered baking industries in the research by Ayub et al. [49]. Regarding the MC, bread is a food product with intermediate moisture [50], typically ranging from 35–42% [51], which also corresponds with our results. Nonetheless, most of the fresh foods are perishable because of their high values for MC [52], as it was also shown in our vegetable samples (>80%). Similarly to our results, Sipahioglu and Barringer [53] estimated the values for MC and aw in carrot and potato to be 89.97% and 0.996, and 75.19% and 0.990, respectively. All the mentioned findings indicate the suitability of chosen food models (bread, carrot, potato) for in situ antifungal analysis of the EOs investigated.

After testing the ability of the food spoilage fungi (*P. citrinum*, *P. crustosum*, and *P. expansum*) to grow on the selected food models, the antifungal efficacy of REO, NEO and FEO was evaluated. Generally, food products require protection against fungi deratoration during their storage [54]. In bakery products, the microscopic filamentous microorganisms (mainly *Penicillium* spp. and *Aspergillus* spp.) are the most common species causing their spoilage [55,56]. These fungi are able to grow in the bread surface and form a greenish-blue layer [57]. Besides the degraded quality including external appearance, the fungi are responsible for an unpleasant taste and aftertaste formation, and the production of mycotoxins and allergenic compounds. Importantly, these substances may be created even before the growth is visible [55]. Bakery goods usually have a short shelf life (only a few days at room temperature) due to their high aw, as it was confirmed in the previous part of this research. Spoilage of fruits and vegetables caused by fungi during storage is also the major concern affecting their quality and shortening their shelf life [54]. Therefore, the application of EOs in packaging may be the principal choice for satisfying and ecologically demanding ways to extend the shelf life of products without using synthetic preservatives.

The antimicrobial effect of EOs against various types of food spoilage microorganisms, after being applied by direct contact, has been demonstrated on a large scale. However, the vapour phase and volatile components present in EOs have not been thoroughly investigated [58]. The bioactivity of EOs in the vapour phase seems to be an interesting alternative that makes them potentially useful as antimicrobial agents for the preservation of stored fresh products. In effect, promising results in this field of research have been obtained, especially for bacteria and fungi [59–61]. Additionally, some studies have reported that vapour generated by EOs has a greater antimicrobial effect as compared to their liquid form applied by direct contact [62,63]. Moreover, Nadjib et al. [64] found that lipophilic molecules in the aqueous phase associate to form micelles restraining the attachment of EOs to microorganisms, whereas the vapour phase allows for free attachment.

The results of our in situ analysis revealed the antifungal effects of all three tested EOs (REO, NEO, and FEO) against selected *Penicillium* spp.; however, their effectiveness varied depending on diverse factors including the type of EOs, their concentrations, and also the fungi and food model used. Although *Penicillium* strains are known due to their high resistance [65], their mycelial growth on all three food models was suppressed by all EOs used with the strongest effectiveness in their highest concentration (500 μL/L). We suppose that the major compounds found in the EOs concept can primarily contribute to the antifungal effects. However, it is necessary to keep in mind that the overall EO effect is not attributed to just one or only some of its components [66], but it is a result of the synergistic action of all its constituents [67]. The antifungal efficacy of linalool (present in our REO in 57.5%) was also confirmed in the study performed by Xu et al. [68] who exogenously applied linalool (in the concentration of 20.95 μM and 2095 μM) on strawberry fruits infected with *Botrytis cinerea*. The antimicrobial potential of linalool has also been demonstrated in the research of Chang et al. [69] who evaluated the effect of active polyethylene (PE) film containing linalool active components (0–2%) on the microbial shelf life of mozzarella cheese. They found that PE films with a higher linalool content significantly suppressed fungal growth throughout the storage period (30 days). Moreover, the results obtained by Soković et al. [70] suggest that 1,8-cineole and α-pinene (primary constituents in our NEO and FEO compositions, respectively) are very strong antifungal agents with MICs of 2.00 μL/L (*Trichoderma harzianum*) and 7.00 μL/L (*Verticillium fungicola*), and 3.00 μL/L (*Trichoderma harzianum*) and 8.00 μL/L (*Verticillium fungicola*), respectively.

From the findings of all our analyses it can be concluded that REO, NEO, and FEO may be promising constituents with a potential use for extending the shelf life of bread and vegetables in the commercial scale of the food industry.

4. Materials and Methods

4.1. Tested EOs

Rosalina EO (REO; *M. armillaris* subsp. *armillaris*), niaouli EO (NEO; *M. quinquenervia*), and fir EO (FEO; *A. alba*) were extracted by steam distillation of fresh needles and leaves. These EOs were obtained by a commercial producer Hanus Ltd. (Nitra, Slovakia), and were preserved at 4 °C in the laboratory refrigerator until their next use. Importantly, the research complements our knowledge gained in previous similar experiments [33,71]. In this regard, it provides a comprehensive overview of the biological effects of different types of commercial EOs obtained from the same company.

4.2. Determination of EOs Volatile Constituents

The volatile compounds of the EOs were determined using gas chromatography with mass spectrometry (GC-MS), as it was described by Valková et al. [33]. In brief, the analysis was carried out by Agilent Technology 6890N (Agilent Technologies, Santa Clara, CA, USA) coupled to quadrupole mass spectrometer 5975B (Agilent Technologies, Santa Clara, CA, USA). Separation of compounds was carried out using HP-5MS capillary column (30 m × 0.25 mm × 0.25 m). The temperature program was as follows: 60 °C to 150 °C (increasing rate 3 °C/min) and 150 °C to 280 °C (increasing rate 5 °C/min), using helium 5.0 as the carrier gas with a flow rate of 1 mL/min. Samples of essential oils were desolved in pentane, and injection volume was 1μL. The split/splitless injector temperature was set at 280 °C. The investigated samples were injected in the split mode with a split ratio at 40.8:1. Electron-impact mass spectrometric data (EI-MS; 70 eV) were acquired in scan mode over the m/z range 35–550. The mass spectrometry ion source temperature was 230 °C, while the temperature of MS quadrupole was set at 150 °C. Solvent delay time of 3 min. After the separation, the components were identified based on the comparison of their relative retention index and compared with the library mass spectral database (Wiley and NIST databases). The percentage composition of compounds (relative quantity; amounts higher than 0.1%) was measured based on the peak area [33]. The retention indices were experimentally determined by injecton of standard n-alkanes (C_6–C_{34}) under the same chromatographic conditions.

4.3. Evaluation of EOs Antifungal Potential

4.3.1. Fungal Strains and Culture Media

In the current study, three strains of genus *Penicillium* (*P. expansum*, *P. crustosum*, *P. citrinum*), isolated from berry samples of *Vitis vinifera* (growing in vineyards localized in Slovakia), were employed. Consequently, the microscopic filamentous fungi were classified using a reference-based MALDI-TOF MS Biotyper, and validated by comparison with the taxonomic identification using 16S ribosomal RNA (16S rRNA) gene sequences analysis.

To prepare fungal media, the strains were inoculated in Sabouraud Dextrose Agar (SDA; Oxoid, Basingstoke, UK) and incubated for 5 days at 25 °C. Subsequently, small aliquots of the fungi were transferred to test tubes, each containing 3 mL of distilled water. The inoculum concentration was standardized by comparison with the 0.5 McFarland scale (1.5×10^8 CFU/mL).

4.3.2. In Vitro Antifungal Activity of EOs

Evaluation of the in vitro antifungal activity of the EOs was performed using the agar disc diffusion method, according to Valková et al. [33] with minor modifications. For this purpose, an aliquot of 100 μL of culture media was inoculated on SDA. Then, the discs of filter paper (6 mm) were impregnated with 10 μL of each EO sample (in four concentrations: 62.5, 125, 250, and 500 μL/L), and applied on the SDA surfaces. Fungi were incubated aerobically at 25 ± 1 °C for 5 days. After the incubation, diameters of the inhibition zones in mm were measured. The values for inhibitory activity increased in the following manner: weak antifungal activity (5–10 mm) < moderate antifungal activity (10–15 mm) < very strong antifungal activity (zone > 15 mm).

4.3.3. In Situ Antifungal Activity of EOs

All three fungal strains (*P. expansum*, *P. crustosum*, and *P. citrinum*) were used to evaluate the antifungal activity of the EOs in situ.

4.3.4. Food Models

Three frequently consumed food products, i.e., bread, carrot, and potato were applied as substrates for the growth of the fungi. Among them, white bread was developed in the Laboratory of Cereal Technologies (Research Center AgroBioTech, SUA in Nitra) according to the methodology described in the study by Valková et al. [72]. The vegetables were purchased at the local market (Nitra, Slovakia).

4.3.5. Moisture Content and Water Activity of Food Models

To predict the suitability of substrates for fungal growth, moisture content (MC) and water activity (aw) were determined, as reported by Valková et al. [73].

4.3.6. Vapour Contact Assay

The experiment itself was performed as reported by Valková et al. [33]. After cooling, the bread slices with a thickness of 15 mm were transferred into glass jars (Bormioli Rocco, Fidenza, Italy; 500 mL). The inoculum of tested strains was applied by stabbing three times with an injection pin on the bread substrate. Then, a sterile filter paper disc (60 mm) was placed under the jar top, and 100 μL of the EOs in concentrations of 62.5, 125, 250, and 500 μL/L (diluted in ethyl acetate) were applied to it. The control bread was not treated with the EOs. Finally, the jars were hermetically closed and stored in an incubator for 14 days at 25 ± 1 °C. For vegetables (carrot and potato) used as food models, the methodology was slightly modified. Firstly, sliced carrot and potato (5 mm) were placed on the bottom of Petri dishes (PDs), and the inoculum was applied by stabbing one time with an injection pin on the vegetable surface. Further, 10 μL of the EOs (in the same four concentrations) was applied on the sterile filter paper disc (60 mm), then, it was placed at the top of PD. Subsequently, PDs were hermetically closed using parafilm and cultivated at 25 °C for 14 days.

4.3.7. Determination of Fungal Growth Inhibition

In situ fungal growth was determined using stereological methods. In this concept, the volume density (Vv) of visible fungal colonies was firstly established using ImageJ software counting the points of the stereological grid hitting the colonies (P) and those (p) falling to the reference space (growth substrate used: bread, carrot, and potato). The volume density of strain colonies was consequently calculated as follows: Vv (%) = P/p. Finally, the antifungal potential of the EOs was expressed as the percentage of fungal growth inhibition (FGI) according to the formula FGI = [(C − T)/C] × 100, where C and T is the growth of fungal strains (expressed as Vv) in the control and treatment group, respectively [33].

4.4. Data Processing

The data were submitted to one-way analysis of variance (ANOVA) and the means were compared by the Tukey test at 5% of probability using statistical software Prism 8.0.1 (GraphPad Software, San Diego, CA, USA). All analyses were performed in triplicate.

5. Conclusions

The present research was carried out to analyze the presence of volatile profile and antifungal efficacies of commercial REO, NEO, and FEO against *Penicillium* strains isolated from berry fruits. Our findings revealed a variable chemical profile of analyzed EO samples with linalool (REO), 1,8-cineole (NEO), and α-pinene (FEO) as principal compounds of their composition. Regarding the in vitro antifungal activities, the EOs were effective in inhibiting the growth of all *Penicillium* strains (*P. expansum*, *P. citrinum*, and *P. crustosum*) in concentrations more than 125 μL/L, whereas the very strong inhibitory effect (16.00 ± 1.00 mm) was detected for the highest REO concentration against *P. citrinum*. Furthermore, the results from the estimation of MC and aw in the food substrates (bread, carrots, potatoes) showed a good growth potential of microscopic filamentous fungi. Similar trends, as found in in vitro antifungal analyses, were also observed in those performed on food models (in situ), indicating dose-dependent antifungal action of all EOs with the highest mycelial growth inhibition being recorded in their highest concentrations against all *Penicillium* spp. inoculated on all food models employed. In summary, our obtained data suggest that REO, NEO, and FEO have promising perspectives as innovative natural agents for application in the storage of food products (including bakery products and vegetables) to prolong their shelf life. To explain the utilization of our EOs as antifungal additives, we plan to perform a sensory evaluation of the food models investigated to reveal which effective concentrations of the EOs are still acceptable for the product consumers. Moreover, our findings complement our previous studies which contribute to create a more comprehensive overview of the biological properties of diverse, commercially available EOs purchased from the same company, Hanus Ltd. (Nitra, Slovakia).

Author Contributions: Conceptualization, V.V., H.Ď. and M.K.; methodology, V.V., H.Ď., N.L.V., M.V., M.K. (Maciej Kluz) and M.K. (Miroslava Kačániová); software, V.V., M.K. (Miroslava Kačániová); validation, V.V., M.K. (Maciej Kluz) and M.K. (Miroslava Kačániová); formal analysis, V.V., H.Ď., M.K. (Miroslava Kačániová); investigation, V.V., H.Ď., N.L.V., M.V., M.K. (Miroslava Kačániová); resources, M.K. (Miroslava Kačániová); data curation, V.V., M.K. (Miroslava Kačániová); writing—original draft preparation, V.V., H.Ď., M.K. (Miroslava Kačániová); writing—review and editing, V.V., H.Ď.; visualization, V.V.; supervision, M.K. (Miroslava Kačániová); project administration, M.K. (Miroslava Kačániová); funding acquisition, M.K. (Miroslava Kačániová) All authors have read and agreed to the published version of the manuscript.

Funding: This work has been supported by the APVV SK-BY-RD-19-0014 grant "The formulation of novel compositions and properties study of the polysaccharides based edible films and coatings with antimicrobial and antioxidant plant additives."

Institutional Review Board Statement: Not applicable.

Informed Consent Statement: Not applicable.

Data Availability Statement: Not applicable.

Acknowledgments: This work was supported by the grants of the VEGA no. 1/0180/20.

Conflicts of Interest: The authors declare no conflict of interest.

Sample Availability: Not applicable.

References

1. Mollet, B.; Rowland, I. Functional foods: At the frontier between food and pharma. *Curr. Opin. Biotechnol.* **2002**, *5*, 483–485. [CrossRef]
2. Yahia, E.M.; García-Solís, P.; Celis, M.E.M. Contribution of fruits and vegetables to human nutrition and health. In *Postharvest Physiology and Biochemistry of Fruits and Vegetables*; Woodhead Publishing: Sawston, UK, 2019; pp. 19–45.
3. Weegels, P.L. The future of bread in view of its contribution to nutrient intake as a starchy staple food. *Plant Foods Hum. Nutr.* **2019**, *74*, 1–9. [CrossRef] [PubMed]
4. Šimović, M.; Delaš, F.; Gradvol, V.; Kocevski, D.; Pavlović, H. Antifungal effect of eugenol and carvacrol against foodborne pathogens *Aspergillus carbonarius* and *Penicillium roqueforti* in improving safety of fresh-cut watermelon. *J. Intercult. Ethnopharmacol.* **2014**, *3*, 91. [PubMed]
5. Otero, C.; Arredondo, C.; Echeverría-Vega, A.; Gordillo-Fuenzalida, F. *Penicillium* spp. mycotoxins found in food and feed and their health effects. *World Mycotoxin J.* **2020**, *13*, 323–343. [CrossRef]
6. Bakkali, F.; Averbeck, S.; Averbeck, D.; Idaomar, M. Biological effects of essential oils—A review. *Food Chem. Toxicol.* **2008**, *46*, 446–475. [CrossRef]
7. Donadu, M.G.; Peralta-Ruiz, Y.; Usai, D.; Maggio, F.; Molina-Hernandez, J.B.; Rizzo, D.; Bussu, F.; Rubino, S.; Zanetti, S.; Papparela, A.; et al. Colombian essential oil of *Ruta graveolens* against nosocomial antifungal resistant *Candida* strains. *J. Fungi* **2021**, *7*, 383. [CrossRef]
8. Diánez, F.; Santos, M.; Parra, C.; Navarro, M.J.; Blanco, R.; Gea, F.J. Screening of antifungal activity of 12 essential oils against eight pathogenic fungi of vegetables and mushroom. *Lett. Appl. Microbiol.* **2018**, *67*, 400–410. [CrossRef]
9. Serra, E.; Hidalgo-Bastida, L.A.; Verran, J.; Williams, D.; Malic, S. Antifungal activity of commercial essential oils and biocides against *Candida albicans*. *Pathogens* **2018**, *7*, 15. [CrossRef]
10. Donato, R.; Sacco, C.; Pini, G.; Bilia, A.R. Antifungal activity of different essential oils against *Malassezia* pathogenic species. *J. Ethnopharmacol.* **2020**, *249*, 112376. [CrossRef]
11. Bhavaniramya, S.; Vishnupriya, S.; Al-Aboody, M.S.; Vijayakumar, R.; Baskaran, D. Role of essential oils in food safety: Antimicrobial and antioxidant applications. *Grain Oil Sci. Technol.* **2019**, *2*, 49–55. [CrossRef]
12. Cai, R.; Hu, M.; Zhang, Y.; Niu, C.; Yue, T.; Yuan, Y.; Wang, Z. Antifungal activity and mechanism of citral, limonene and eugenol against *Zygosaccharomyces rouxii*. *LWT* **2019**, *106*, 50–56. [CrossRef]
13. Butnariu, M. Plants as Source of Essential Oils and Perfumery Applications. *Bioprospect. Plant Biodivers. Ind. Mol.* **2021**, *2021*, 261–292.
14. Aziz, Z.A.; Ahmad, A.; Setapar, S.H.M.; Karakucuk, A.; Azim, M.M.; Lokhat, D.; Rafatullah, M.; Ganash, M.; Kamal, M.A.; Ashraf, G.M. Essential oils: Extraction techniques, pharmaceutical and therapeutic potential-a review. *Curr. Drug Metab.* **2018**, *19*, 1100–1110. [CrossRef] [PubMed]
15. Abdellatif, F.; Hassani, A. Chemical composition of the essential oils from leaves of *Melissa officinalis* extracted by hydrodistillation, steam distillation, organic solvent and microwave hydrodistillation. *J. Mater. Environ. Sci.* **2015**, *6*, 207–213.
16. Sharifi-Rad, J.; Sureda, A.; Tenore, G.C.; Daglia, M.; Sharifi-Rad, M.; Valussi, M.; Tundis, R.; Sharifi-Rad, M.; Loizzo, M.R.; Ademiluyi, A.O.; et al. Biological activities of essential oils: From plant chemoecology to traditional healing systems. *Molecules* **2017**, *22*, 70. [CrossRef]
17. Contreras-Moreno, B.Z. Chemical Composition of Essential Oil of Genus Pimenta (*Myrtaceae*). In *Potential of Essential Oils*; El-Shemy, H., Ed.; IntechOpen Limited: London, UK, 2018; p. 21.
18. Farag, R.S.; Shalaby, A.S.; El-Baroty, G.A.; Ibrahim, N.A.; Ali, M.A.; Hassan, E.M. Chemical and biological evaluation of the essential oils of different Melaleuca species. *Phytother. Res.* **2004**, *18*, 30–35. [CrossRef]
19. Lee, Y.S.; Kim, J.; Shin, S.C.; Lee, S.G.; Park, I.K. Antifungal activity of Myrtaceae essential oils and their components against three phytopathogenic fungi. *Flavour Fragr. J.* **2008**, *23*, 23–28. [CrossRef]
20. Yang, S.A.; Jeon, S.K.; Lee, E.J.; Im, N.K.; Jhee, K.H.; Lee, S.P.; Lee, I.S. Radical scavenging activity of the essential oil of silver fir (*Abies alba*). *J. Clin. Biochem. Nutr.* **2009**, *44*, 253–259. [CrossRef]
21. Marjanovic-Balaban, Z.; Cvjetkovic, V.G.; Stanojevic, L.; Stanojevic, J.; Nikolic, L.; Danilovic, B. Quality testing of industrially produced essential oil of fir (*Abies alba* L.) from the Republic of Srpska. *J. Essent. Oil-Bear. Plants* **2020**, *23*, 503–513. [CrossRef]
22. Pérez Zamora, C.M.; Torres, C.A.; Nuñez, M.B. Antimicrobial activity and chemical composition of essential oils from Verbenaceae species growing in South America. *Molecules* **2018**, *23*, 544. [CrossRef]
23. Barbosa, L.C.A.; Silva, C.J.; Teixeira, R.R.; Meira, R.M.S.A.; Pinheiro, A.L. Chemistry and biological activities of essential oils from *Melaleuca* L. species. *Agric. Conspec. Sci.* **2013**, *78*, 11–23.
24. Brophy, J.J.; Doran, J.C. Geographic variation in oil characteristics in *Melaleuca ericifolia*. *J. Essent. Oil Res.* **2014**, *16*, 4–8. [CrossRef]
25. Zhao, Q.; Bowles, E.J.; Zhang, H.Y. Antioxidant activities of eleven Australian essential oils. *Nat. Prod. Commun.* **2008**, *3*, 1934578X0800300531. [CrossRef]
26. Ireland, B.F.; Hibbert, D.B.; Goldsack, R.J.; Doran, J.C.; Brophy, J.J. Chemical variation in the leaf essential oil of *Melaleuca quinquenervia* (Cav.) ST Blake. *Biochem. Syst. Ecol.* **2002**, *30*, 457–470. [CrossRef]
27. Chalchat, J.C.; Sidibé, L.; Maksimovic, Z.A.; Petrovic, S.D.; Gorunovic, M.S. Essential oil of *Abies alba* Mill., Pinaceae, from the pilot production in Montenegro. *J. Essent. Oil Res.* **2001**, *13*, 288–289. [CrossRef]

28. Wajs, A.; Urbańska, J.; Zaleśkiewicz, E.; Bonikowski, R. Composition of essential oil from seeds and cones of *Abies alba*. *Nat. Prod. Commun.* **2010**, *5*, 1934578X1000500830. [CrossRef]
29. Zeneli, G.; Tsitsimpikou, C.; Petrakis, P.V.; Naxakis, G.; Habili, D.; Roussis, V. Foliar and cortex oleoresin variability of silver fir (*Abies alba* Mill.) in Albania. *Z. Für Nat. C* **2001**, *56*, 531–539. [CrossRef]
30. Roussis, V.; Couladis, M.; Tzakou, O.; Loukis, A.; Petrakis, P.V.; Dukic, N.M.; Jancic, R. A comparative study on the needle volatile constituents of three *Abies* species grown in South Balkans. *J. Essent. Oil Res.* **2000**, *12*, 41–46. [CrossRef]
31. Tsasi, G.; Danalatos, G.; Tomou, E.M.; Sakadani, E.; Bethanis, M.; Kolokouris, A.; Ćirić, A.; Soković, M.; Skaltsa, H. Chemical composition and antimicrobial activity of the essential oil of *Abies cephalonica* Loudon from Mount Ainos (Kefalonia, Greece). *J. Essent. Oil Res.* **2022**, *34*, 143–147. [CrossRef]
32. Felšöciová, S.; Vukovic, N.; Jeżowski, P.; Kačániová, M. Antifungal activity of selected volatile essential oils against *Penicillium* sp. *Open Life Sci.* **2020**, *15*, 511–521. [CrossRef]
33. Valková, V.; Ďúranová, H.; Galovičová, L.; Borotová, P.; Vukovic, N.L.; Vukic, M.; Kačániová, M. *Cymbopogon citratus* Essential Oil: Its Application as an Antimicrobial Agent in Food Preservation. *Agronomy* **2022**, *12*, 155. [CrossRef]
34. Dias, I.J.; Trajano, E.R.I.S.; Castro, R.D.; Ferreira, G.L.S.; Medeiros, H.C.M.; Gomes, D.Q.C. Antifungal activity of linalool in cases of *Candida* spp. isolated from individuals with oral candidiasis. *Braz. J. Biol.* **2017**, *78*, 368–374. [CrossRef] [PubMed]
35. Guo, F.; Chen, Q.; Liang, Q.; Zhang, M.; Chen, W.; Chen, H.; Chen, W. Antimicrobial activity and proposed action mechanism of linalool against *Pseudomonas fluorescens*. *Front. Microbiol.* **2021**, *49*, 562094. [CrossRef] [PubMed]
36. Guo, F.; Liang, Q.; Zhang, M.; Chen, W.; Chen, H.; Yun, Y.; Chen, W. Antibacterial activity and mechanism of linalool against *Shewanella putrefaciens*. *Molecules* **2021**, *26*, 245. [CrossRef]
37. Vilela, G.R.; de Almeida, G.S.; D'Arce, M.A.B.R.; Moraes, M.H.D.; Brito, J.O.; da Silva, M.F.D.G.; Silva, S.C.; Piedade, S.M.S.; Calori-Domingues, D.A.; da Gloria, E.M. Activity of essential oil and its major compound, 1,8-cineole, from *Eucalyptus globulus* Labill., against the storage fungi *Aspergillus flavus* Link and *Aspergillus parasiticus* Speare. *J. Stored Prod. Res.* **2009**, *45*, 108–111. [CrossRef]
38. Tančinová, D.; Foltinová, D.; Mašková, Z.; Štefániková, J.; Árvay, J. Effect of essential oils of myrtaceae plants on the *Penicillium commune*. *Potravinarstvo* **2019**, *13*, 604–613. [CrossRef]
39. Li, L.; Li, Z.W.; Yin, Z.Q.; Wei, Q.; Jia, R.Y.; Zhou, L.J.; Yu, W. Antibacterial activity of leaf essential oil and its constituents from *Cinnamomum longepaniculatum*. *Int. J. Clin. Exp.* **2014**, *7*, 1721.
40. Yu, D.; Wang, J.; Shao, X.; Xu, F.; Wang, H. Antifungal modes of action of tea tree oil and its two characteristic components against *Botrytis cinerea*. *J. Appl. Microbiol.* **2015**, *119*, 1253–1262. [CrossRef]
41. Nikolova, M.T.; Yordanov, P.; Slavov, S.; Berkov, S. Antifungal activity of plant extracts against phytopathogenic fungi. *J. Biosci. Biotechnol. Discv.* **2017**, *6*, 155–161.
42. Salamon, I.; Kryvtsova, M.; Bucko, D.; Tarawneh, A.H. Chemical characterization and antimicrobial activity of some essential oils after their industrial large-scale distillation. *J. Microbiol. Biotechnol. Food Sci.* **2021**, *2021*, 984–988. [CrossRef]
43. Cox, S.D.; Mann, C.M.; Markham, J.L.; Bell, H.C.; Gustafson, J.E.; Warmington, J.R.; Wyllie, S.G. The mode of antimicrobial action of the essential oil of *Melaleuca alternifolia* (tea tree oil). *J. Appl. Microbiol.* **2000**, *88*, 170–175. [CrossRef] [PubMed]
44. Mathlouthi, M. Water content, water activity, water structure and the stability of foodstuffs. *Food Control* **2001**, *12*, 409–417. [CrossRef]
45. Chirife, J.; del Pilar Buera, M.; Labuza, T.P. Water activity, water glass dynamics, and the control of microbiological growth in foods. *Crit. Rev. Food Sci. Nutr.* **1996**, *36*, 465–513. [CrossRef] [PubMed]
46. Achata, E.; Esquerre, C.; O'Donnell, C.; Gowen, A. A study on the application of near infrared hyperspectral chemical imaging for monitoring moisture content and water activity in low moisture systems. *Molecules* **2015**, *20*, 2611–2621. [CrossRef]
47. Parasoglou, P.; Parrott, E.P.J.; Zeitler, J.A.; Rasburn, J.; Powell, H.; Gladden, L.F.; Johns, M.L. Quantitative moisture content detection in food wafers. In Proceedings of the 34th International Conference on Infrared, Millimeter, and Terahertz Waves, IRMMW-THz 2009, Busan, Korea, 21–25 September 2009; pp. 1–2.
48. Roos, Y.H.; Jouppila, K.; Söderholm, E.S. Crystallization of amorphous food components and polymers. In *Water Management in the Design and Distribution of Quality Foods: ISOPOW, 7*; CRP Press: New York, NY, USA, 2009; pp. 429–451.
49. Ayub, M.; Wahab, S.; Durrani, Y. Effect of water activity (aw), moisture content and total microbial count on the overall quality of bread. *Int. J. Agric. Biol.* **2003**, *5*, 273–278.
50. Lahlali, R.; Serrhini, M.N.; Jijakli, M.H. Studying and modelling the combined effect of temperature and water activity on the growth rate of *P. expansum*. *Int. J. Food Microbiol.* **2005**, *103*, 315–322. [CrossRef]
51. Day, L. Cereal Food Production with Low Salt. In *Encyclopedia of Food Grains*; Elsevier: Amsterdam, The Netherlands, 2016; pp. 396–402.
52. Dagnas, S.; Gougouli, M.; Onno, B.; Koutsoumanis, K.P.; Membré, J.M. Quantifying the effect of water activity and storage temperature on single spore lag times of three moulds isolated from spoiled bakery products. *Int. J. Food Microbiol.* **2017**, *240*, 75–84. [CrossRef]
53. Sipahioglu, O.; Barringer, S.A. Dielectric properties of vegetables and fruits as a function of temperature, ash, and moisture content. *J. Food Sci.* **2003**, *68*, 234–239. [CrossRef]
54. Tejeswini, M.G.; Sowmya, H.V.; Swarnalatha, S.P.; Negi, P.S. Antifungal activity of essential oils and their combinations in in vitro and in vivo conditions. *Arch. Phytopathol. Pflanzenschutz* **2014**, *47*, 564–570. [CrossRef]

55. Lund, F.; Filtenbory, O.; Westall, S.; Frisvad, J.C. Associated mycoflora of rye bread. *Lett. Appl. Microbiol.* **1996**, *23*, 213–217. [CrossRef]
56. Garcia, M.V.; Bernardi, A.O.; Copetti, M.V. The fungal problem in bread production: Insights of causes, consequences, and control methods. *Curr. Opin.* **2019**, *29*, 1–6. [CrossRef]
57. Pitt, J.I.; Hocking, A.D. *Fungi and Food Spoilage*; Springer: New York, NY, USA, 2019; Volume 519, p. 388.
58. Reyes-Jurado, F.; Navarro-Cruz, A.R.; Ochoa-Velasco, C.E.; Palou, E.; López-Malo, A.; Ávila-Sosa, R. Essential oils in vapor phase as alternative antimicrobials: A review. *Crit. Rev. Food Sci. Nutr.* **2020**, *60*, 1641–1650. [CrossRef] [PubMed]
59. Moreira, M.R.; Ponce, A.G.; Del Valle, C.E.; Roura, S.I. Inhibitory parameters of essential oils to reduce a foodborne pathogen. *LWT* **2005**, *38*, 565–570. [CrossRef]
60. Sacchetti, G.; Maietti, S.; Muzzoli, M.; Scaglianti, M.; Manfredini, S.; Radice, M.; Bruni, R. Comparative evaluation of 11 essential oils of different origin as functional antioxidants, antiradicals and antimicrobials in foods. *Food Chem.* **2005**, *91*, 621–632. [CrossRef]
61. Murbach Teles Andrade, B.F.; Nunes Barbosa, L.; da Silva Probst, I.; Fernandes Júnior, A. Antimicrobial activity of essential oils. *J. Essent. Oil Res.* **2014**, *26*, 34–40. [CrossRef]
62. Tullio, V.; Nostro, A.; Mandras, N.; Dugo, P.; Banche, G.; Cannatelli, M.A.; Cuffini, A.M.; Alonzo, V.; Carlone, N.A. Antifungal activity of essential oils against filamentous fungi determined by broth microdilution and vapour contact methods. *J. Appl. Microbiol.* **2007**, *102*, 1544–1550. [CrossRef]
63. Fisher, K.; Phillips, C. Potential antimicrobial uses of essential oils in food: Is citrus the answer? *Trends Food Sci. Technol.* **2008**, *19*, 156–164. [CrossRef]
64. Nadjib, B.M.; Amine, F.M.; Abdelkrim, K.; Fairouz, S.; Maamar, M. Liquid and vapour phase antibacterial activity of *Eucalyptus globulus* essential oil = susceptibility of selected respiratory tract pathogens. *Am. J. Infect. Dis.* **2014**, *10*, 105. [CrossRef]
65. Samson, R.A.; Houbraken, J.; Thrane, U.; Frisvad, J.C.; Andersen, B. *Food and Indoor Fungi*; CBS–KNAW Fungal Biodiversity Centre: Utrecht, The Netherlands, 2010; p. 390.
66. Wang, W.; Wu, N.; Zu, Y.G.; Fu, Y.J. Antioxidative activity of Rosmarinus officinalis L. essential oil compared to its main components. *Food Chem.* **2008**, *108*, 1019–1022. [CrossRef]
67. Guimarães de Lima, L.G.; Cardoso das Graças, M.; de Sousa, P.E.; de Andrade, J.; Vieira, S.S. Antioxidant and fungitoxic activities of the lemongrass essential oil and citral. *Rev. Cienc. Agron.* **2011**, *42*, 464–472.
68. Xu, Y.; Tong, Z.; Zhang, X.; Wang, Y.; Fang, W.; Li, L.; Luo, Z. Unveiling the mechanisms for the plant volatile organic compound linalool to control gray mold on strawberry fruits. *J. Agric. Food Chem.* **2019**, *67*, 9265–9276. [CrossRef] [PubMed]
69. Chang, S.; Mohammadi Nafchi, A.; Baghaie, H. Development of an active packaging based on polyethylene containing linalool or thymol for mozzarella cheese. *Food Sci. Nutr.* **2021**, *9*, 3732–3739. [CrossRef]
70. Soković, M.; Van Griensven, L.J. Antimicrobial activity of essential oils and their components against the three major pathogens of the cultivated button mushroom, *Agaricus bisporus*. *Eur. J. Plant Pathol.* **2006**, *116*, 211–224. [CrossRef]
71. Valková, V.; Ďúranová, H.; Galovičová, L.; Vukovic, N.L.; Vukic, M.; Kačániová, M. In Vitro antimicrobial activity of lavender, mint, and rosemary essential oils and the effect of their vapours on growth of *Penicillium* spp. in a bread model system. *Molecules* **2021**, *26*, 3859. [CrossRef] [PubMed]
72. Valková, V.; Ďúranová, H.; Štefániková, J.; Miškeje, M.; Tokár, M.; Gabríny, L.; Kowalczewski, P.L.; Kačániová, M. Wheat bread with grape seeds micropowder: Impact on dough rheology and bread properties. *Appl. Rheol.* **2020**, *30*, 138–150. [CrossRef]
73. Valková, V.; Ďúranová, H.; Galovičová, L.; Ivanišová, E.; Kačániová, M. Evaluation of vapor-phase antifungal activities of selected plant essential oils against fungal strains growing on bread food model. *Potravinarstvo* **2020**, *15*, 210–217. [CrossRef]

molecules

Article

Antibacterial Activity of *Thymus vulgaris* L. Essential Oil Vapours and Their GC/MS Analysis Using Solid-Phase Microextraction and Syringe Headspace Sampling Techniques

Julien Antih [1], Marketa Houdkova [1], Klara Urbanova [2] and Ladislav Kokoska [1,*]

[1] Department of Crop Sciences and Agroforestry, Faculty of Tropical Agrisciences, Czech University of Life Sciences Prague, Kamycka 129, 165 21 Prague-Suchdol, Czech Republic; antih@ftz.czu.cz (J.A.); houdkovam@ftz.czu.cz (M.H.)

[2] Department of Sustainable Technologies, Faculty of Tropical Agrisciences, Czech University of Life Sciences Prague, Kamycka 129, 165 21 Prague-Suchdol, Czech Republic; urbanovak@ftz.czu.cz

* Correspondence: kokoska@ftz.czu.cz; Tel.: +420-224-382-180

Abstract: While the inhalation of *Thymus vulgaris* L. essential oil (EO) is commonly approved for the treatment of mild respiratory infections, there is still a lack of data regarding the antimicrobial activity and chemical composition of its vapours. The antibacterial activity of the three *T. vulgaris* EOs against respiratory pathogens, including *Haemophilus influenzae*, *Staphylococcus aureus*, and *Streptococcus pyogenes*, was assessed in both liquid and vapour phases using the broth microdilution volatilisation (BMV) method. With the aim of optimising a protocol for the characterisation of EO vapours, their chemical profiles were determined using two headspace sampling techniques coupled with GC/MS: solid-phase microextraction (HS-SPME) and syringe headspace sampling technique (HS-GTS). All EO sample vapours exhibited antibacterial activity with minimum inhibitory concentrations (MIC) ranging from 512 to 1024 µg/mL. According to the sampling technique used, results showed a different distribution of volatile compounds. Notably, thymol was found in lower amounts in the headspace—peak percentage areas below 5.27% (HS-SPME) and 0.60% (HS-GTS)—than in EOs (max. 48.65%), suggesting that its antimicrobial effect is higher in vapour. Furthermore, both headspace sampling techniques were proved to be complementary for the analysis of EO vapours, whereas HS-SPME yielded more accurate qualitative results and HS-GTS proved a better technique for quantitative analysis.

Keywords: antimicrobial activity; broth microdilution; headspace analysis; respiratory infections; thyme; vapour phase

Citation: Antih, J.; Houdkova, M.; Urbanova, K.; Kokoska, L. Antibacterial Activity of *Thymus vulgaris* L. Essential Oil Vapours and Their GC/MS Analysis Using Solid-Phase Microextraction and Syringe Headspace Sampling Techniques. *Molecules* **2021**, *26*, 6553. https://doi.org/10.3390/molecules26216553

Academic Editor: Daniela Rigano

Received: 27 August 2021
Accepted: 26 October 2021
Published: 29 October 2021

Publisher's Note: MDPI stays neutral with regard to jurisdictional claims in published maps and institutional affiliations.

Copyright: © 2021 by the authors. Licensee MDPI, Basel, Switzerland. This article is an open access article distributed under the terms and conditions of the Creative Commons Attribution (CC BY) license (https://creativecommons.org/licenses/by/4.0/).

1. Introduction

Pneumonia is an acute lower respiratory infection (ALRI) mainly caused by bacterial pathogens, including Gram-positive bacteria such as *Staphylococcus aureus*, *Streptococcus pneumoniae*, and *Streptococcus pyogenes*, or Gram-negative strains such as *Haemophilus influenzae* [1–3]. For decades, they have been amongst the major causes of morbidity and mortality worldwide; this is particularly significant in low-income countries where vulnerable populations such as children under five years old, the elderly, and the immunocompromised population are at higher risk [4]. In 2019, ALRIs remained the world's most deadly communicable diseases with nearly 2.6 million deaths and accounted for 15% of deaths in children under 5 years of age [5–7]. To reduce the fatality of these infections, the recommended treatment is the administration of a full course of appropriate antibiotics. Currently, nebulised antibiotics offer significant advantages over intravenous and oral therapies; amongst others, inhalation therapy allows to deliver high drug concentrations and to have more efficient and faster action while reducing the risk of side effects compared to systemic administrations [8]. Today, such treatments are already included in care protocols, especially for cystic fibrosis. For example, nebulised tobramycin, colistin, and aztreonam lysine are commonly used in Europe as inhaled antibiotics against chronic *Pseudomonas aeruginosa* infections [9]. However, difficulties with implementing optimal nebulisation techniques and the lack of robust clinical data have limited the widespread adoption of aerosol treatments [10–12]. Among other reasons, this can be attributed to both particle-related and patient-related factors. For instance, efficient delivery of aerosolised antibiotic particles in the lower parts of the lungs rests on the aerodynamic behaviour of the particles: larger particles preferably accumulate in the oropharyngeal area, while smaller particles deposit in the lower airways. Consequently, this alters drug delivery in the lungs as small particles carry fewer active substances [13]. Furthermore, the distribution of nebulised particles depends on the breathing manoeuvre used by the patient: rapid and forceful inhalations will see an increased deposition of drug in the upper airways, whereas slow and deep inhalations deposit particles in the lower part of the respiratory system [14]. As a result, drug delivery is also affected by the quality of inhalations which hinge on the patient's age (especially children and elderly), the severity of the respiratory disease, or even the physical capability to perform a breathing manoeuvre [15].

In this context, plant-derived preparations, and more specifically, essential oils (EOs), are interesting alternatives in the development of novel antimicrobial volatile agents [16,17]. EOs' typical physicochemical feature is high volatility at room temperature. Therefore, compared to aerosolised antibiotics, EO vapours are easy to inhale without requiring any specific breathing technique. At the same time, this characteristic allows for a uniform distribution of EOs' active substances, at a significant concentration, in the lower section of the lungs [18]. As an example, a recent clinical trial study has already demonstrated that a *Thymus vulgaris* L. EO inhalation therapy would improve the respiratory conditions in patients with ventilator-associated pneumonia. Among other benefits, this treatment would facilitate the clearance of the mucous membrane, improve gas exchange in alveoli, and thus increase the oxygen saturation in patients' blood [19]. Furthermore, EOs are complex mixtures containing a broad spectrum of chemically diverse compounds that work in synergy to provide their antibacterial properties. Their composition includes hydrocarbons (terpenes and terpenoids) and oxygenated compounds (phenols, aldehydes, alcohols, ketones, esters, and lactones) as well as sulphur or nitrogen substances such as isothiocyanates and pyrazines synthesised by certain plant families [20–22]. Because of their well-studied chemical compositions, EOs can be used for medicinal purposes if their long-standing usage and experience can be demonstrated [23]. Amongst the nearly 30 EOs approved by the Committee on Herbal Medicinal Products (HMPC) of the European Medicines Agency (EMA), the essential oil of *T. vulgaris* is one of the most important plant species used traditionally to treat respiratory disorders, including cough, bronchitis, laryngitis, and other respiratory congestions [24]. Moreover, several reports are validating the in vitro antibacterial activities of this EO on respiratory pathogens such as *S. aureus* and *P. aeruginosa*. Today, its composition is well known and comprises components such as thymol methyl ether, *p*-cymene, 1,8 cineol, and α-pinene, but it is widely believed that its antimicrobial activity is related to its content in two phenolic monoterpenes: thymol and carvacrol [25–27]. *T. vulgaris* is also used in complex herbal formulations. For instance, Bronchipret (Bionorica, Neumarkt, Germany) is an over-the-counter oral treatment consisting mainly of *Hedera helix* L. and *T. vulgaris* extracts. This herbal product is recommended in acute inflammation of the respiratory tract associated with cough and thick mucus [7]. Although recent studies suggest that the EO of *T. vulgaris* produce better results in antimicrobial activities in the vapour phase than their liquid phase [17,28,29], there is still a lack of available data regarding its growth-inhibitory effect against pneumonia-causing bacteria [28].

As opposed to well-established methods for the testing of antimicrobial susceptibility in both solid (i.e., agar disc diffusion) and liquid (broth microdilution) media, there is no standardised antimicrobial assay to study EO vapour phases according to, for example, the Clinical and Laboratory Standards Institute (CLSI) [30–32]. Today, disc volatilisation is the most common method to examine the in vitro growth inhibition of EOs' volatile agents [33]; however, similar methods have already been developed and used (e.g., dressing model volatilisation test [34], or airtight apparatus disc volatilisation methods [28]). Unfortunately, such tests have limitations: they are generally not designed for high-throughput screening, some only allow the evaluation of a single concentration of each sample, and usually only provide qualitative data [35]. Furthermore, the lack of standardisation of methods has led to difficulties not only in interpreting but also in comparing antimicrobial growth results. For instance, from all studies examining *T. vulgaris* EO vapour antimicrobial activity, it is not unusual to observe results expressed in different ways, including unit volume of air [28] or even various definitions of minimum inhibitory concentrations (MICs) [36,37].

The antimicrobial properties of EO vapours are determined by the relative volatility and antimicrobial effects of their compounds [38]. It is therefore essential to understand their chemical composition, something that has not been well-explored yet [39]. Nowadays, to examine the content of EOs vapour, static headspace extraction coupled with gas chromatography–mass spectrometry (SHE-GC/MS) offers more advantages compared to traditional liquid extraction of samples. It is a simple, rapid, and solventless technique that requires a small amount of sample and allows the analysis of highly volatile compounds [40]. Consequently, studies examining the antimicrobial activity of EO vapours are also focused on their chemical composition using SHE-GC/MS. As an example, several studies have explored the chemical composition of *T. vulgaris*'s volatile agents using different SHE techniques, including solid-phase microextraction (HS-SPME) [37,41,42], which is currently the preferred technique to examine complex, volatile mixtures in laboratories [43]. This method uses a fused silica fibre coated with a stationary phase that is directly exposed to the sample headspace. After absorption of the analytes to the fibre coating, the sample is transferred into a GC injection port for thermal desorption [40]. HS-SPME sampling is not only a sensitive technique due to the concentration achieved by the fibre but also a selective one thanks to the different coating material available; repeatability is also one of its assets when used with a dedicated autosampler [40,44,45]. However, using fibre coating suffers limitations; for example, it is not uniformly sensitive; therefore, competitive adsorption between volatile agents for the limited number of active sites can be observed. Similarly, selectivity will be different depending on the coating polymer used [46,47]. As a result, GC peak areas might not reflect the exact compounds'; composition and proportion in the headspace. That is why other approaches can be considered for analysing the EO vapour profile, including gas tight syringe headspace sampling (HS-GTS). This technique is the most convenient and inexpensive way to sample highly volatile compounds from the headspace of a closed vessel [40,48]. The gas syringe with a pressure-lock valve is inserted into the headspace, and a fraction of its volume is removed. The gas sample, locked into the syringe, is then transferred and injected into the GC inlet. Despite the above-mentioned advantages, the potential of this method for headspace sampling of EO vapours has not fully been exploited yet. Intending to identify EO vapours with the potential to inhibit the growth of bacteria causing pneumonia, we performed a series of preliminary experiments with herbal products approved by the HMPC of the EMA for the treatment of infectious cough and cold [23]. It encompassed the assessment of the in vitro growth-inhibitory effect of *Pimpinella anisum* L. seed, *Eucalyptus globulus* Labill. leaf, *Thymus vulgaris* L. aerial part, *Mentha* × *piperita* L. leaf, and *Foeniculum vulgare* Mill. seed EOs using the broth microdilution volatilisation (BMV) method against *H. influenzae*, *S. aureus*, and *S. pyogenes*. This novel method was recently developed by our team [35]; it is a simple, rapid, and simultaneous technique that allows the assessment of EOs antibacterial activities at different concentrations in both liquid and vapour phases. As a result of these exploratory tests, *T. vulgaris* EOs was selected for further evaluation due to the lowest MICs it generated (J. Antih, M. Houdkova, and L. Kokoska, unpublished data). We therefore investigated the in vitro growth-inhibitory effect of *T. vulgaris* EOs in both liquid and vapour phases against the above-mentioned respiratory pathogens and compared their chemical compositions using GC-MS. Then, with the aim of optimising a protocol for the chemical characterisation of EO vapours, we performed a time series of headspace analyses comparing both HS-SPME and HS-GTS sampling techniques.

2. Results

2.1. Antimicrobial Activity

In this study, essential oil samples from dry plant material of *T. vulgaris* from three different suppliers were tested against three standard bacterial strains associated with respiratory infections using the BMV method (Table 1). All EOs presented a certain degree of antibacterial efficacy ranging from 512 to 1024 µg/mL in both liquid and vapour phases. Supplier C's EO was the most active, with the lowest MICs value of 512 µg/mL in both liquid and vapour phases for the three bacteria strains. Similarly, *S. pyogenes* and *H. influenzae* growth were more affected by the EO of supplier B than *S. aureus* with MICs at 512 µg/mL and 1024 µg/mL in both broth and agar, respectively. On the contrary, the least effective EO source was from supplier A: results showed mild efficacy against *H. influenzae* (512 µg/mL), whereas a weaker inhibitory effect of 1024 µg/mL was observed against both *S. aureus* and *S. pyogenes*. Likewise, each EO affected the growth of *S. pyogenes*, *H. influenzae*, and *S. aureus* similarly in both liquid and vapour phases. No discrepancy between broth and agar results was observed on the tested strains. *H. influenzae* was the most susceptible bacterial strain (MICs = 512 µg/mL for all EOs tested) followed by *S. pyogenes* (MICs = 512 µg/mL supplier B and C and MIC = 1024 µg/mL for supplier C), while *S. aureus* was the least sensitive (MIC = 512 µg/mL supplier C and MICs = 1024 µg/mL for supplier A and B).

Table 1. In vitro growth-inhibitory effect of *Thymus vulgaris* L. essential oils in liquid and vapour phases against respiratory infection bacteria.

Supplier	Bacterium/Growth Medium/Minimum Inhibitory Concentration								
	Staphylococcus aureus			*Streptococcus pyogenes*			*Haemophilus influenzae*		
	Broth	Agar		Broth	Agar		Broth	Agar	
	(µg/mL)	(µg/mL)	(µg/cm^3)	(µg/mL)	(µg/mL)	(µg/cm^3)	(µg/mL)	(µg/mL)	(µg/cm^3)
A	1024	1024	256	1024	1024	256	512	512	128
B	1024	1024	256	512	512	128	512	512	128
C	512	512	128	512	512	128	512	512	128
				Positive antibiotic control					
Oxacillin	0.25	>2	>0.5	NT	NT	NT	NT	NT	NT
Amoxicilin	NT	NT	NT	0.06	>2	>0.5	NT	NT	NT
Ampicilin	NT	NT	NT	NT	NT	NT	0.5	>16	>4

NT = Not tested.

2.2. Chemical Analysis of EOs

In this investigation, *T. vulgaris* EOs from three different suppliers (A, B, and C) were extracted with respective yield values of 0.73, 1.23 and 1.25%. All EOs presented a *strong* herbaceous *fragrance* while being of different shades of orange colour. The complete chemical analyses of all samples are provided in Table 2 as well as in Figure 1a,b. Using the HP-5MS column, 54 compounds were identified in EOs of suppliers A and B, whereas 62 components were found in supplier C's EO, representing 99.55, 99.65, and 99.62% of their respective total constituents, respectively. Similarly, using the DB-HeavyWAX column, 44, 43, and 36 components were determined, which constituted 99.30, 99.68, and 99.61% of the volatile oils, respectively. In the three samples analysed, monoterpenoids represented by thymol (phenolic monoterpene) followed by sesquiterpenoids mainly represented by β-caryophyllene were the two dominant groups of volatile agents identified. To a lesser extent, other groups were identified, such as oxygenated phenylpropanoids, oxygenated aliphatics, furanoids, and oxygenated diterpenes.

Table 2. Chemical composition of *Thymus vulgaris* L. essential oils obtained from 3 different commercial suppliers.

RI [a]		Compound	Cl. [b]	Supplier/Column/Content (%)						Identification [e]	
				A		B		C [c]			
Obs.	Lit			HP-5MS	DB-WAX	HP-5MS	DB-WAX	HP-5MS	DB-WAX	HP-5MS	DB-WAX
761	783 [d]	Methyl α-methylbutanoate	OA	0.08	- [g]	0.08	-	0.05 ± 0.01	-	GC/MS	-
923	924	α-Thujene	MH	0.10	0.08	0.10	0.13	0.26 ± 0.02	0.24 ± 0.06	RI, GC/MS	GC/MS
929	939	α-Pinene	MH	0.62	0.38	0.61	0.62	0.43 ± 0.03	0.38 ± 0.07	RI, GC/MS	GC/MS
944	945	Camphene	MH	0.44	0.26	0.43	0.43	0.19 ± 0.01	0.18 ± 0.03	RI, GC/MS	GC/MS
972	969	Sabinene	MH	0.09	0.05	0.08	0.09	0.10 ± 0.01	0.09 ± 0.02	RI, GC/MS	GC/MS
977	979	1-octen-3-ol	OA	0.31	0.20	0.32	0.30	0.35 ± 0.02	0.35 ± 0.06	RI, GC/MS	GC/MS
989	988	β-Myrcene	MH	0.31	0.21	0.31	0.35	0.58 ± 0.04	0.58 ± 0.11	RI, GC/MS	GC/MS
994	996	3-Octanol	OA	0.05	-	tr. [f]	-	tr.	-	RI, GC/MS	-
1002	1002	α-Phellandrene	MH	0.06	-	0.06	0.05	0.10 ± 0.01	0.08 ± 0.02	RI, GC/MS	GC/MS
1007	1008	3-Carene	MH	-	-	-	-	tr.	-	RI, GC/MS	-
1014	1014	α-Terpinene	MH	0.62	-	0.61	-	1.07 ± 0.07	-	RI, GC/MS	-
1025	1020	*p*-Cymene	AH	22.15	16.43	22.03	25.35	12.75 ± 0.84	12.90 ± 2.55	RI, GC/MS	GC/MS
1027	1031	D-Limonene	MH	-	0.16	-	0.25	0.41 ± 0.03	0.23 ± 0.04	RI, GC/MS	GC/MS
1028	1023	*m*-Cymene	AH	0.46	-	0.47	-	-	0.18 ± 0.17	RI, GC/MS	GC/MS
1030	1032	1,8-Cineole	OM	0.69	0.51	0.70	0.76	0.39 ± 0.03	0.47 ± 0.10	RI, GC/MS	GC/MS
1058	1054	γ-Terpinene	MH	2.82	2.02	2.78	2.90	5.22 ± 0.33	5.40 ± 1.11	RI, GC/MS	GC/MS
1065	1068	*cis*-thujanol	OM	0.17	0.06	0.18	0.07	0.37 ± 0.03	0.14 ± 0.03	RI, GC/MS	GC/MS
1071	1078	*cis*-Linalool oxide	F	tr.	-	tr.	-	-	-	RI, GC/MS	-
1087	1086	Terpinolene	MH	0.18	-	0.17	-	0.11 ± 0.01	-	RI, GC/MS	-
1096	1102	*trans*-thujanol	OM	-	0.09	-	0.13	tr.	0.37 ± 0.08	RI, GC/MS	GC/MS
1100	1095	Linalool	OM	1.80	1.08	1.82	1.77	1.53 ± 0.07	1.38 ± 0.24	RI, GC/MS	GC/MS
1149	1141	Camphor	OM	0.32	0.23	0.18	0.34	0.06 ± 0.01	0.16 ± 0.03	RI, GC/MS	GC/MS
1165	1165	Endo-borneol	OM	1.14	-	1.15	-	0.63 ± 0.03	-	RI, GC/MS	-
1176	1174	Terpinen-4-ol	OM	0.61	-	0.61	-	0.62 ± 0.04	-	RI, GC/MS	-
1186	1180	m-Cymen-8-ol	OM	0.05	-	-	-	-	-	RI, GC/MS	-
1190	1183	*p*-Cymen-8-ol	OM	0.14	0.11	0.12	0.16	-	-	RI, GC/MS	GC/MS
1196	1189	α-Terpineol	OM	0.14	-	0.14	-	0.16 ± 0.02	-	RI, GC/MS	-
1204	1195	*cis*-Dihydrocarvone	OM	tr.	-	tr.	-	tr.	-	RI, GC/MS	-
1235	1235	Thymol methyl ether	OPM	1.15	2.11	1.15	3.01	0.76 ± 0.05	3.17 ± 0.67	RI, GC/MS	GC/MS
1244	1241	Carvacrol methyl ether	OPM	0.91	0.86	0.91	1.35	0.66 ± 0.04	1.09 ± 0.23	RI, GC/MS	GC/MS
1256	1242	Carvone	OM	1.39	1.05	1.40	1.38	1.01 ± 0.05	1.05 ± 0.24	RI, GC/MS	GC/MS
1269	1255	Geraniol	MH	0.05	-	0.05	0.09	0.05 ± 0.01	tr.	RI, GC/MS	GC/MS

Table 2. *Cont.*

RI [a]		Compound	Cl. [b]	Supplier/Column/Content (%)						Identification [e]	
				A		B		C [c]			
Obs.	Lit			HP-5MS	DB-WAX	HP-5MS	DB-WAX	HP-5MS	DB-WAX	HP-5MS	DB-WAX
1286	1282	Anethole	OPP	0.22	0.21	0.23	0.30	0.11 ± 0.06	0.07 ± 0.02	RI, GC/MS	GC/MS
1304	1290	Thymol	PM	38.42	29.52	38.93	41.31	48.09 ± 2.53	48.65 ± 10.80	RI, GC/MS	GC/MS
1310	1298	Carvacrol	PM	10.61	3.84	10.85	5.59	10.92 ± 3.50	3.60 ± 0.78	RI, GC/MS	GC/MS
1361	1357	Estragole	OPP	-	-	0.07	0.05	tr.	0.09 ± 0.02	RI, GC/MS	GC/MS
1379	1374	α-Copaene	SH	0.21	-	0.21	-	0.09 ± 0.01	-	RI, GC/MS	-
1388	1387	β-Bourbonene	SH	0.07	tr.	0.07	-	tr.	-	RI, GC/MS	GC/MS
1424	1418	β-Caryophyllene	SH	1.79	-	1.79	-	2.33 ± 0.13	-	RI, GC/MS	-
1433	1446 [d]	Isogermacrene D	SH	-	-	-	-	tr.	-	GC/MS	-
1458	1452	α-Humulene	SH	0.09	tr.	0.08	0.07	0.11 ± 0.01	0.12 ± 0.02	RI, GC/MS	GC/MS
1475	1475	Geranyl propionate	OM	0.06	0.06	0.10	0.14	0.10 ± 0.01	0.11 ± 0.03	RI, GC/MS	GC/MS
1480	1478	γ-Muurolene	SH	0.19	0.15	0.21	0.22	0.21 ± 0.02	0.21 ± 0.04	RI, GC/MS	GC/MS
1483	1493	α-Amorphene	SH	-	-	tr.	-	tr.	-	RI, GC/MS	-
1498	1495	Valencene	SH	0.09	-	0.09	-	0.11 ± 0.01	-	RI, GC/MS	-
1503	1499	α-Muurolene	SH	0.10	-	0.10	-	0.09 ± 0.01	-	RI, GC/MS	-
1510	1509	β-Bisabolene	SH	0.06	-	0.06	-	0.06 ± 0.01	-	RI, GC/MS	-
1518	1513	γ-Cadinene	SH	0.38	0.43	0.39	0.58	0.33 ± 0.02	0.73 ± 0.16	RI, GC/MS	GC/MS
1527	1524	δ-Cadinene	SH	0.48	-	0.48	-	0.50 ± 0.03	-	RI, GC/MS	-
1561	1564	Nerolidol	OS	tr.	-	-	-	tr.	-	RI, GC/MS	-
1585	1578	Spathulenol	OS	-	-	tr.	-	tr.	-	RI, GC/MS	-
1591	1581	Caryophyllene oxide	OS	1.31	0.92	1.32	1.24	0.68 ± 0.03	0.66 ± 0.15	RI, GC/MS	GC/MS
1617	1606	Humulene epoxide II	OS	0.06	tr.	0.06	0.05	tr.	-	RI, GC/MS	GC/MS
1621	1627	Epicubenol	OS	0.06	tr.	0.06	-	tr.	tr.	RI, GC/MS	GC/MS
1628	1630	γ-Eudesmol	OS	0.13	0.10	0.13	0.13	0.13 ± 0.01	0.13 ± 0.03	RI, GC/MS	GC/MS
1634	1642	Cubenol	OS	tr.	-	-	0.09	tr.	-	RI, GC/MS	GC/MS
1648	1640	α-epi-Cadinol	OS	0.48	tr.	0.49	-	0.36 ± 0.03	0.05 ± 0.05	RI, GC/MS	GC/MS
1652	1645	δ-Cadinol	OS	-	0.06	-	-	tr.	tr.	RI, GC/MS	GC/MS
1661	1653	α-Cadinol	OS	0.05	-	0.05	-	0.07 ± 0.02	-	RI, GC/MS	-
1682	1677	Cadalene	SH	0.12	-	0.14	-	tr.	-	RI, GC/MS	-
1844	1844	Perhydrofarnesyl acetone	OM	tr.	-	tr.	-	tr.	-	RI, GC/MS	-
[h]	1001	4-Carene	MH	-	-	-	0.64	-	1.05 ± 0.23	-	GC/MS
[h]	NA	4-Pentenyl butyrate	OM	-	tr.	-	0.06	-	-	-	GC/MS
[h]	1351	α-Cubebene	SH	-	0.06	-	0.10	-	0.07 ± 0.01	-	GC/MS

Table 2. *Cont.*

RI [a]		Compound	Cl. [b]	Supplier/Column/Content (%)						Identification [e]	
				A		B		C [c]			
Obs.	Lit			HP-5MS	DB-WAX	HP-5MS	DB-WAX	HP-5MS	DB-WAX	HP-5MS	DB-WAX
h	1290	Isobornyl acetate	OM	-	0.08	-	0.11	-	-	-	GC/MS
h	1521	Calamenene	SH	-	-	-	0.19	-	tr.	-	GC/MS
h	1156	Isoborneol	OM	-	0.86	-	1.17	-	0.74 ± 0.16	-	GC/MS
h	1455	Aromandendrene	SH	-	tr.	-	0.05	-	tr.	-	GC/MS
h	NA	Lavandulyl butyrate	OM	-	0.05	-	0.07	-	tr.	-	GC/MS
h	1372	*p*-Cymen-7-ol	OM	-	tr.	-	0.05	-	-	-	GC/MS
h	2105	Phytol	OD	-	0.07	-	0.23	-	tr.	-	GC/MS
		Total identified (%)		99.55	99.30	99.65	99.68	99.62	99.61		

[a] RI = retention indices. Obs. = retention indices determined relative to a homologous series of *n*-alkanes (C8–C40) on an HP-5MS column. Lit. = literature RI values [49,50]; [b] Cl = class; AH—aromatic hydrocarbon, F—furanoid, MH—monoterpene hydrocarbon, OA—oxygenated aliphatic, OD—oxygenated diterpene, OM—oxygenated monoterpene, OPM—oxygenated phenolic monoterpene, OPP—oxygenated phenylpropanoid, PM—phenolic monoterpene, OS—oxygenated sesquiterpene, SH—sesquiterpene hydrocarbon. [c] Relative peak area percentage as mean of three measurements ± standard deviation. [d] Literature RI values [51]. [e] Identification method: GC/MS = mass spectrum was identical to that of National Institute of Standards and Technology Library (ver. 2.0.f); RI = the retention index was matching literature database. [f] tr. = traces, relative peak area < 0.05%. [g] - = not detected. [h] Retention indices were not calculated for compounds determined by DB-WAX column.

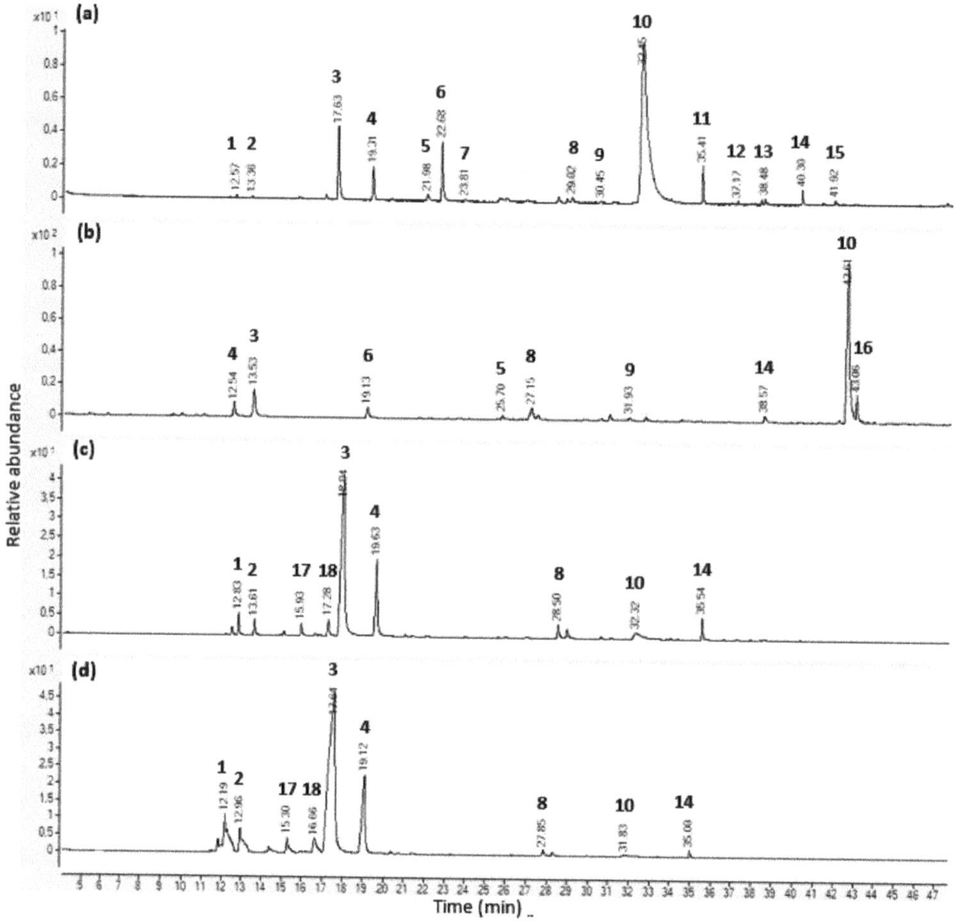

Figure 1. GC-MS chromatograms of *Thymus vulgaris* L. essential oil obtained from commercial source C and headspace analysis of its vapour. (**a**) Chromatogram on an HP-5MS column; (**b**) chromatogram on a DB-HeavyWAX column; (**c**) chromatogram of the headspace after a 9-h incubation period using solid-phase microextraction sampling technique; (**d**) chromatogram of the headspace after a 9-h incubation period using gas tight syringe sampling technique. Peak number and compound names: 1. α-pinene, 2. camphene, 3. *p*-cymene, 4. γ-terpinene, 5. linalool, 6. methyl octanoate, 7. Camphor, 8. Thymol methyl ether, 9. Carvone, 10. thymol. 11. β-caryophyllene, 12. γ-Muurolene, 13. δ-cadinene, 14. caryophyllene oxide, 15. α-epi-cadinol, 16. Carvacrol, 17. β-myrcene, 18. α-terpinene.

In all *T. vulgaris* EOs, thymol was the most abundant compound. When considering both HP-5MS/DB-HeavyWAX columns, supplier C's EO percentage values were 48.09/48.65% in contrast to suppliers A and B results—i.e., 38.42/29.52% and 38.93/41.31%, respectively. Likewise, *p*-cymene was the second most abundant constituent in the three EO samples. While in sample C, the peak percentage area represented 12.75/12.90%, its value was twice as high in both supplier A and B EOs—22.15/16.43% and 22.03/25.35%. Carvacrol was the third abounding compound in A, B, and C samples, with percentage values of 10.61/3.84%, 10.85/5.59%, and 10.92/3.60%, respectively. Although the content of γ-terpinene was relatively high in the three samples, the amount detected in EOs from suppliers A and B (2.82/2.02% and 2.78/2.90%) were significantly lower than in supplier C

EO (5.22/5.40%). Eventually, in EO samples A, B, and C, other compounds were detected in amounts lower than 1.80/1.08%, 1.82/1.77%, and 1.53/1.38%, respectively. Furthermore, ten compounds, including 4-carene, 4-pentenyl butyrate, α-cubebene, isobornyl acetate, calamenene, isoborneol, aromandendrene, lavandulyl butyrate, *p*-cymen-7-ol, and phytol, were only detected by the DB-HeavyWAX column at an amount lower than 0.86% in supplier A, 1.17% in supplier B, and 1.05% in supplier C.

2.3. Chemical Analysis of EOs' Vapour Phase

In this study, the sampling of the headspace above a mixture of *T. vulgaris* EO and MH broth has been carried out using two different methods of extraction, i.e., HS-SPME and HS-GTS extractions. Headspace chemical compositions were measured every 3 h during a 12-h experiment using the HP-5MS column. Complete analyses are provided in Tables 3 and 4, as well as Figure 1c,d. Using HS-SPME extraction, a total of 40, 38, and 43 volatile compounds were identified in the samples of suppliers A, B, and C, respectively. This represented 99.85, 99.89, and 99.63% of their respective total constituents at 12 hrs. In contrast, a significantly lower number of compounds was detected when using the HS-GTS extraction method. While 32 constituents were found in EO sample C, only 26 components in samples A and B were found, which accounted for 99.87, 99.95, and 99.71% of their total contents at time 12 h. Regardless of the extraction method used and the three samples analysed, monoterpenoids followed by sesquiterpene hydrocarbons were by far the two most predominant chemical groups of volatile compounds identified in the headspace. Other chemical categories present in minor amounts (oxygenated phenylpropanoids, oxygenated aliphatics, and furanoids) were identified using HS-SPME extraction only. On the other hand, cyclic ethers such as furan derivatives were only present in samples obtained by HS-GTS extraction.

Using the HS-SPME extraction method, the most abundant volatile substance across all *T. vulgaris* was *p*-cymene. Its percentage values during the whole experiment were rather similar in the headspace of samples A, B, and C, ranging from 54.57 to 58.61%, 69.91 to 74.50%, and 58.38 to 67.21%, respectively. Likewise, the second most abundant compound was γ-terpinene. While in sample A the peak percentage area was between 19.74 and 20.45% during the 12 h experiment, these values were lower in both supplier C and B, ranging from 12.37 to 16.18% and 7.07 to 9.30%, respectively. Thymol was the third abounding compound in A, B, and C samples during the entire 12-h period, with percentage values ranging from 2.61 to 3.36%, 2.13 to 5.25%, and 3.72 to 5.27%, respectively. Similarly using HS-GTS extraction, *p*-cymene (47.05 to 50.73% for sample A, 52.28 to 57.41% for sample B, and 44.80 to 49.28% for sample C), as well as γ-terpinene (13.40 to 17.45%, 5.76 to 6.99%, and 9.54 to 11.85% for sample A, B, and C, respectively), were the first two most abundant compounds in the headspace for all EO samples and during the full experiment. However, the third volatile substance detected in a significant amount was α-pinene, with percentage values over the 12-h experiment between 5.74 and 6.83%, 6.34 and 8.98%, and 6.90 and 7.96% for EO samples A, B, and C, respectively. Other differences have been observed when comparing the two sampling methods. Firstly, the number of compounds detected using HS-SPME was higher than when using HS-GTS extraction—i.e., on average, 41 versus 28 compounds. Then, the chemical analysis showed that when using HS-SPME extraction, a larger number of sesquiterpene hydrocarbons (14 components) were found in amounts lower than 3.09%, 1.59%, and 2.94% in EO samples A, B, and C, respectively. In contrast, using the HS-GTS method, only six compounds were found, including α-copaene, β-bourbonene, β-caryophyllene, γ-muurolene, γ-cadinene, and δ-cadinene, at amounts lower than 0.38% in sample A, 0.28% in sample B, and 0.58% in sample C. Similarly, percentage values of phenolic monoterpenes and derivatives were considerably higher using HS-SPME sampling method (overtime for samples A, B, and C, lower than 3.14%, 5.25%, and 5.27%, respectively) than HS-GTS extraction (values lower than 0.67%, 0.49%, and 0.58%, respectively). Eventually, despite the above-mentioned discrepancies, the headspace analysis of both sampling methods showed that there were no significant changes in the chemical composition in the vapour of the three EO samples of *T. vulgaris* over time.

Table 3. Chemical composition of the headspace above a mixture of Mueller–Hinton broth and *Thymus vulgaris* L. essential oils at a concentration of 512 μg/mL over a 12-h period using solid-phase microextraction sampling technique.

RI [a]		Compounds	Cl. [b]	Supplier/Time (h)/Content (%)														
				A					B					C [c]				
Obs.	Lit.			0	3	6	9	12	0	3	6	9	12	0	3	6	9	12
921	921	Tricyclene	MH	0.05	0.05	0.05	0.05	0.05	tr. [e]	tr.	0.05	0.07	0.07	0.08 ± 0.02	0.08 ± 0.01	0.07 ± 0.02	0.10 ± 0.01	0.08 ± 0.02
927	924	α-Thujene	MH	0.95	0.91	0.83	0.84	0.93	0.97	0.79	0.98	1.10	1.07	1.01 ± 0.05	1.00 ± 0.03	0.93 ± 0.03	0.94 ± 0.13	0.94 ± 0.05
933	939	α-Pinene	MH	2.37	2.25	2.18	2.11	2.37	1.97	1.60	2.23	2.45	2.41	2.73 ± 0.18	2.69 ± 0.13	2.57 ± 0.26	2.82 ± 0.30	2.70 ± 0.26
949	945	Camphene	MH	1.47	1.42	1.37	1.33	1.51	0.85	0.83	1.58	1.69	1.67	1.58 ± 0.56	1.56 ± 0.53	1.52 ± 0.58	2.13 ± 0.16	1.82 ± 0.57
976	969	Sabinene	MH	tr.	tr.	- [f]	tr.	tr.	tr.	tr.	0.05	0.06	0.06	tr.	tr.	tr.	0.07 ± 0.00	0.06 ± 0.02
978	974	β-Pinene	MH	0.49	0.47	0.42	0.45	0.48	0.43	0.40	0.49	0.51	0.53	0.45 ± 0.01	0.49 ± 0.06	0.48 ± 0.07	0.56 ± 0.04	0.51 ± 0.06
991	988	3-Octanone	OA	tr.	tr.	-	tr.	tr.	-	-	tr.	tr.	tr.	-	-	-	-	-
995	988	β-Myrcene	MH	2.76	2.70	2.63	2.48	2.72	1.53	1.45	1.79	1.88	1.88	1.62 ± 0.32	1.53 ± 0.18	1.51 ± 0.29	1.83 ± 0.03	1.62 ± 0.26
1009	1002	α-Phellandrene	MH	0.36	0.35	0.34	0.33	0.35	0.19	0.19	0.20	0.20	0.20	0.26 ± 0.03	0.26 ± 0.02	0.25 ± 0.03	0.28 ± 0.01	0.24 ± 0.03
1013	1008	3-Carene	MH	0.21	0.21	0.21	0.20	0.22	0.09	0.08	0.13	0.14	0.14	0.11 ± 0.04	0.10 ± 0.03	0.10 ± 0.03	0.14 ± 0.01	0.12 ± 0.03
1021	1014	α-Terpinene	MH	3.76	3.72	3.61	3.58	3.72	1.61	1.58	2.34	2.37	2.38	2.20 ± 0.66	2.17 ± 0.64	2.14 ± 0.65	2.92 ± 0.05	2.46 ± 0.64
1036	1020	*p*-Cymene	AH	58.61	56.24	55.41	54.57	56.35	74.22	74.50	71.03	71.16	69.91	67.21 ± 7.03	66.38 ± 7.46	65.89 ± 6.38	58.38 ± 0.23	60.62 ± 6.81
1055	1023	*m*-Cymene	AH	tr.	0.05	0.05	tr.	tr.	-	-	-	-	-	-	-	-	tr.	tr.
1067	1054	γ-Terpinene	MH	19.7	20.45	20.30	20.23	20.18	7.07	7.23	9.30	9.18	9.17	12.37 ± 2.86	12.51 ± 2.64	12.67 ± 3.02	16.18 ± 0.16	14.14 ± 2.88
1094	1086	Terpinolene	MH	0.31	0.33	0.31	0.32	0.32	0.10	0.11	0.24	0.24	0.24	0.16 ± 0.10	0.16 ± 0.10	0.17 ± 0.10	0.27 ± 0.01	0.21 ± 0.09
1101	1089	*p*-Cymenene	MH	0.20	0.25	0.19	0.21	0.22	0.15	0.14	0.19	0.21	0.20	0.12 ± 0.06	0.10 ± 0.04	0.09 ± 0.02	0.14 ± 0.01	0.11 ± 0.03
1115	1095	Linalool	OM	0.27	0.26	0.25	0.26	0.33	0.13	0.10	0.28	0.34	0.34	0.23 ± 0.21	0.16 ± 0.12	0.16 ± 0.13	0.29 ± 0.03	0.19 ± 0.10
1153	1141	Camphor	OM	0.05	tr.	tr.	0.05	0.05	tr.	tr.	0.06	0.06	0.06	tr.	tr.	0.06 ± 0.04	0.09 ± 0.01	0.08 ± 0.04
1186	1165	Endo-borneol	OM	tr.	tr.	tr.	tr.	tr.	0.08	0.07	tr.	0.05	0.05	0.11 ± 0.02	0.09 ± 0.01	0.08 ± 0.01	0.07 ± 0.01	0.08 ± 0.01
1192	1174	Terpinen-4-ol	OM	0.06	0.08	0.08	0.10	0.10	tr.	tr.	0.08	0.08	0.09	tr.	tr.	0.06 ± 0.05	0.10 ± 0.01	0.09 ± 0.04
1213	1195	Estragole	OPP	tr.	tr.	tr.	tr.	tr.	-	-	-	-	tr.	0.10 ± 0.06	0.13 ± 0.09	0.11 ± 0.04	0.17 ± 0.03	0.18 ± 0.06
1247	1235	Thymol methyl ether	OPM	1.78	2.48	2.68	2.82	2.44	3.05	3.77	2.68	2.24	2.58	1.83 ± 0.13	2.41 ± 0.17	2.77 ± 0.41	2.61 ± 0.17	2.91 ± 0.34
1257	1241	Carvacrol methyl ether	OPM	1.15	1.52	1.68	1.68	1.51	1.67	2.02	1.55	1.40	1.49	1.08 ± 0.12	1.37 ± 0.11	1.50 ± 0.14	1.54 ± 0.03	1.69 ± 0.11
1293	1287	Bornyl acetate	OM	0.17	0.24	0.29	0.33	0.21	0.05	0.08	0.15	0.11	0.14	tr.	0.15 ± 0.08	0.15 ± 0.06	0.25 ± 0.04	0.24 ± 0.11
1306	1284	Anethol	OPP	tr.	tr.	tr.	0.05	tr.	-	-	-	-	-	tr.	tr.	0.08 ± 0.00	0.12 ± 0.02	0.12 ± 0.03
1339	1290	Thymol	PM	2.61	2.80	3.14	3.36	3.12	5.25	3.50	2.13	2.64	3.15	5.27 ± 1.27	4.57 ± 0.81	4.62 ± 0.84	3.72 ± 0.07	5.05 ± 0.45
1368	1298	Carvacrol	PM	tr.	tr.	tr.	tr.	tr.	-	-	-	-	-	-	tr.	tr.	tr.	tr.
1381	1372	*p*-Cymen-7-ol	OM	tr.	tr.	tr.	tr.	tr.	-	-	tr.	tr.	tr.	-	-	-	tr.	tr.

Table 3. *Cont.*

RI [a]		Compounds	Cl. [b]	Supplier/Time (h)/Content (%)														
				A					B					C [c]				
Obs.	Lit.			0	3	6	9	12	0	3	6	9	12	0	3	6	9	12
1386	1374	α-Copaene	SH	0.09	0.13	0.17	0.19	0.12	tr.	tr.	0.11	0.07	0.11	tr.	tr.	0.07 ± 0.08	0.16 ± 0.01	0.19 ± 0.05
1395	1387	β-Bourbonene	SH	tr.	tr.	0.05	0.07	tr.	tr.	tr.	tr.	tr.	0.05	tr.	tr.	tr.	0.09 ± 0.01	0.10 ± 0.01
1434	1418	β-Caryophyllene	SH	1.59	2.14	2.74	3.09	1.94	0.26	0.32	1.59	1.13	1.49	0.77 ± 0.82	1.19 ± 1.26	1.28 ± 1.29	2.94 ± 0.32	2.40 ± 1.59
1442	1446 [d]	isogermacrene D	SH	tr.	0.06	0.09	tr.	tr.	-	-	tr.	tr.	tr.	-	tr.	tr.	0.08 ± 0.06	tr.
1469	1452	α-Humulene	SH	tr.	tr.	0.07	0.07	tr.	-	tr.	tr.	tr.	tr.	tr.	tr.	tr.	0.07 ± 0.01	0.06 ± 0.03
1475	1465	*cis*-muurola-4(14),5-diene	SH	-	tr.	tr.	tr.	-	-	-	tr.	-	tr.	-	-	-	tr.	tr.
1482	1475	Geranyl propionate	OM	tr.	tr.	tr.	tr.	-	-	-	-	-	-	-	-	-	-	tr.
1491	1478	γ-Muurolene	SH	0.06	0.08	0.12	0.14	0.07	tr.	tr.	0.06	tr.	0.06	tr.	0.07 ± 0.04	0.08 ± 0.05	0.12 ± 0.01	0.11 ± 0.05
1509	1480	Germacrene D	SH	tr.	tr.	tr.	tr.	-	-	-	tr.	-	-	-	-	-	tr.	tr.
1512	1491	Valencene	SH	-	tr.	tr.	tr.	-	-	-	-	-	-	-	-	-	tr.	tr.
1515	1499	α-Muurolene	SH	tr.	tr.	tr.	tr.	tr.	-	-	tr.	tr.	tr.	-	-	-	tr.	tr.
1520	1509	β-Bisabolene	SH	tr.	tr.	tr.	tr.	tr.	-	-	tr.	-	tr.	-	-	-	tr.	tr.
1532	1513	γ-Cadinene	SH	0.08	0.08	0.12	0.14	0.07	tr.	tr.	0.07	tr.	0.07	tr.	0.10 ± 0.04	0.08 ± 0.03	0.12 ± 0.02	0.12 ± 0.05
1538	1524	δ-Cadinene	SH	0.1	0.1	0.15	0.17	0.09	tr.	tr.	0.05	tr.	0.05	tr.	0.07 ± 0.04	0.08 ± 0.04	0.12 ± 0.02	0.11 ± 0.05
1541	1521	Calamenene	SH	tr.	tr.	-	0.07	tr.	tr.	0.05	0.05	0.03	0.05	tr.	0.07 ± 0.01	0.06 ± 0.01	0.07 ± 0.01	0.07 ± 0.01
1605	1581	Caryophyllene oxide	OS	tr.	tr.	tr.	tr.	tr.	-	-	-	tr.	tr.	-	-	tr.	tr.	tr.
		Total identified (%)		99.68	99.83	99.91	99.77	99.85	99.89	99.18	99.72	99.64	99.89	99.33	99.87	99.87	99.74	99.63

[a] RI = retention indices. Obs. = retention indices determined relative to a homologous series of n-alkanes (C8-C40) on an HP-5MS column. Lit. = literature Ri values [49,50]; [b] Cl = class; AH—aromatic hydrocarbon, MH—monoterpene hydrocarbon, OA—oxygenated aliphatic, OM—oxygenated monoterpene, OPM—oxygenated phenolic monoterpene, OPP—oxygenated phenylpropanoid, PM—phenolic monoterpene, OS—oxygenated sesquiterpene, SH—sesquiterpene hydrocarbon. [c] Relative peak area percentage as mean of three measurements ± standard deviation. [d] Literature RI values from [51]. [e] tr. = traces, relative peak area < 0.05%. [f] - = not detected.

Table 4. Chemical composition of the headspace above a mixture of Mueller–Hinton broth and *Thymus vulgaris* L. essential oils at a concentration of 512 μg/mL over a 12-h period using gas tight syringe headspace sampling technique.

RI [a]		Compounds	Cl. [b]	Supplier/Time (h)/Content (%)														
				A					B					C [c]				
Obs.	Lit.			0	3	6	9	12	0	3	6	9	12	0	3	6	9	12
-	NA	2-Ethyl furan	F	tr. [d]	tr.	tr.	- [e]	tr.	0.09	0.09	0.09	0.07	0.04	0.09 ± 0.05	0.07 ± 0.01	0.08 ± 0.04	0.08 ± 0.02	0.07 ± 0.00
778	NA	Methyl α-methylbutyrate	OA	0.08	0.10	0.13	0.08	0.12	0.22	0.17	0.23	0.16	0.11	0.10 ± 0.07	0.08 ± 0.02	0.10 ± 0.04	0.07 ± 0.01	0.09 ± 0.00
912	921	Tricyclene	MH	0.11	0.20	0.16	0.12	0.19	0.17	0.15	0.06	0.20	0.18	0.19 ± 0.01	0.17 ± 0.02	0.19 ± 0.01	0.20 ± 0.04	0.24 ± 0.00
918	924	α-Thujene	MH	2.01	2.17	2.03	1.75	2.23	2.09	2.03	2.11	2.38	2.34	1.97 ± 0.24	1.89 ± 0.07	1.95 ± 0.15	1.91 ± 0.17	2.30 ± 0.00
924	939	α-Pinene	MH	6.19	5.70	6.83	4.63	5.74	8.98	7.75	7.84	6.79	6.34	7.96 ± 2.05	6.94 ± 1.11	6.90 ± 1.73	7.20 ± 0.33	7.11 ± 0.00
939	945	Camphene	MH	3.40	3.21	3.57	2.69	3.28	4.59	4.04	4.07	3.81	3.71	4.59 ± 0.91	4.22 ± 0.35	4.16 ± 0.76	4.24 ± 0.03	4.45 ± 0.00
953	969	Sabinene	MH	-	-	-	0.07	-	-	-	-	-	-	tr.	tr.	0.05 ± 0.02	0.13 ± 0.19	0.06 ± 0.00
967	980	β-Pinene	MH	0.91	0.83	0.85	0.78	0.87	1.07	0.88	0.99	0.89	0.81	0.75 ± 0.37	0.94 ± 0.04	0.91 ± 0.11	0.74 ± 0.25	0.81 ± 0.00
984	988	β-Myrcene	MH	2.88	2.80	3.14	3.13	2.88	1.88	1.89	1.98	1.90	2.06	1.82 ± 0.06	1.96 ± 0.24	1.88 ± 0.08	1.63 ± 0.15	2.09 ± 0.00
998	1002	α-Phellandrene	MH	0.31	0.35	0.23	0.20	0.35	0.05	0.11	0.11	0.10	0.07	0.13 ± 0.05	0.23 ± 0.06	0.20 ± 0.10	0.19 ± 0.03	0.22 ± 0.00
1002	1008	3-Carene	MH	0.28	0.25	0.36	0.13	0.26	0.30	0.12	0.11	0.08	0.05	0.14 ± 0.10	0.10 ± 0.05	0.13 ± 0.08	0.13 ± 0.05	0.16 ± 0.00
1010	1014	α-Terpinene	MH	3.64	3.57	2.57	4.17	3.80	2.21	2.06	1.93	2.12	2.34	2.57 ± 0.18	2.87 ± 0.50	2.77 ± 0.33	2.41 ± 0.18	2.88 ± 0.00
1028	1020	*p*-Cymene	AH	45.93	47.05	47.48	50.73	48.78	52.28	54.29	56.81	57.30	57.41	44.80 ± 1.03	49.19 ± 2.67	49.20 ± 3.00	46.10 ± 2.10	49.28 ± 0.00
1044	1023	*m*-Cymene	AH	tr.	tr.	0.27	0.28	tr.	-	-	-	-	-	0.07 ± 0.07	0.09 ± 0.06	tr.	0.06 ± 0.08	tr.
1058	1054	γ-Terpinene	MH	13.40	13.90	15.46	17.54	14.80	5.76	6.22	6.58	6.69	6.99	9.54 ± 1.00	11.77 ± 1.52	11.82 ± 1.70	10.32 ± 0.68	11.85 ± 0.00
1083	1086	Terpinolene	MH	0.18	0.19	0.23	0.26	0.20	0.10	0.12	0.14	0.14	0.14	0.13 ± 0.02	0.18 ± 0.04	0.18 ± 0.03	0.14 ± 0.02	0.17 ± 0.00
1089	1089	*p*-Cymenene	MH	0.10	0.09	0.13	0.14	0.09	0.08	0.09	0.10	0.09	0.09	0.06 ± 0.00	0.08 ± 0.01	0.07 ± 0.01	0.06 ± 0.02	0.06 ± 0.00
1103	1095	Linalool	OM	0.09	0.08	0.15	0.16	0.10	0.08	0.06	0.12	0.11	0.10	0.09 ± 0.03	0.09 ± 0.03	0.09 ± 0.01	0.06 ± 0.03	0.08 ± 0.00
1133	1141	Camphor	OM	tr.	tr.	tr.	tr.	tr.	tr.	tr.	tr.	tr.	tr.	tr.	tr.	tr.	tr.	tr.
1171	NA	2-Ethyl-5-methylfuran	F	tr.	tr.	tr.	tr.	tr.	tr.	-	tr.	-	tr.	tr.	tr.	tr.	tr.	tr.
1180	1174	Terpinen-4-ol	OM	tr.	tr.	tr.	tr.	tr.	tr.	tr.	tr.	tr.	tr.	tr.	tr.	tr.	tr.	tr.
1200	1195	Estragole	OPP	-	-	tr.	tr.	-	-	-	-	-	-	tr.	tr.	tr.	tr.	tr.
1234	1235	Thymol methyl ether	OPM	0.34	0.34	0.62	0.67	0.43	0.24	0.39	0.45	0.43	0.49	0.31 ± 0.12	0.53 ± 0.15	0.57 ± 0.12	0.43 ± 0.08	0.50 ± 0.14
1244	1241	Carvacrol methyl ether	OPM	0.21	0.19	0.38	0.40	0.25	0.13	0.23	0.25	0.25	0.28	0.19 ± 0.08	0.30 ± 0.09	0.33 ± 0.08	0.24 ± 0.05	0.29 ± 0.08
1281	1287	Bornyl acetate	OM	tr.	tr.	0.05	0.05	tr.	-	tr.	tr.	tr.	tr.	tr.	tr.	tr.	tr.	tr.
1328	1290	Thymol	PM	0.50	0.27	0.58	0.60	0.28	0.25	0.10	0.28	0.26	0.22	0.46 ± 0.31	0.51 ± 0.36	0.55 ± 0.30	0.25 ± 0.10	0.58 ± 0.37
1372	1374	α-Copaene	SH	tr.	tr.	tr.	tr.	tr.	-	tr.	tr.	-	tr.	tr.	tr.	tr.	tr.	tr.
1381	1387	β-Bourbonene	SH	-	-	tr.	tr.	-	-	-	-	-	-	-	tr.	tr.	-	tr.

Table 4. *Cont.*

RI [a]		Compounds	Cl. [b]	Supplier/Time (h)/Content (%)														
				A					B					C [c]				
Obs.	Lit.			0	3	6	9	12	0	3	6	9	12	0	3	6	9	12
1417	1418	β-Caryophyllene	SH	0.17	0.17	0.38	0.34	0.19	0.07	0.11	0.14	0.13	0.15	0.21 ± 0.11	0.37 ± 0.15	0.41 ± 0.00	0.28 ± 0.08	0.39 ± 0.20
1475	1478	γ-Muurolene	SH	-	-	tr.	tr.	-	-	-	-	-	tr.	tr.	tr.	tr.	tr.	tr.
1513	1513	γ-Cadinene	SH	-	-	tr.	tr.	-	-	-	-	-	-	-	tr.	tr.	-	tr.
1521	1524	δ-Cadinene	SH	-	-	tr.	tr.	-	-	-	-	-	-	-	tr.	tr.	-	tr.
		Total identified (%)		99.95	99.80	99.94	99.94	99.95	99.98	99.86	99.90	99.92	99.71	99.87	99.83	99.90	99.90	99.87

[a] RI = retention indices. Obs. = retention indices determined relative to a homologous series of *n*-alkanes (C8–C40) on an HP-5MS column. Lit. = literature Ri values [49–51]. [b] Cl = Class; AH—aromatic hydrocarbon, MH—monoterpene hydrocarbon, OA—oxygenated aliphatic, OM—oxygenated monoterpene, OPM—oxygenated phenolic monoterpene, OPP—oxygenated phenylpropanoid, PM—phenolic monoterpene, OS—oxygenated sesquiterpene, SH—sesquiterpene hydrocarbon. [c] Relative peak area percentage as mean of three measurements ± standard deviation. [d] tr. = traces, relative peak area < 0.05%. [e] - = not detected.

3. Discussion

As reported by Houdkova and Kokoska [52], several assays have previously been developed for the evaluation of the antibacterial activity of volatile plant compounds in the vapour phase. However, there is still a lack of standardised methods, something that makes any interpretation and comparison difficult [52]. For instance, three different tests have been identified to investigate *T. vulgaris* EO vapours activity, and results are described differently according to each author. A study led by Inouye et al. [28] assessed the growth-inhibitory effects of two *T. vulgaris* EO vapours using the airtight box disc volatilisation method. Introducing the minimum inhibitory dose (MID) expressed in mg/L air, they determined that the vapour of carvacrol chemotype EOs were more active against Gram-negative *H. influenzae* than against Gram-positive strains such as *S. pyogenes* and *S. aureus* (MIDs of 3.13, 6.25, and 12.5 mg/L air, respectively). In this study, we reported similar results: *H. influenzae* was more susceptible—i.e., MIC = 512 µg/mL or 128 µg/cm^3 considering the volume of the entire well—to the vapour of our three thymol chemotype EOs than the *S. pyogenes*, and *S. aureus* (MICs comprised in between 1024–512 µg/mL or 256–128 µg/cm^3 for both bacteria strains). It is largely admitted that Gram-negative bacteria are more resistant to EOs than Gram-positive ones [53,54]. This weak antibacterial activity was attributed to the presence of hydrophilic polysaccharide chains in the outer membrane structure, preventing hydrophobic EOs from reaching the bacteria cell membrane [55]. One reason that could explain the higher susceptibility of *H. influenzae* to EOs would be the more hydrophobic nature of its outer membrane composed of oligosaccharide shorter chains [28]. This was confirmed by Reyes-Jurado et al. [56], who demonstrated that compounds including *p*-cymene, linalool, and thymol were able to disintegrate such outer membrane structures. Furthermore, various studies have subsequently assessed the growth-inhibitory effects of *T. vulgaris* EO vapour using vapour diffusion assay developed by Lopez et al. [37]. Nedorostova et al. [36] reported a MIC of 17 µL/L against *S. aureus*, which is, according to the author, equivalent to the result of Inouye et al. [28] against the same bacterium (MID = 12.5 mg/L of air). Similarly, Kloucek et al. [57] have observed that the vapour of *T. vulgaris* EO consisting mainly of geraniol had a MIC = 125 µL/L against *S. aureus* as well. In contrast to these findings, MIC values recorded in our study were usually higher—i.e., 1024–512 µg/mL (or 256–128 µg/cm^3 considering the volume of the entire well). The discrepancy in results could be firstly explained by the quality of the EO samples used and their chemical compositions [28], but also by the disparate strains of bacteria used in the different antimicrobial assays [58]. Most importantly, this variation may also be attributed to the diverse methods used to explore the antimicrobial effect of *T. vulgaris* EO vapour, allowing various interpretations of the MIC [35,37,57]. That is why the BMV assay presented here is a powerful alternative to the previously developed techniques. While their designs only enable the assessment of EO vapour at a single concentration [57], the BMV assay is conceived to evaluate the EOs' in vitro growth-inhibitory effect in liquid and gaseous phases simultaneously and at different concentrations, something that allows fast comparison of MIC values in both liquid and solid media. Based on broth microdilution [59] and disc volatilisation (DV) assays [33], the BMV experiments are conducted on standard 96-well immune plates, which offer several advantages such as cost and labour efficiency [57,60]: microplates are standard laboratory equipment that are commonly available and compared, for example, to the special airtight experimental apparatus used in certain methods such as in Seo et al. [61]. Furthermore, microplates can also be employed in fully automated workstations, unlike Petri dishes used in DV assays which therefore suffer from a lack of repeatability [62]. Other studies developed assays using microplates for detecting volatile substances antimicrobial activity, such as the vapour-phase-mediated patch assay of Feyaerts et al. [63]. However, contrary to the BMV assay, their designs only allow to determine relative microbial inhibition values, and their main limitation lies in only providing qualitative results [64]. Eventually, another asset of our assay is that modifications can be easily implemented for new applications. For instance, Netopilova et al. [17] modified the test to explore the combinatory effects of volatile substances using a chequerboard design and thus determine fractional inhibitory concentration (FIC) indices, something not possible with other methods. Overall, these features allow our method to be suitable for high-throughput screening and thus be a simple, fast, and reliable assay as well as providing reproducible and quantitative results [65]. Nevertheless, despite its numerous advantages, serially produced microplates are not designed for volatile substance testing, something that contributes to weaknesses shown by our assay. For example, only a limited volume of agar can be pipetted into each flange of the lid; this could impact the bacterial growth and thus affect the results. Most importantly, when testing EO vapours, the BMV assay also faces specific problems linked to the physico-chemical nature of EOs and, more particularly, their high volatility, viscosity, and hydrophobicity [52]. Firstly, volatility allows a loss of active substances by evaporation that can happen at different steps of the protocol, including during sample handling and experiment preparation, complications that are shared by all tests assessing EO properties [66,67]. More specifically to volatilisation assays, the level of vapour transition from the matrix into which the EO is included and its distribution into the inner atmosphere of the well during the experiment are two critical factors that may affect the results. For instance, the matrix influences both the intensity and the speed of the evaporation [35]. In our assay, a broth medium was used, which according to Orchard et al.'s [68] observations, would slow the level of vapour transition during the experiment and thus affect the bacterial growth. Similarly, as described by Reyes-Jurado et al. [69], the hydrophobicity and viscosity properties of the EO may also cause its uneven distribution through the broth medium, something that could also alter the distribution of the volatile agents into the well's atmosphere. That is why the concentrations in the vapour phase used in our experiment should only be considered as indicative values.

The antibacterial properties of *T. vulgaris* are mainly attributed to the chemical composition of its EO, which has already been extensively studied. Its major constituents are mainly monoterpenoids, such as carvacrol, thymol, γ-terpinene, and *p*-cymene, but also sesquiterpenoids such as β-caryophyllene. Within the same species, the variation of proportions of these compounds defines the EO chemotype, which is named after the predominant constituent identified [70]. Our chemical analyses showed that thymol was the most abundant constituent within our samples, followed by *p*-cymene and carvacrol. This characterises our three EOs as thymol chemotypes. These findings are in accordance with several studies previously published. For instance, Schmidt et al. [71], Grosso et al. [72], and Nikolić et al. [25] reported thymol as the major component of their *T. vulgaris* thymol chemotype EOs (peak percentage area of 38.8%, 41.6%, and 49.1%, respectively); *p*-cymene was the second most abundant with 24.0%, 28.9%, and 20.0%, respectively. However, instead of carvacrol, γ-terpinene was detected as the third most abundant constituent (9.5%, 5.1%, and 4.2%, respectively). The variations in yields and concentrations of volatile compounds in our samples could be attributed to several factors occurring at different stages, from the growing conditions of the plants, the harvesting period to the storage conditions of the plant materials by the commercial suppliers [18,73]. For instance, Nezhadali et al. [56] showed that *T. vulgaris* harvested in the same location but at different stages of the plant growth resulted in different yields of EOs: the highest oil yield (1.39%) was obtained during the flowering period, whereas the lowest yield (0.83%) corresponded to the fruiting stage. Similarly, he reported that the concentration of thymol between those periods dropped from 63.01% to 38.23% of the total EO content. Subsequently, the characterisation of the EOs was carried out by GC/MS using two capillary columns of different polarities. The concomitant use of a polar column (DB-HeavyWAX) along with a non-polar (HP-5MS) allows revealing overlapped signal peaks and thus improves the identification of the separated compounds. Hudaib et al. [74], who analysed the chemical profile of *T. vulgaris* EO by GC/MC using the same approach, demonstrated that the polar column helped to enhance the resolution between compounds co-eluting on a non-polar column such as the couples (α-thujene, α-pinene) or (sabinene, β-pinene). This has also been observed in our study with, for instance, the couple (*p*-cymene-limonene). Moreover, as described by Fan et al. [75], while the main constituents of an EO are equally detected by both columns, a fraction is identified by either of them: in our study, 23 compounds out of 75 were detected using HP-5MS only—i.e., representing on average 5.82% of the compounds identified within the three samples—and 10 compounds (1.88% on average) with DB-HeavyWAX. The difference in detection recorded would be the result of the different polarity and material of the columns used [76]. Overall, these results suggest that complementing a non-polar with a polar column provides a more precise picture of the EO analysed than if displayed individually.

The characterisation of *T. vulgaris* EO vapours was carried out using two different sampling methods: HS-SPME and HS-GTS. In recent years, HS-SPME has become the preferred laboratory method for identifying EOs volatile compounds: it is a simple, fast, cost-effective, selective, and sensitive method that provides high-quality results [40,42,60]. With the aim to simulate the conditions of the antimicrobial susceptibility testing experiments performed in this study, the EO samples were prepared identically to the most active EO during the BMV assay (i.e., incubation temperature was at 37 °C and EOs dissolved in Mueller–Hinton (MH) broth medium concentrated at 512 μg/mL). In addition, we used a mixed coating material (DVB/CAR/PDMS) that gives better extraction yields for both polar and non-polar volatile constituents than simple fibre coatings [41,77]. As a result, our investigation revealed that although *p*-cymene and γ-terpinene were abundant in the headspace, the amount of thymol extracted by the coated fibre was unusually low (peak percentage area lower than 5.27% across the three EO samples). This observation is in contradiction with what was described in previously published research. For instance, Lugo-Estrada et al. [41], Soleimani et al. [78], and Nezhadali et al. [79], who investigated *T. vulgaris* EO vapour composition, all reported that thymol was the most abundant constituent of the headspace (peak percentage area of 34.28%, 28.50%, and 45.45%, respectively). As described by Adam et al. [80], efficient extractions of EO volatile compounds depend on optimised experimental parameters such as the selection of the fibre coating material and the incubation temperature of the EO sample. The lack of thymol could, therefore, be explained by several reasons. Firstly, it is the selectivity and sensitivity of the DVB/CAR/PDMS coating. Although it proved to be the most universal assembly for the isolation of compounds with diverse physico-chemical properties [81], Soleimani et al. [77] demonstrated in their comparative study that phenolic monoterpenes and, more particularly, thymol, were better extracted by a simple polydimethylsiloxane (PDMS) fibre than using mixed coating materials. Furthermore, as the transfer rate of volatile agents toward the fibre increases with the incubation temperature of the sample [81], the temperature chosen could have potentially limited the extraction of thymol in our investigation. For example, in their research work, Nezhadali et al. [82] compared the HS-SPME extraction efficiency of *T. vulgaris* EO main volatile agents at 25 °C and 50 °C using a water-based matrix. The result showed that the amount of thymol at higher temperatures is almost twice as high as at lower temperatures (73.09% and 45.45%, respectively). This suggests that different experimental conditions using the HS-SPME technique can yield different distributions of EO volatile compounds, and therefore, the result may not necessarily illustrate the actual chemical composition of the headspace above the EO samples. In contrast, the HS-GTS technique provided a different perspective on the constituency of the headspace, perhaps closer to the real distribution of volatile compounds in the vial at equilibrium [83]. Despite this, the results showed a peak percentage area for thymol lower than 0.60% across the three EO samples—even below the levels observed with the HS-SPME method. Using identical experimental conditions with both methods has demonstrated unusually low concentration levels of thymol. The possible explanation may therefore lie in the matrix in which the EO was inserted. As previously mentioned, the hydrophobic nature of volatile compounds worsens their solubility in water-based media (here the MH broth medium), which may reduce their dilution capability and result in an unequal distribution throughout the medium, something that may also affect their distribution into the well's atmosphere [69].

The HS-GTS method also yielded interesting results with other components; for instance, the amount of α-pinene obtained with HS-GTS sampling (peak percentage area lower than 7.96% across the three EO samples) was at least three times higher than the amount extracted with HS-SPME (peak percentage area lower than 2.82%). Despite a lack of academic studies examining *T. vulgaris* EO using HS-GTS, these findings are consistent with the data available in other publications. Coleman et al. [84], for instance, compared the distribution of volatile constituents in the headspace above a sample of *Juniperus virginiana* EO using both HS-SPME and HS-GTS techniques. He observed that α-pinene dominated the headspace with 88.0% of the total composition when using HS-GTS, whereas it only exhibited 32.4% with the HS-SPME technique. According to the author, this difference may be explained by the sorption behaviour of α-pinene over time when using a coated fibre assembly. In fact, he demonstrated that the percentage composition of volatile compounds changes significantly with the fibre exposure time, with α-pinene levels dramatically descending with greater time exposure, while high molecular weight components increased with the sampling time.

When comparing the chemical profiles of *T. vulgaris* EO in the liquid phase with the headspace analyses, we observed significant differences between both phases. The EO liquid phase contained a greater number of substances with a variable distribution, whereas the headspace composition detected a smaller number of compounds represented mostly by highly volatile substances. This phenomenon is in accordance with what was already described in previous research [83]. Nevertheless, the three EO samples showed identical concentrations of antimicrobial activities during BMV experiments in both liquid and vapour phases. This raises the question as to the levels of the active concentration of the EO compounds in the vapour, and more particularly thymol as its main antimicrobial constituent [57]. As a phenolic compound, thymol is very stable, moderately soluble in water and of low volatility, and was detected in a lower amount in the headspace—i.e., peak area percentage lower than 5.27% (HS-SPME) and 0.60% (HS-GTS)—than in the sample of EO diluted in the broth (max. 48.65%). By contrast, α-pinene, which is highly volatile and extremely insoluble in water, followed the opposite trend (peak area percentage lower than 2.82% (HS-SPME), 7.84% (HS-GTS), and 0.62% in the sample of EO diluted in the broth). This was observed in previous studies and is explained by the difference in volatility: when the EO is introduced into a closed environment, volatile compounds start to diffuse at different rates according to their molecular weight until they reach equilibrium [69]. Despite its slower diffusion rate, we could possibly argue that the low levels of thymol detected in the headspace was sufficient to generate the same antimicrobial activity as its amount in the EO concentrated at 512 μg/mL. To compare, Wang et al. [84] observed that the active concentration of thymol in its vapour phase samples against oral pathogens was between 100–400 μg/mL. Although the correlation between the concentration of essential oil and thymol were not the subject of this study, the results strongly suggest that the effect of antimicrobial constituents of *T. vulgaris* EO (e.g., of thymol) is higher in vapour than in the liquid phase. To further support this, an additional investigation focused on thymol behaviour in vapour would be needed.

4. Materials and Methods

4.1. Plant Material and Sample Preparation

Dried aerial part bulk of *Thymus vulgaris* L. were randomly purchased at three local spice stores and e-shops (supplier A = Kralovství chuti s.r.o., Prague, Czech Republic; supplier B = Byliny Mikes s.r.o., Číčenice, Czech Republic; supplier C = Lbros s.r.o., Vrchlabí, Czech Republic). Subsequently, plant materials were ground and homogenised using a Grindomix apparatus (GM 100 Retsch, Haan, Germany). The residual moisture content was evaluated gravimetrically at 130 °C for 1 h by Scaltec SMO 01 Analyzer (Scaltec Instruments, Gottingen, Germany) in triplicates and results were expressed as arithmetic averages (15.29%, 14.54%, and 13.13% for suppliers A, B, and C, respectively).

4.2. Hydrodistillation of the Essential Oils

EOs were extracted by hydrodistillation of the ground material following the indication provided by the European Pharmacopoeia [85]: 100 g of ground plant materials placed in 1 L of distilled water was distilled for 3 h using Clevenger-type apparatus (Merci, Brno, Czech Republic). Since hydrodistillation is one of the commonly used methods for commercial production of *Thymus vulgaris* L. EO, the properties of samples prepared in our study should be similar to those commercially available [86]. Eventually, the extracted EOs were stored in sealed glass vials at 4 °C until further handling.

4.3. Bacterial Strains and Culture Media

In this study, the following standard strains of the American Type Culture Collection (ATCC) were used: *H. influenzae* ATCC 49247, *S. aureus* ATCC 29213, and *S. pyogenes* ATCC 19615. Both cultivation and assay media (broth/agar) were MH, complemented with Haemophilus Test Medium and defibrinated horse blood for *H. influenzae*, MH only for *S. aureus*, and Brain Heart Infusion when working with *S. pyogenes*. The pH of broths was equilibrated to a final value of 7.6 using Trizma base (Sigma-Aldrich, Prague, Czech Republic). All microbial strains, growth media, and additions were purchased from Oxoid (Basingstoke, Hampshire, UK).

Stock cultures of bacterial strains were cultivated in appropriate media at 37 °C for 24 h prior to the testing, and the bacterial suspension turbidity was adjusted to 0.5 McFarland standard using Densi-La-Meter II (Lachema, Brno, Czech Republic) to reach the final concentration of 10^7 CFU/mL. Ampicillin, oxacillin, and amoxicillin were purchased from Sigma-Aldrich (Prague, Czech Republic) and assayed as positive antibiotic controls [30].

4.4. Antimicrobial Assay

The in vitro growth-inhibitory effect of EOs was assessed using BVM method that allows simultaneous assessment of EOs antibacterial activities at different concentrations in both liquid and vapour phases [4,35]. Experiments were performed using standard 96-well microtiter plates (well volume = 400 µL) covered by tight-fitting lids with flanges designed to reduce evaporation (SPL Life Sciences, Naechon-Myeon, Korea). Initially, 30 µL of agar was pipetted into every flange on the lid, except the outermost flanges, and inoculated with 5 µL of bacterial suspension after solidification of the agar. Subsequently, each EO sample was dissolved in dimethyl sulfoxide (DMSO) (Sigma-Aldrich, Prague, Czech Republic) at a maximum concentration of 1% and diluted in an appropriate broth medium. Seven two-fold serially diluted concentrations of samples starting from 1024 µg/mL were prepared for all EOs. The final volume in each well was 100 µL. The plates were subsequently inoculated with a bacterial suspension using a 96-pin multi-blot replicator (National Institute of Public Health, Prague, Czech Republic). The wells containing inoculated and non-inoculated broth were prepared as growth and purity controls simultaneously. The outermost wells were left empty to prevent the edge effect. Eventually, clamps (Lux Tool, Prague, Czech Republic) were used for fastening the plate and lid together with handmade wooden pads (size 8.5 × 13 × 2 mm) and the microtiter plates were incubated at 37 °C for 24 h. The MICs were evaluated by visual assessment of bacterial growth after colouring of metabolically active bacterial colonies with thiazolyl blue tetrazolium bromide dye (MTT) at a concentration of 600 µg/mL (Sigma-Aldrich, Prague, Czech Republic), when the interface of colour change from yellow to purple (relative to that of colours in control wells) was recorded in broth and agar. The MIC values were determined as the lowest concentrations that inhibited bacterial growth compared with the compound-free control and expressed in µg/mL (1024, 512, 256,128, 64, 32, 16, and 8 µg/mL, respectively). In the case of the vapour phase, considering a uniform distribution of the volatile compounds in the liquid and gaseous phase, these concentrations can be expressed as weight of volatile agent per volume unit of a well; therefore, MIC values would be 256, 128, 64, 32, 16, 8, 4, and 2 µg/cm^3, respectively). The DMSO used as the negative control at a concentration of 1% did not inhibit any of the strains tested either in broth or agar media. All experiments were set in triplicates in three independent measurements, and results were expressed as median/modal MICs values. According to the widely accepted norm in MIC testing, the mode and median were used for the final value calculation when triplicate endpoints were within the two- and three-dilution ranges, respectively.

4.5. Chemical Analysis of EOs

For the characterisation of the EOs, GC/MS analysis was performed using the dual-column/dual-detector gas chromatograph Agilent GC-7890B system (Agilent Technologies, Santa Clara, CA, USA). equipped with autosampler Agilent 7693, two columns, a fused-silica HP-5MS column (30 m × 0.25 mm, film thickness 0.25 μm, Agilent 19091s-433) and a DB-HeavyWAX (30 m × 0.25 mm, film thickness 0.25 μm, Agilent 122–7132), and a flame ionisation detector (FID) coupled with single quadrupole mass selective detector Agilent MSD-5977B. The operational parameters were the following: helium as carrier gas at 1 mL/min, injector temperature 250 °C for both columns. The oven temperature was raised for both columns after 3 min from 50 to 280 °C. Initially, the heating velocity was 3 °C/min until the system reached a temperature of 120 °C. Subsequently, the velocity increased to 5 °C/min until a temperature of 250 °C, and after 5 min holding time, the heating speed reached 15 °C/min until obtaining a temperature of 280 °C. Heating was followed by an isothermic period of 20 min. The EO samples were diluted in n-hexane for GC/MS (Merck KGaA, Darmstadt, Germany) at the concentration 20 μL/mL. One microliter of the solution was injected in split mode in a split ratio 1:30. The mass detector was set to the following conditions: ionisation energy 70 eV, ion source temperature 230 °C, scan time 1 s, mass range 40–600 m/z.

The identification of constituents was based on comparison of their retention indices (RIs), retention times (RT), and spectra with the National Institute of Standards and Technology Library ver. 2.0.f and the available literature [57]. The RIs were calculated for compounds separated by the HP-5MS column using the retention times of n-alkanes series ranging from C8 to C40 (Sigma-Aldrich, Prague, Czech Republic). For each EO analysed, the final number of compounds was calculated as the sum of components simultaneously identified using both columns and the remaining constituents identified by individual columns only. Quantitative data are expressed as relative percentage content of constituents determined by the FID.

4.6. Chemical Analysis of EOs' Vapour Phase

The analysis of the chemical composition of the headspace above a mixture of MH broth and *T. vulgaris* EO at a concentration of 512 μg/mL (i.e., the lowest MIC value obtained from the BMV assay) was performed using two different sampling techniques: HS-SPME and HS-GTS. Regardless of the sampling method used, for each experiment, a set of five samples were prepared, and a volume of 2 mL of the above-mentioned mixture was introduced into a 4 mL glass vial. Except for the first sample (t = 0 h), all EO samples were placed into an oven set at a temperature of 37 °C for incubation until their analysis at 3, 6, 9, and 12 h.

In HS-SPME, the headspace sampling was achieve using a fibre assembly coated with a 50/30 μm mixed layer of divinylbenzene/carboxen/polydimehylsiloxane (DVB/CAR/PDMS—SUPELCO, Bellefonte, PA, USA). When equilibrium was reached between the mixture and the headspace, the needle of the HS-SPME holder was inserted into the vial, and the coated fibre was exposed to the headspace for 15 min for adsorption of the volatile compounds. The needle was subsequently removed, inserted into the GC injector port, and set in splitless mode, where the desorption of analytes occurred. The injector temperature was set at 250 °C, and the fibre was left into the injector for the whole analysis until the next measure.

As for HS-GTS, however, the sampling technique was carried out using a 2.5 mL SampleLock gas tight syringe (Hamilton Bonaduz AG, Bonaduz, Switzerland), including a twist valve lock and a positive rear plunger stop to prevent sample loss. At equilibrium, with the valve of the syringe closed, the needle was passed through the vial septum and inserted until reaching the middle of the headspace. The valve was then opened, and a 2.5 mL sample was collected. Afterwards, the valve was closed again; the syringe was removed from the vial and inserted into the GC injector at a similar temperature of 250 °C but set in splitless mode. Finally, the valve was opened one more time to inject the headspace sample, and the syringe was immediately removed.

For both sampling methods, measurements were repeated every 3 h during a 12-h incubation period. Furthermore, analyses were performed on the HP-5MS column with similar operational parameters as described in Section 4.5 for GC/MS analyses, and their quantification was expressed as relative percentage content of constituents determined by the FID.

4.7. Statistical Analysis

The chemical analysis of the EO sample from supplier C (the most active EO based on the results of the antimicrobial assay) was performed in triplicate, including the chromatographic analysis of its liquid-phase and the headspace analysis using both extraction techniques (HS-SPME and HS-GTS). Relative peak area percentages were expressed as mean average of these three measurements ± standard deviation. For all EO samples' chemical profiles to be compared with one another, chemical analysis of EO samples from suppliers A and B was carried out in one replication only.

5. Conclusions

To summarise, this study reports the antibacterial activity of three *T. vulgaris* EOs hydrodistilled from commercial samples of different origins against three standard bacterial strains associated with respiratory infections, namely, *H. influenzae*, *S. aureus*, and *S. pyogenes*, in both liquid and vapour phases when assayed using the BVM method. While all bacterial strains were sensitive to *T. vulgaris* EO vapours to a certain degree, Gram-negative strains of *H. influenzae* showed the highest susceptibility. The GC/MS analysis identified the EO samples as a thymol chemotype, whose major constituents were monoterpenoids such as thymol, carvacrol linalool, γ-terpinene, and *p*-cymene, but also sesquiterpenoids represented by β-caryophyllene. In opposition, the chemical analysis of the headspace reported fewer compounds in the vapour with a predominance of *p*-cymene, γ-terpinene, and α-pinene, whereas the amount of thymol was unusually low. As for which of the two headspace sampling techniques could prove more valuable for the chemical analysis of EO vapours, results show that both methods are rather complementary and interdependent: HS-SPME with optimised experimental conditions may yield more accurate results when aiming for qualitative aspects, while HS-GTS could provide more accurate data representing the true headspace distribution of the EO volatile agents, therefore proving to be a better technique when aiming for quantitative analysis. Further research, however, is needed to corroborate this argument. Overall, the potential of the procedures examined in this study could be further exploited to better assess the benefits of EO volatile compounds and their applications in the healthcare and pharmaceutical industries. Results of this study also suggest a potential of *T. vulgaris* EO for application in inhalation therapy against respiratory infections; however, a further pharmacological evaluation will be necessary in order to verify its potential practical use.

Author Contributions: Conceptualization, L.K.; Formal analysis, J.A.; Investigation, J.A.; Supervision, M.H., K.U. and L.K.; Writing—original draft, J.A.; Writing—review & editing, M.H., K.U. and L.K. All authors have read and agreed to the published version of the manuscript.

Funding: This work was supported by the Czech University of Life Sciences Prague (IGA 20213109) and the METROFOOD-CZ research infrastructure project (MEYS Grant No: LM2018100), including access to its facilities.

Institutional Review Board Statement: Not applicable.

Informed Consent Statement: Not applicable.

Data Availability Statement: Not applicable.

Conflicts of Interest: The authors declare no conflict of interest.

Sample Availability: Not available.

References

1. Nguyen, T.K.P.; Tran, T.H.; Roberts, C.L.; Graham, S.M.; Marais, B.J. Child pneumonia—Focus on the Western Pacific Region. *Paediatr. Respir. Rev.* **2017**, *21*, 102–110. [CrossRef] [PubMed]
2. Pneumonia: The Forgotten Killer of Children. Available online: https://apps.who.int/iris/handle/10665/43640 (accessed on 5 January 2021).
3. Baron, S. *Medical Microbiology*, 4th ed.; University of Texas Medical Branch at Galveston: Galveston, TX, USA, 1996.
4. Houdkova, M.; Urbanova, K.; Doskocil, I.; Rondevaldova, J.; Novy, P.; Nguon, S.; Chrun, R.; Kokoska, L. In vitro growth-inhibitory effect of Cambodian essential oils against pneumonia causing bacteria in liquid and vapour phase and their toxicity to lung fibroblasts. *S. Afr. J. Bot.* **2018**, *118*, 85–97. [CrossRef]
5. World Health Organization—Forum of International Respiratory Societies. *The Global Impact of Respiratory Disease*, 2nd ed.; WHO: Geneva, Switzerland, 2017.
6. Wang, X.; Li, Y.; O'Brien, K.L.; Madhi, S.A.; Widdowson, M.-A.; Byass, P.; Omer, S.B.; Abbas, Q.; Ali, A.; Amu, A.; et al. Global burden of respiratory infections associated with seasonal influenza in children under 5 years in 2018: A systematic review and modelling study. *Lancet Glob. Health* **2020**, *8*, 497–510. [CrossRef]
7. Kokoska, L.; Kloucek, P.; Leuner, O.; Novy, P. Plant-derived products as antibacterial and antifungal agents in human health care. *Curr. Med. Chem.* **2019**, *26*, 5501–5541. [CrossRef]
8. Borghardt, J.M.; Kloft, C.; Sharma, A. Inhaled therapy in respiratory disease: The complex interplay of pulmonary kinetic processes. *Can. Respir. J.* **2018**, *2018*, 2732017. [CrossRef] [PubMed]
9. Döring, G.; Hoiby, N. Consensus study group early intervention and prevention of lung disease in cystic fibrosis: A European consensus. *J. Cyst. Fibros.* **2004**, *3*, 67–91. [CrossRef] [PubMed]
10. de Pablo, E.; Fernández-García, R.; Ballesteros, M.P.; Torrado, J.J.; Serrano, D.R. Nebulised antibiotherapy: Conventional versus nanotechnology-based approaches, is targeting at a nano scale a difficult subject? *Ann. Transl. Med.* **2017**, *5*, 448. [CrossRef]
11. Schreiber, M.P.; Shorr, A.F. Inhaled antibiotics for the treatment of pneumonia. *Curr. Opin. Pulm. Med.* **2019**, *25*, 289–293. [CrossRef] [PubMed]
12. Ehrmann, S.; Chastre, J.; Diot, P.; Lu, Q. Nebulized antibiotics in mechanically ventilated patients: A challenge for translational research from technology to clinical care. *Ann. Intensive Care* **2017**, *7*, 78. [CrossRef] [PubMed]
13. Tiddens, H.A.W.M.; Bos, A.C.; Mouton, J.W.; Devadason, S.; Janssens, H.M. Inhaled antibiotics: Dry or wet? *Eur. Clin. Respir. J.* **2014**, *44*, 1308–1318. [CrossRef] [PubMed]
14. de Jongh, F.H.C.; Rinkel, M.J.G.; Hoeijmakers, H.W.M. Aerosol deposition in the upper airways of a child. *J. Aerosol Med.* **2006**, *19*, 279–289. [CrossRef] [PubMed]
15. Laube, B.L.; Jashnani, R.; Dalby, R.N.; Zeitlin, P.L. Targeting aerosol deposition in patients with cystic fibrosis: Effects of alterations in particle size and inspiratory flow rate. *Chest* **2000**, *118*, 1069–1076. [CrossRef]
16. Heinrich, M. *Fundamentals of Pharmacognosy and Phytotherapy*, 2nd ed.; Elsevier: Edinburgh, UK, 2012.
17. Netopilova, M.; Houdkova, M.; Urbanova, K.; Rondevaldova, J.; Damme, P.; Kokoska, L. In vitro antimicrobial combinatory effect of *Cinnamomum cassia* essential oil with 8-hydroxyquinoline against *Staphylococcus aureus* in liquid and vapour phase. *J. Appl. Microbiol.* **2020**, *129*, 906–915. [CrossRef] [PubMed]
18. Başer, K.H.C.; Buchbauer, G. *Handbook of Essential Oils: Science, Technology, and Applications*, 3rd ed.; CRC Press: Boca Raton, FL, USA, 2020.
19. Ghahremani-Chabok, A.; Bagheri-Nesami, M.; Shorofi, S.A.; Mousavinasab, S.N.; Gholipour-Baradari, A.; Saeedi, M. The effects of *Thymus vulgaris* inhalation therapy on airway status and oxygen saturation of patients under mechanical ventilation: A randomized clinical trial. *Adv. Integr. Med.* **2020**, *8*, 92–100. [CrossRef]
20. Bakkali, F.; Averbeck, S.; Averbeck, D.; Idaomar, M. Biological effects of essential oils—A review. *Food Chem. Toxicol.* **2008**, *46*, 446–475. [CrossRef] [PubMed]
21. Pisoschi, A.; Pop, A.; Cecilia, G.; Turcuş, V.; Olah, N.-K.; Mathe, E. An overview of natural antimicrobials role in food. *Eur. J. Med. Chem.* **2017**, *143*, 922–935. [CrossRef] [PubMed]
22. Tongnuanchan, P.; Benjakul, S. Essential oils: Extraction, bioactivities, and their uses for food preservation. *J. Food Sci.* **2014**, *79*, R1231–R1249. [CrossRef] [PubMed]
23. HMPC Monographs: Overview of Recommendations for the Uses of Herbal Medicinal Products in the Paediatric Population. Available online: https://www.ema.europa.eu/en/documents/other/hmpc-monographs-overview-recommendations-uses-herbal-medicinal-products-paediatric-population_en.pdf (accessed on 6 January 2021).
24. Assessment Report on Thymus vulgaris L., Thymus zygis L., Aetheroleum. Available online: https://www.ema.europa.eu/en/documents/herbal-report/draft-assessment-report-thymus-vulgaris-l-thymus-zygis-l-aetheroleum-revision-1_en.pdf (accessed on 6 January 2021).

25. Nikolić, M.; Glamočlija, J.; Ferreira, I.C.F.R.; Calhelha, R.C.; Fernandes, Â.; Marković, T.; Marković, D.; Giweli, A.; Soković, M. Chemical composition, antimicrobial, antioxidant and antitumor activity of *Thymus serpyllum* L., *Thymus algeriensis* Boiss. and Reut and *Thymus vulgaris* L. essential oils. *Ind. Crop. Prod.* **2014**, *52*, 183–190. [CrossRef]
26. Pina-Vaz, C.; Rodrigues, A.G.; Pinto, E.; Costa-de-Oliveira, S.; Tavares, C.; Salgueiro, L.; Cavaleiro, C.; Gonçalves, M.; de Oliveira, M. Antifungal activity of *Thymus* oils and their major compounds. *J. Eur. Acad. Dermatol. Venereol.* **2004**, *18*, 73–78. [CrossRef] [PubMed]
27. Fani, M.; Kohanteb, J. In vitro antimicrobial activity of *Thymus vulgaris* essential oil against major oral pathogens. *J. Evid. Based Complement. Altern. Med.* **2017**, *22*, 660–666. [CrossRef] [PubMed]
28. Inouye, S.; Takizawa, T.; Yamaguchi, H. Antibacterial activity of essential oils and their major constituents against respiratory tract pathogens by gaseous contact. *J. Antimicrob. Chemother.* **2001**, *47*, 565–573. [CrossRef]
29. Tullio, V.; Nostro, A.; Mandras, N.; Dugo, P.; Banche, G.; Cannatelli, M.A.; Cuffini, A.M.; Alonzo, V.; Carlone, N.A. Antifungal activity of essential oils against filamentous fungi determined by broth microdilution and vapour contact methods. *J. Appl. Microbiol.* **2007**, *102*, 1544–1550. [CrossRef] [PubMed]
30. M100Ed30 | Performance Standards for Antimicrobial Susceptibility Testing, 30th ed. Available online: https://clsi.org/standards/products/microbiology/documents/m100/ (accessed on 2 February 2021).
31. Borugă, O.; Jianu, C.; Mişcă, C.; Goleţ, I.; Gruia, A.; Horhat, F. *Thymus vulgaris* essential oil: Chemical composition and antimicrobial activity. *J. Med. Life* **2014**, *7*, 56–60. [PubMed]
32. Ács, K.; Balázs, V.L.; Kocsis, B.; Bencsik, T.; Böszörményi, A.; Horváth, G. Antibacterial activity evaluation of selected essential oils in liquid and vapor phase on respiratory tract pathogens. *BMC Complement. Altern. Med.* **2018**, *18*, 227. [CrossRef]
33. Bueno, J. Models of evaluation of antimicrobial activity of essential oils in vapour phase: A promising use in healthcare decontamination. *Nat. Volatiles Essent. Oils* **2015**, *2*, 16–29.
34. Edwards-Jones, V.; Buck, R.; Shawcross, S.G.; Dawson, M.M.; Dunn, K. The effect of essential oils on methicillin-resistant *Staphylococcus aureus* using a dressing model. *Burns* **2004**, *30*, 772–777. [CrossRef]
35. Houdkova, M.; Rondevaldova, J.; Doskocil, I.; Kokoska, L. Evaluation of antibacterial potential and toxicity of plant volatile compounds using new broth microdilution volatilization method and modified MTT assay. *Fitoterapia* **2017**, *118*, 56–62. [CrossRef] [PubMed]
36. Nedorostova, L.; Kloucek, P.; Kokoska, L.; Stolcova, M.; Pulkrabek, J. Antimicrobial properties of selected essential oils in vapour phase against foodborne bacteria. *Food Control* **2009**, *20*, 157–160. [CrossRef]
37. López, P.; Sánchez, C.; Batlle, R.; Nerín, C. Solid- and vapor-phase antimicrobial activities of six essential oils: Susceptibility of selected foodborne bacterial and fungal strains. *J. Agric. Food Chem.* **2005**, *53*, 6939–6946. [CrossRef]
38. Laird, K.; Phillips, C. Vapour Phase: A potential future use for essential oils as antimicrobials?: Essential oil vapours and their antimicrobial activity. *Lett. Appl. Microbiol.* **2012**, *54*, 169–174. [CrossRef] [PubMed]
39. Zhang, W.; Liang, X. Headspace gas chromatography-mass spectrometry for volatile components analysis in *Ipomoea cairica* (L.) sweet leaves: Natural deep eutectic solvents as green extraction and dilution matrix. *Foods* **2019**, *8*, 205. [CrossRef] [PubMed]
40. Ross, C.F. Headspace analysis. In *Comprehensive Sampling and Sample Preparation*; Elsevier: Amsterdam, The Netherlands, 2012; pp. 27–50.
41. Lugo-Estrada, L.; Galindo-Rodríguez, S.A.; Pérez-López, L.A.; de Torres, N.W.; Álvarez-Román, R. Headspace–Solid-phase microextraction gas chromatography method to quantify *Thymus vulgaris* essential oil in polymeric nanoparticles. *Pharmacogn. Mag.* **2019**, *15*, 473.
42. López, P.; Sánchez, C.; Batlle, R.; Nerín, C. Vapor-phase activities of cinnamon, thyme, and oregano essential oils and key constituents against foodborne microorganisms. *J. Agric. Food Chem.* **2007**, *55*, 4348–4356. [CrossRef] [PubMed]
43. Reyes-Jurado, F.; Navarro-Cruz, A.R.; Ochoa-Velasco, C.E.; Palou, E.; López-Malo, A.; Ávila-Sosa, R. Essential oils in vapor phase as alternative antimicrobials: A review. *Food Sci. Nutr.* **2020**, *60*, 1641–1650. [CrossRef]
44. Snow, N.H.; Bullock, G.P. Novel techniques for enhancing sensitivity in static headspace extraction-gas chromatography. *J. Chromatogr. A* **2010**, *1217*, 2726–2735. [CrossRef] [PubMed]
45. Bicchi, C. Essential oils | gas chromatography. In *Encyclopedia of Separation Science*; Elsevier: Amsterdam, The Netherlands, 2000; pp. 2744–2755.
46. Murray, R.A. Limitations to the use of solid-phase microextraction for quantitation of mixtures of volatile organic sulfur compounds. *Anal. Chem.* **2001**, *73*, 1646–1649. [CrossRef]
47. Regti, A.; Kassimi, A.E.; Laamari, M.R.; Haddad, M.E. Competitive adsorption and optimization of binary mixture of textile dyes: A factorial design analysis. *J. Assoc. Arab Univ. Basic Appl. Sci.* **2017**, *24*, 1–9. [CrossRef]
48. Kolb, B.; Ettre, L.S. *Static Headspace-Gas Chromatography: Theory and Practice*, 2nd ed.; Wiley: Hoboken, NJ, USA, 2006.
49. Adams, R.P. *Identification of Essential Oil Components by Gas Chromatography/Mass Spectorscopy*, 4th ed.; Allured Publishing Corp: Carol Stream, IL, USA, 2007.
50. National Institute of Standards and Technology. Available online: https://www.nist.gov/ (accessed on 15 December 2020).
51. Satyal, P.; Murray, B.; McFeeters, R.; Setzer, W. Essential oil characterization of *Thymus vulgaris* from various geographical locations. *Foods* **2016**, *5*, 70. [CrossRef] [PubMed]
52. Houdkova, M.; Kokoska, L. Volatile antimicrobial agents and In vitro methods for evaluating their activity in the vapour phase: A review. *Planta Med.* **2020**, *86*, 822–857. [CrossRef] [PubMed]

53. Smith-Palmer, A.; Stewart, J.; Fyfe, L. Antimicrobial properties of plant essential oils and essences against five important food-borne pathogens. *Lett. Appl. Microbiol.* **1998**, *26*, 118–122. [CrossRef] [PubMed]
54. Mann, C.M.; Cox, S.D.; Markham, J.L. The outer membrane of *Pseudomonas aeruginosa* NCTC 6749 contributes to its tolerance to the essential oil of *Melaleuca alternifolia* (tea tree oil). *Lett. Appl. Microbiol.* **2000**, *30*, 294–297. [CrossRef] [PubMed]
55. Reyes-Jurado, F.; Cervantes-Rincón, T.; Bach, H.; López-Malo, A.; Palou, E. Antimicrobial activity of Mexican oregano (*Lippia berlandieri*), Thyme (*Thymus vulgaris*), and Mustard (*Brassica nigra*) essential oils in gaseous phase. *Ind. Crops Prod.* **2019**, *131*, 90–95. [CrossRef]
56. Nezhadali, A.; Nabavi, M.; Rajabian Noghundar, M.; Akbarpour, M.; Pourali, P.; Amini, F. Chemical variation of leaf essential oil at different stages of plant growth and in vitro antibacterial activity of *Thymus vulgaris Lamiaceae*, from Iran. *Beni-Suef Univ. J. Basic Appl. Sci.* **2014**, *3*, 87–92. [CrossRef]
57. Kloucek, P.; Smid, J.; Frankova, A.; Kokoska, L.; Valterová, I.; Pavela, R. Fast screening method for assessment of antimicrobial activity of essential oils in vapor phase. *Food Res. Int.* **2012**, *47*, 161–165. [CrossRef]
58. Muthaiyan, A.; Biswas, D.; Crandall, P.G.; Wilkinson, B.J.; Ricke, S.C. Application of orange essential oil as an antistaphylococcal agent in a dressing model. *BMC Complement. Altern. Med.* **2012**, *12*, 125. [CrossRef] [PubMed]
59. Wiegand, I.; Hilpert, K.; Hancock, R.E.W. Agar and broth dilution methods to determine the minimal inhibitory concentration (MIC) of antimicrobial substances. *Nat. Protoc.* **2008**, *3*, 163–175. [CrossRef] [PubMed]
60. Mottaleb, M.A.; Meziani, M.J.; Rafiq Islam, M. Solid-phase microextraction (SPME) and its application to natural products. In *Encyclopedia of Analytical Chemistry: Applications, Theory and Instrumentation*; Wiley: Hoboken, NJ, USA, 2006; pp. 105–127.
61. Seo, H.S.; Beuchat, L.R.; Kim, H.; Ryu, J.H. Development of an experimental apparatus and protocol for determining antimicrobial activities of gaseous plant essential oils. *Int. J. Food Microbiol.* **2015**, *215*, 95–100. [CrossRef] [PubMed]
62. Jorgensen, J.H.; Turnidge, J.D. Susceptibility test methods: Dilution and disk diffusion methods. In *Manual of Clinical Microbiology*; Jorgensen, J.H., Carroll, K.C., Funke, G., Pfaller, M.A., Landry, M.L., Richter, S.S., Warnock, D.W., Eds.; ASM Press: Washington, DC, USA, 2015; pp. 1253–1273.
63. Feyaerts, A.F.; Mathé, L.; Luyten, W.; Tournu, H.; Dyck, K.V.; Broekx, L.; Dijck, P.V. Assay and recommendations for the detection of vapour-phase-mediated antimicrobial activities. *Flavour Fragr. J.* **2017**, *32*, 347–353. [CrossRef]
64. Amat, S.; Baines, D.; Alexander, T.W. A vapour phase assay for evaluating the antimicrobial activities of essential oils against bovine respiratory bacterial pathogens. *Lett. Appl. Microbiol.* **2017**, *65*, 489–495. [CrossRef] [PubMed]
65. Casey, J.T.; O'Cleirigh, C.; Walsh, P.K.; O'Shea, D.G. Development of a robust microtiter plate-based assay method for assessment of bioactivity. *J. Microbiol. Methods* **2004**, *58*, 327–334. [CrossRef] [PubMed]
66. Fisher, K.; Phillips, C. Potential antimicrobial uses of essential oils in food: Is citrus the answer? *Trends Food Sci. Technol.* **2008**, *19*, 156–164. [CrossRef]
67. Kalemba, D.; Kunicka, A. Antibacterial and antifungal properties of essential oils. *Curr. Med. Chem.* **2003**, *10*, 813–829. [CrossRef] [PubMed]
68. Orchard, A.; van Vuuren, S. Commercial essential oils as potential antimicrobials to treat skin diseases. *Evid. Based Complement. Alternat. Med.* **2017**, *2017*, 4517971. [CrossRef]
69. Reyes-Jurado, F.; Franco-Vega, A.; Ramírez-Corona, N.; Palou, E.; López-Malo, A. Essential oils: Antimicrobial activities, extraction methods, and their modeling. *Food Eng. Rev.* **2015**, *7*, 275–297. [CrossRef]
70. Leyva-López, N.; Gutiérrez-Grijalva, E.P.; Vazquez-Olivo, G.; Heredia, J.B. Heredia Essential oils of oregano: Biological activity beyond their antimicrobial properties. *Molecules* **2017**, *22*, 989. [CrossRef]
71. Schmidt, E.; Wanner, J.; Höferl, M.; Jirovetz, L.; Buchbauer, G.; Gochev, V.; Girova, T.; Stoyanova, A.; Geissler, M. Chemical composition, olfactory analysis and antibacterial activity of *Thymus vulgaris* chemotypes geraniol, 4-Thujanol/Terpinen-4-ol, thymol and linalool cultivated in southern France. *Nat. Prod. Commun.* **2012**, *7*, 1934578X1200700833. [CrossRef]
72. Grosso, C.; Figueiredo, A.C.; Burillo, J.; Mainar, A.M.; Urieta, J.S.; Barroso, J.G.; Coelho, J.A.; Palavra, A.M.F. Composition and antioxidant activity of *Thymus vulgaris* volatiles: Comparison between supercritical fluid extraction and hydrodistillation. *J. Sep. Sci.* **2010**, *33*, 2211–2218. [CrossRef] [PubMed]
73. Figueiredo, A.C.; Barroso, J.G.; Pedro, L.G.; Scheffer, J.J.C. Factors affecting secondary metabolite production in plants: Volatile components and essential oils. *Flavour Fragr. J.* **2008**, *23*, 213–226. [CrossRef]
74. Hudaib, M.; Speroni, E.; Di Pietra, A.M.; Cavrini, V. GC/MS evaluation of thyme (*Thymus vulgaris* L.) oil composition and variations during the vegetative cycle. *J. Pharm. Biomed.* **2002**, *29*, 691–700. [CrossRef]
75. Fan, S.; Chang, J.; Zong, Y.; Hu, G.; Jia, J. GC/MS Analysis of the composition of the essential oil from *Dendranthema indicum* vararomaticum using three extraction methods and two columns. *Molecules* **2018**, *23*, 576. [CrossRef] [PubMed]
76. Netopilova, M.; Houdkova, M.; Urbanova, K.; Rondevaldova, J.; Kokoska, L. Validation of qualitative broth volatilization checkerboard method for testing of essential oils: Dual-column GC–FID/MS analysis and in vitro combinatory antimicrobial effect of *Origanum vulgare* and *Thymus vulgaris* against *Staphylococcus aureus* in liquid and vapor phases. *Plants* **2021**, *10*, 393.
77. Adam, M.; Dobiáš, P.; Pavlíková, P.; Ventura, K. Comparison of solid-phase and single-drop microextractions for headspace analysis of herbal essential oils. *Open Chem.* **2009**, *7*, 303–311. [CrossRef]
78. Soleimani, M.; Daryasari, A.P.; Ghorbani, A.; Hejri, O.M.; Mazaheri, R. Analysis of the volatile compounds in *Thymus vulgaris* L. using improved HS-SPME-GC/MS and comparison with conventional methods. *J. Essent. Oil-Bear. Plants* **2014**, *17*, 1233–1240. [CrossRef]

79. Nezhadali, A.; Akbarpour, M.; Shirvan, B.Z.; Mousavi, M. Comparison of volatile organic compounds of *Thymus vulgaris* using hydrodistillation and headspace solid phase microextraction gas chromatography mass spectrometry. *J. Chin. Chem. Soc.* **2010**, *57*, 40–43. [CrossRef]
80. Leggio, A.; Leotta, V.; Belsito, E.L.; Di Gioia, M.L.; Romio, E.; Santoro, I.; Taverna, D.; Sindona, G.; Liguori, A. Aromatherapy: Composition of the gaseous phase at equilibrium with liquid bergamot essential oil. *Chem. Cent. J.* **2017**, *11*, 111. [CrossRef] [PubMed]
81. Coleman, W.M.; Lawrence, B.M. A Comparison of selected analytical approaches to the analysis of an essential oil. *Flavour Fragr. J.* **1997**, *12*, 1–8. [CrossRef]
82. Cui, S.; Tan, S.; Ouyang, G.; Jiang, S.; Pawliszyn, J. Headspace solid-phase microextraction gas chromatography–mass spectrometry analysis of *Eupatorium odoratum* extract as an oviposition Repellent. *J. Chromatogr. B* **2009**, *877*, 1901–1906. [CrossRef] [PubMed]
83. Chialva, F.; Gabri, G.; Liddle, P.A.P.; Ulian, F. Qualitative evaluation of aromatic herbs by direct headspace GC analysis. Applications of the method and comparison with the traditional analysis of essential oils. *J. High. Resolut. Chromatogr.* **1982**, *5*, 182–188. [CrossRef]
84. Wang, T.-H.; Hsia, S.-M.; Wu, C.-H.; Ko, S.-Y.; Chen, M.Y.; Shih, Y.-H.; Shieh, T.-M.; Chuang, L.-C.; Wu, C.-Y. Evaluation of the antibacterial potential of liquid and vapor phase phenolic essential oil compounds against oral microorganisms. *PLoS ONE* **2016**, *11*, e0163147. [CrossRef]
85. EDQM (Council of Europe). *European Pharmacopoeia*, 10th ed.; EDQM: Strasbourg, France, 2019.
86. Thymi Aetheroleum. Available online: https://www.ema.europa.eu/en/medicines/herbal/thymi-aetheroleum (accessed on 16 February 2021).

molecules

MDPI

Article

New Broth Macrodilution Volatilization Method for Antibacterial Susceptibility Testing of Volatile Agents and Evaluation of Their Toxicity Using Modified MTT Assay In Vitro

Marketa Houdkova [1], Aishwarya Chaure [1], Ivo Doskocil [2], Jaroslav Havlik [3] and Ladislav Kokoska [1,*]

[1] Department of Crop Sciences and Agroforestry, Faculty of Tropical AgriSciences, Czech University of Life Sciences Prague, 16500 Prague, Czech Republic; houdkovam@ftz.czu.cz (M.H.); chaure@ftz.czu.cz (A.C.)
[2] Department of Microbiology, Nutrition and Dietetics, Faculty of Agrobiology, Food and Natural Resources, Czech University of Life Sciences Prague, 16500 Prague, Czech Republic; doskocil@af.czu.cz
[3] Department of Food Science, Faculty of Agrobiology, Food and Natural Resources, Czech University of Life Sciences Prague, 16500 Prague, Czech Republic; havlik@af.czu.cz
* Correspondence: kokoska@ftz.czu.cz; Tel.: +420-224382180

Citation: Houdkova, M.; Chaure, A.; Doskocil, I.; Havlik, J.; Kokoska, L. New Broth Macrodilution Volatilization Method for Antibacterial Susceptibility Testing of Volatile Agents and Evaluation of Their Toxicity Using Modified MTT Assay In Vitro. *Molecules* **2021**, *26*, 4179. https://doi.org/10.3390/molecules26144179

Academic Editor: Domenico Montesano

Received: 28 May 2021
Accepted: 5 July 2021
Published: 9 July 2021

Publisher's Note: MDPI stays neutral with regard to jurisdictional claims in published maps and institutional affiliations.

Copyright: © 2021 by the authors. Licensee MDPI, Basel, Switzerland. This article is an open access article distributed under the terms and conditions of the Creative Commons Attribution (CC BY) license (https://creativecommons.org/licenses/by/4.0/).

Abstract: In this study, a new broth macrodilution volatilization method for the simple and rapid determination of the antibacterial effect of volatile agents simultaneously in the liquid and vapor phase was designed with the aim to assess their therapeutic potential for the development of new inhalation preparations. The antibacterial activity of plant volatiles (β-thujaplicin, thymohydroquinone, thymoquinone) was evaluated against bacteria associated with respiratory infections (*Haemophilus influenzae, Staphylococcus aureus, Streptococcus pneumoniae, Streptococcus pyogenes*) and their cytotoxicity was determined using a modified thiazolyl blue tetrazolium bromide assay against normal lung fibroblasts. Thymohydroquinone and thymoquinone possessed the highest antibacterial activity against *H. influenzae*, with minimum inhibitory concentrations of 4 and 8 μg/mL in the liquid and vapor phases, respectively. Although all compounds exhibited cytotoxic effects on lung cells, therapeutic indices (TIs) suggested their potential use in the treatment of respiratory infections, which was especially evident for thymohydroquinone (TI > 34.13). The results demonstrate the applicability of the broth macrodilution volatilization assay, which combines the principles of broth microdilution volatilization and standard broth macrodilution methods. This assay enables rapid, simple, cost- and labor-effective screening of volatile compounds and overcomes the limitations of assays currently used for screening of antimicrobial activity in the vapor phase.

Keywords: antimicrobial; cytotoxicity; macrodilution method; respiratory infections; β-thujaplicin; thymohydroquinone; thymoquinone; vapor phase; volatile compound

1. Introduction

Bacterial infections of the lower respiratory tract, such as pneumonia and bronchitis, are some of the leading global causes of death, especially among children under five years of age and elderly people [1,2]. Typical causative bacterial species of respiratory infections include *Haemophilus influenzae, Staphylococcus aureus, Streptococcus pneumoniae,* and *Streptococcus pyogenes* [3]. Moreover, co-infection with viruses (e.g., severe acute respiratory syndrome coronavirus) can significantly increase the morbidity and mortality rates [4]. Inhalation therapy represents the appropriate way to treat respiratory disorders. If a medication is inhaled, it is directly delivered into the airways to the site of infection. This gives a relatively faster onset of action, while using a lower dose of active agent, which consequently causes fewer side effects in districts where its action is not needed [5,6]. Although antibiotic treatment is generally considered as effective against most infective strains, it is increasingly failing due to some limitations including allergies, bacterial resistance, inadequate penetration in lung tissues, and undesirable adverse effects [7]. Several types of devices for the delivery of inhaled medications to the lungs have been invented, such as nebulizers, pressurized metered-dose and dry-powder inhalers; however, they all face various limitations, including a short-life because of pulmonary clearance, enzymatic degradation, fast systemic absorption, and poor bioavailability of bioactive agents at the target site [8]. Moreover, the efficiency of the inhalable therapies may be affected by the deposition of the aerolized particles in the oropharyngeal region and upper airways while the deposition of drugs in the lungs can be reduced due to the inappropriate size of its droplets or due to specific respiratory tract anatomy, and proper operation, especially in children and elderly patients [9,10].

Since plant-derived antimicrobial agents have been traditionally used in the prevention and treatment of respiratory infections, they are considered promising sources of novel chemical scaffolds for the development of new drugs against bacteria causing respiratory diseases [11]. For this reason, the biological potential of natural substances, including volatile plant-derived compounds has been intensively studied in recent years with the aim to find potential alternatives to conventional synthetic antimicrobial agents [12,13]. For example, β-thujaplicin, a monoterpenoid isolated from the wood of the Cupressaceae species, and thymohydroquinone, a monoterpenoid phenol occurring in the Ranunculaceae and Lamiaceae families, have been reported to exhibit antimicrobial properties, including effects against bacteria causing respiratory infections [14–17]. Due to the high vapor pressure of volatile plant-derived products at the ambient temperature, they have the benefit of bioactivity in the vapor phase [18], and thus, plant volatiles have great potential for the development of novel preparations for inhalation [19,20]. For example, thymoquinone, a benzoquinone occurring in the seeds of *Nigella sativa* (Ranunculaceae), has been recorded to possess a relatively strong antibacterial effect in the vapor phase against pathogens that cause pneumonia [21]. Although plant-derived products are generally considered as relatively safe, the toxicological evaluation of volatile agents is necessary to confirm their non-toxicity for their practical application in inhalation therapy [22]. Despite the availability of several lung cell in vitro models, there is an urgent need for the development of more appropriate non-animal methods of inhalation toxicity, particularly for predicting effects in humans [23].

In vitro screening is typically the first step in the process of discovery of new antimicrobial drugs, including those derived from plant volatiles. However, the susceptibility testing of microorganisms to volatile agents using standard methods, such as broth dilution and disk diffusion assays, is a challenging task because of their specific physico-chemical properties, including high volatility, hydrophobicity, and viscosity [24]. The main problem is that their hydrophobic nature worsens the solubility of these compounds in water-based media (e.g., agar, broth) and their volatility increases the risk of loss of active substances via evaporation during sample handling, experiment preparation, and incubation. In addition, the transition of the vapors can affect the microplate assay results, as described in our previous studies [25,26]. This is even more complicated in the case of antimicrobial vapors testing. In contrast to well-established methods for antimicrobial susceptibility testing on solid (agar disc diffusion) and liquid (broth dilution) media [27–32], there are no standardized methods for the determination of microbial sensitivity to volatile compounds in the vapor phase. In recent years, several methods for the testing of the antimicrobial effects of volatile plant-derived products in the vapor phase have been developed. However, most of them have some specific limitations, such not being designed for high-throughput screening, and some of them need special equipment that is not commonly available [33]. Recently, we proposed a broth microdilution volatilization assay [21] based on the principles of broth microdilution and disc volatilization methods, which is suitable for high-throughput screening of volatile compounds simultaneously in the liquid and vapor phase. This method can also be easily used for determination of the antibacterial effects of essential oil vapors [34,35]. Although the broth microdilution volatilization method is fast, simple and labor-effective, it has several weaknesses. For example, clamps and wooden pads are required for a better sealing and fixing the microtiter plate and its lid together. Moreover, the limited volume of agar that is applied on the lid can affect the growth of the microorganisms tested.

With the aim to overcome the above mentioned drawbacks of previously developed methods used for testing of volatile antimicrobial agents in the vapor phase, we designed a novel macrodilution volatilization assay that combines the principles of broth microdilution volatilization [21] and standard broth macrodilution [27] methods. The validity of the method for susceptibility testing of bacterial pathogens causing respiratory infections was evaluated using three antimicrobial phytochemicals, namely β-thujaplicin, thymohydroquinone, and thymoquinone (Figure 1). In addition, the cytotoxicity of these compounds was analyzed using a modified thiazolyl blue tetrazolium bromide (MTT) cytotoxicity assay to assess their safety for use in the treatment of respiratory infections.

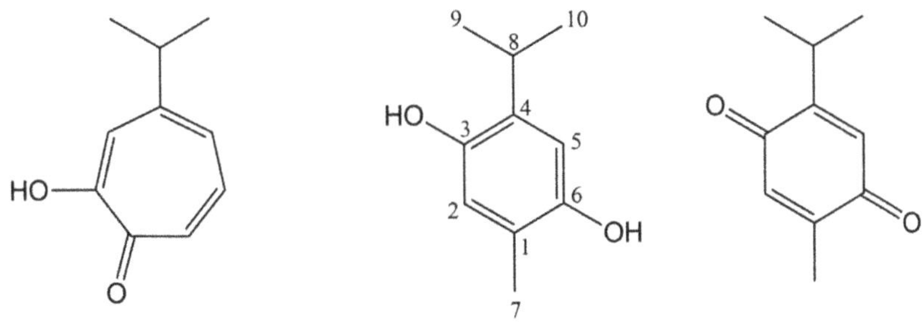

ß-thujaplicin thymohydroquinone thymoquinone

Figure 1. Chemical structures of plant volatile compounds tested.

2. Results and Discussion

2.1. Antimicrobial Activity

The results of the antibacterial effect of plant-derived volatiles against respiratory pathogens in the liquid and vapor phases assessed using the broth macrodilution volatilization method are listed in Table 1. All the compounds tested exhibited a certain degree of growth-inhibitory effect in both the liquid and vapor phase and their effectiveness varied in the ranges 4–64 μg/mL and 8–1024 μg/mL in the broth and agar media, respectively. To the best of our knowledge, this is the first report on the antibacterial activity of thymohydroquinone in the vapor phase. Moreover, the antimicrobial susceptibility of *H. influenzae* to β-thujaplicin and thymohydroquinone, *S. pyogenes* to thymohydroquinone and thymoquinone, and *S. pneumoniae* to thymohydroquinone was also described for the first time in this study.

In the liquid phase, the lowest minimal inhibitory concentration (MIC) values were observed for thymohydroquinone and thymoquinone against *H. influenzae* (4 μg/mL), followed by the growth-inhibitory effects against *S. aureus*, *S. pneumoniae*, and *S. pyogenes* with respective MIC values of 8, 16, and 32 μg/mL. β-thujaplicin exhibited the same level of antibacterial activity against all bacterial strains tested with a MIC of 64 μg/mL. As well as in the broth, thymohydroquinone and thymoquinone were found to be the most active antibacterial agents against *H. influenzae* in the vapor phase with a MIC of 8 μg/mL. In addition, both compounds effectively inhibited the growth of *S. aureus* and *S. pyogenes* on agar medium at the same concentrations, 16 and 32 μg/mL, respectively. However, their activity differed against *S. pneumoniae*, where a lower MIC value was detected for thymoquinone (16 μg/mL) than for thymohydroquinone (32 μg/mL). In the case of β-thujaplicin, a moderate antibacterial efficacy was recorded against *H. influenzae* and *S. aureus* (512 μg/mL) and a low antibacterial efficacy against *S. pneumoniae* and *S. pyogenes*.

Table 1. Antibacterial activity of plant volatile compounds in the liquid and vapor phases against respiratory pathogens.

Plant Volatile Compound	Bacterium/Minimal Inhibitory Concentration												$\bar{x}$-MIC
	Haemophilus influenzae			*Staphylococcus aureus*			*Streptococcus pneumoniae*			*Streptococcus pyogenes*			
	Broth	Agar		Broth	Agar		Broth	Agar		Broth	Agar		
	(μg/mL)	(μg/mL)	(μg/cm³)	(μg/mL)	(μg/mL)	(μg/cm³)	(μg/mL)	(μg/mL)	(μg/cm³)	(μg/mL)	(μg/mL)	(μg/cm³)	
β-thujaplicin	64	512	320	64	512	320	64	1024	640	64	1024	640	64
thymohydroquinone	4	8	5	8	16	10	16	32	20	32	32	20	15
thymoquinone	4	8	5	8	16	10	16	16	10	32	32	20	15
positive antibiotic control	1 [a]	n.d.	n.d.	0.5 [b]	n.d.	n.d.	0.25 [c]	n.d.	n.d.	0.25 [d]	n.d.	n.d.	-

$\bar{x}$-MIC: mean value of minimal inhibitory concentrations in broth medium, positive antibiotic control: [a] ampicillin, [b] oxacillin, [c] amoxicillin, [d] tetracycline, n.d.: not detected.

In general, the results of antibacterial potential of plant-derived volatiles obtained by our novel broth macrodilution volatilization assay correspond with those observed by other authors using standard broth and agar dilution methods. The MICs previously determined by Domon et al., Morita et al., and Inoue et al. [17,36,37] for β-thujaplicin against various strains of *S. aureus* in the range 12.5–160 μg/mL were close to the MIC detected in our study. However, a much higher antimicrobial efficacy of this compound was recorded against *S. pneumoniae* and *S. pyogenes* with MICs of 0.3–1 μg/mL in previously published studies [17]. Other authors [25,38] observed the growth-inhibitory effect of thymoquinone against *S. aureus* ATCC 29213 and ATCC 25923 with the MIC value of 8 μg/mL, which is the same as that in our study. In the case of thymohydroquinone, a fifty-times-higher MIC value of 400 μg/mL was detected against *S. aureus* ATCC 25923 [39]. As it has previously been observed, various bacterial strains with different susceptibilities to antibacterial agents [40], as well as different antimicrobial assays used for the testing of volatiles [41], may be responsible for the variability of the results obtained by other authors. In contrast to a number of published papers on the antibacterial potential of plant-derived volatile compounds in the liquid phase, literature on the activity of their vapors is limited. Wang et al. [42] recorded a certain level of growth-inhibitory effect of β-thujaplicin against oral microorganisms in the gaseous phase and Inouye et al. [43] described a very potent vapor activity of thymoquinone against *Trichophyton mentagrophytes*; both by using the disc volatilization method. Our previous results on the antimicrobial activity of thymoquinone against respiratory pathogens *H. influenzae*, *S. aureus*, and *S. pneumoniae* obtained by using the broth microdilution volatilization assay with respective MIC values of 8, 16, and 16 μg/mL in the liquid phase and 8, 16, and 32 μg/mL in the vapor phase [21] correspond to this current study and demonstrate the validity of our newly developed method. In general, the compounds tested in this study showed higher or equal antibacterial effects in the liquid phase than in the vapor phase. The largest differences were observed in the case of β-thujaplicin when sixteen times lower MIC values were detected in broth than on the agar against *S. pneumonieae* and *S. pyogenes*.

The above-described results demonstrate the validity of our novel broth macrodilution volatilization assay, which combines the principles of broth microdilution volatilization [21] and standard macrodilution methods [27] and overcomes some of their drawbacks. In comparison with previously established liquid matrix volatilization techniques, the broth macrodilution volatilization method is performed in commercially available microtubes, which can be tightly closed with a snap cap, and therefore, it does not require specialized equipment, such as wooden pads and clamps to prevent the losses of the active agents by evaporation. A variable number of sample concentrations tested in one experiment is another benefit of this method, while the design is not limited by the quantity of wells, as in the case of microplate-based assays. Whereas a higher amount of appropriate media can be applied in microtubes and their caps, which are suitable for the cultivation of slower growing microorganisms (e.g., fungi) that require longer incubation time to produce enough growth for MIC determination [44]. Despite the obvious benefits of our novel antimicrobial susceptibility test based on microtubes, it has some specific limits, such as a lower potential for the automation (e.g., use of automated pipetting platform and reader) and the need for an extra step for the preparation of an appropriate amount of serially diluted concentrations of samples in the test tubes. Due to the transition of antimicrobial compounds between the liquid and vapor phases and their possible losses during the experiment preparation, the final concentration should be considered as indicative only. For that reason, the concentrations of the samples tested in the vapor phase were expressed as the weight of the volatile agent per volume unit of a microtube, that is, 640, 320, 160, 80, 40, 20, 10, and 5 μg/cm^3 for 1024, 512, 128. 64, 32, 16, 8, and 4 μg/mL, respectively. Nevertheless, the real quantity of volatile agents evaporated from the broth should be determined, e.g., using a combination of solid-phase microextraction and gas chromatography analysis [45].

2.2. *Cytotoxicity*

The results of the modified MTT assay performed on lung fibroblast cells are summarized in Table 2. The cytotoxic effect of twelve two-fold serially diluted concentrations of compounds tested are displayed in Figure 2. *β*-Thujaplicin and thymohydroquinone exhibited moderate toxicity to the lung cells with respective half maximal inhibitory concentration (IC_{50}) values of 4.15 and 2.64 μg/mL. Thymoquinone was evaluated as toxic with an IC_{50} value of 1.21 μg/mL. In the case of an 80% inhibitory concentration of proliferation (IC_{80}) determination, the lowest cytotoxic effect was observed for thymohydroquinone followed by β-thujaplicin with IC_{80} values >12.00 and 214.85 μg/mL, respectively. Similar to IC_{50}, the lowest IC_{80} value was recorded for thymoquinone (IC_{80} = 15.00 μg/mL). Although according to the WHO [46], all the compounds were classified as toxic and moderately toxic, the therapeutic index (TI) calculated to compare their antibacterial and cytotoxic effects indicate that thymohydroquinone (TI > 34.13) can be a safe and effective antibacterial agent for inhalation therapy.

Table 2. Cytotoxicity of plant volatile compounds to the normal lung fibroblast cells MRC-5.

Samples	$IC_{50} \pm SD$ (μg/mL)	$IC_{80} \pm SD$ (μg/mL)	TI
Plant volatile compound			
β-thujaplicin	4.15 ± 0.45	214.85 ± 9.71	3.36
thymohydroquinone	2.64 ± 0.33	>512.00	>34.13
thymoquinone	1.21 ± 0.24	15.00 ± 4.46	1.00
Positive control			
vinorelbin	0.54 ± 0.26	>10	-

IC_{50}: half maximal inhibitory concentration of proliferation in μg/mL, IC_{80}: 80% inhibitory concentration of proliferation in μg/mL, SD: standard deviation, TI: therapeutic index (TI = $IC_{80}/\overline{x}$-MIC).

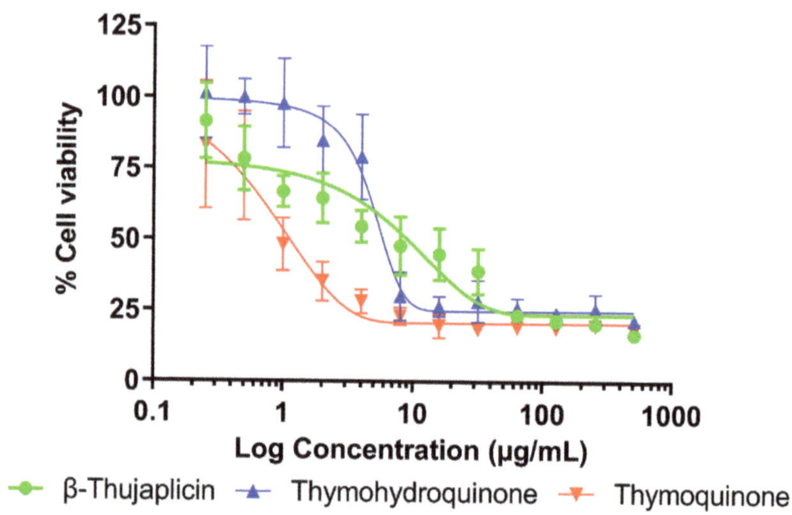

Figure 2. Cytotoxic activity of twelve two-fold serially diluted concentrations (0.25–512 μg/mL) of plant volatile compounds to lung fibroblast cells tested by using MTT assay performed in microtiter plates sealed with vapor barrier EVA Capmat.

Series of assays examining the toxic potential of thymoquinone have been performed including our previous study [21], which observed that this compound has a very similar cytotoxic effect on healthy human lung MRC-5 cell lines with a IC_{50} value of 1.70 μg/mL, while Gurung et al. [47] found that thymoquinone did not alter the viability of normal IMR-90 lung fibroblasts at a concentration of 32.84 μg/mL. In contrast to the well-documented toxicity of thymoquinone, there are only few data available on the safety of other compounds tested in normal human lung fibroblasts. In a study of Ivankovic et al. [48], thymohydroquinone exhibited a certain level of cell growth inhibition against mouse fibroblasts at concentrations of 10 and 100 μg/mL (21% and 63%, respectively). A toxic effect of β-thujaplicin was detected against various types of human lung cancer tissues, with IC_{50} values of 0.26–12.32 μg/mL [49–51]. Regarding the relatively high toxicity of compounds tested in lung cell cultures, their practical application for inhalation therapy to treat respiratory infection seems to be limited. However, their specific structural and technological modifications reducing toxicity are possible. For example, liposomal encapsulation of thymoquinone was found to be effective in decreasing its toxic effects [52].

3. Materials and Methods

3.1. Chemicals and Reagents

The following plant-derived volatile antibacterial agents were assayed: β-thujaplicin (99%, CAS 499-44-5), thymohydroquinone (98%, CAS 2217-60-9), and thymoquinone (99%, CAS 490-91-5). Amoxicillin (90%, CAS 26787-78-0), ampicillin (84.5%, CAS 69-52-3), oxacillin (86.3%, CAS 7240-38-2), and tetracycline (98–102%, CAS 60-54-8) were used as positive antibiotic controls. The chemicals used for antimicrobial susceptibility testing were as follows: dimethyl sulfoxide (DMSO, CAS 67-68-5), thiazolyl blue tetrazolium bromide dye (MTT, CAS 298-93-1), and Tween 20 (CAS 9005-64-5). With the exception of thymohydroquinone prepared by reduction of thymoquinone according to a method described further, all other chemicals were obtained from Sigma-Aldrich (Prague, Czech Republic).

The following chemicals were used for thymohydroquinone preparation and characterization: acetic acid (CAS 64-19-7) purchased from Merck (Darmstadt, Germany), then deuterium oxide containing 3-(trimethylsilyl) propionic-2,2,3,3-d_4 acid sodium salt (CAS 7789-20-0), vanillin (CAS 121-33-5) and zinc (CAS 7440-66-6) purchased from Sigma-Aldrich (Praha, Czech Republic), and others obtained from Penta (Praha, Czech Republic), namely chloroform (CAS 67-66-3), ethanol (96%, CAS 64-17-5), hexane (CAS 110-54-3), hydrochloric acid (35%, CAS 7647-01-0), and sulfuric acid (CAS 7664-93-9).

3.2. Thymohydroquinone Preparation and Characterization

Thymohydroquinone was prepared by the reduction of thymoquinone using acetic acid in the presence of zinc powder as a catalyst according to the method previously described by Tesarova et al. [53]. Briefly, 50 mg of thymoquinone (Sigma Aldrich, Praha, Czech Republic) was dissolved in 3 mL of concentrated acetic acid (99%) and 5 g of zinc powder was subsequently added. The reaction mixture was stirred for 4 h, monitored by thin layer chromatography (TLC) under laboratory conditions, and then filtered. The TLC was performed on TLC Silica gel 60 F254 (Merck, Darmstadt, Germany) plates with chloroform and ethanol in a ratio of 9:1 as the mobile phase. Vanillin dissolved in sulfuric acid in ethanol (1%) was used as a visualizing agent. The liquid fraction was evaporated under vacuum in a rotary evaporator (Rotavapor R-210, Buchi, Flawil, Switzerland). The solid residue was dissolved in water, acidified by the addition of hydrochloric acid (to 5% *v/v*) and yellow residues of thymoquinone were removed by repeated extraction with hexane. Thymohydroquinone was then extracted three times with distilled diethyl ether and evaporated under a stream of N_2. For further purification, sublimation at 168 °C was used and colorless needle crystals of thymohydroquinone were stored in the dark.

The identity and purity of obtained thymohydroquinone were confirmed using gas chromatography/mass spectrometry analysis (GC/MS) and a nuclear magnetic resonance (NMR). The mass spectra of thymohydroquinone were recorded by Agilent GC-7890B and MSD-5977B (Agilent Technologies, Santa Clara, CA, USA) equipped with a fused-silica HP-5MS column (30 m × 0.25 mm, film thickness 0.25 μm, Agilent 19091s-433) and a flame ionization detector coupled with single quadrupole mass selective detector Agilent MSD-5977B (Agilent Technologies, Santa Clara, CA, USA). Operational parameters were: helium as a carrier gas at 1 mL/min, injector temperature 250 °C. The oven temperature was raised from 50 to 280 °C. Thymohydroquinone was diluted in *n*-hexane for GC/MS at a concentration of 20 µg/mL. One microliter of solution was injected in a split mode (split ratio 1:50). The mass detector was set to the following conditions: ionization energy 70 eV, ion source temperature 230 °C, scan time 1 s, mass range 40–600 *m*/*z*. Identification of the sample was based on the comparison of its retention index (RI), retention time (RT) and mass spectra with the National Institute of Standards and Technology Library ver. 2.0.f (National Institute of Standards and Technology, USA) [54].

The ^{1}H-NMR spectra of thymohydroquinone standard (2 mg/mL) were recorded on a Bruker Avance III HD BBFO (Bruker BioSpin GmbH, Rheinstetten, Germany), operating at 500 MHz for ^{1}H-NMR and 126 MHz for ^{13}C-NMR using *noesypr1d* pulse sequence, at 25 °C. The sample was dissolved in H_2O containing 10% deuterium oxide, at pH 7.4, and including 3-(trimethylsilyl) propionic-2,2,3,3-d_4 acid sodium salt (99%) as an internal standard. Supplementary evidence was given by ^{13}C-NMR, HSQC, HMBC and COSY experiments. The experimental chemical shifts in ^{1}H and ^{13}C-NMR spectra of thymohydroquinone closely matched the theoretically predicted chemical shifts obtained using www.nmrdb.org (accessed on 21 June 2021) [55,56].

The signals were confirmed and assigned after inspection of the 1D and 2D spectra and GC/MS data and were as follows: *Thymohydroquinone*. White crystals; RI: 1520, MS, *m*/*z* (rel. int.): 166 (M^+, 43%), 151 (100), 152 (14), 77 (8), 123 (7), 95 (7); ^{1}H-NMR (H_2O/D_2O 9:1, 500 MHz): δ 6.79 (s, 1H, 2-H), 6.73 (s, 1H, 5-H), 3.14 (sept, 1H, J = 6.9 Hz, 8-H), 2.13 (s, 3H, 7-H), 1.17 (d, 6H, J = 6.9 Hz, 9,10-H); ^{13}C-NMR (H_2O/D_2O 9:1, 126 MHz): δ 147.5 (C-6), 145.8 (C-3), 134.4 (C-4), 123.3 (C-1), 118.3 (C-2), 113.2 (C-5), 26.0 (C-8), 22.2 (C-9), 22.2 (C-10), 14.9 (C-7).

3.3. Bacterial Strains and Culture Media

The following four bacterial standard strains from the American Type Culture Collection (ATCC, Manassas, VA, USA) were used: *Haemophilus influenzae* ATCC 49247, *Staphylococcus aureus* ATCC 29213, *Streptococcus pneumoniae* ATCC 49619, and *Streptococcus pyogenes* ATCC 19615. Cultivation and assay media (broth/agar) were Mueller–Hinton (MH) complemented by Haemophilus Tested Medium (*H. influenzae*), MH (*S. aureus*), and Brain Heart Infusion (*S. pneumoniae* and *S. pyogenes*). The pH of the broths was equilibrated to a final value of 7.6 using Trizma base (Sigma-Aldrich, Praha, Czech Republic). All microbial strains and cultivation media were purchased from Oxoid (Basingstoke, UK).

Stock cultures of bacterial strains were cultivated in broth medium at 37 °C for 24 h prior to testing. For the preparation of inoculum, the turbidity of the bacterial suspension was adjusted to 0.5 McFarland standard using a Densi-La-Meter II (Lachema, Brno, Czech Republic) to obtain a final concentration of 10^8 CFU/mL.

3.4. Cell Cultures

Lung fibroblast cell line MRC-5, obtained from ATCC, was propagated in Eagle's Minimum Essential Medium (EMEM) supplemented with 10% fetal bovine serum (FBS), 2 mM glutamine, 10 µL/mL non-essential amino acids, and 1% penicillin–streptomycin solution (10,000 units/mL of penicillin and 10 mg/mL of streptomycin); all these components were purchased from Sigma-Aldrich. The cells were pre-incubated in 96-well microtiter plates at a density of 2.5×10^3 cells per well for 24 h at 37 °C in a humidified incubator in an atmosphere of 5% CO_2 in air.

3.5. Antimicrobial Assay

The antibacterial potential of volatile plant-derived compounds in the liquid and vapor phases was determined using a newly developed broth macrodilution volatilization method performed in standard 2 mL microtubes with snap caps (Eppendorf, Hamburg, Germany). Initially, each sample of compound was dissolved in DMSO at maximum concentration of 1% and diluted in the appropriate broth medium. With the aim to prepare a sufficient amount of stock solutions of the compounds assayed, six two-fold serially diluted concentrations of samples were prepared in 15 mL test tubes closed with plugs to avoid the losses of active compounds by evaporation (Gama Group, Ceske Budejovice, Czech Republic). The concentration of β-thujaplicin started from 1024 μg/mL and from 128 μg/mL for thymohydroquinone and thymoquninone. In the second step, 90 μL of melted agar was pipetted into rims on the caps (Figure 3a) and inoculated with 5 μL of bacterial suspension after agar solidification. Subsequently, the appropriate concentrations of each sample previously prepared in test tubes were pipetted into microtubes in a final volume of 1500 μL. Then, the microtubes were inoculated with 10 μL of bacterial suspension and closed properly (Figure 3b). Microtubes containing inoculated and non-inoculated media were prepared as growth and purity controls simultaneously. After incubation at 37 °C for 24 h, the MICs were evaluated by the visual assessment of bacterial growth after coloring metabolically active bacterial colonies with MTT dye. The respective volumes of 30 and 375 μL of MTT at a concentration of 600 μg/mL were pipetted into the caps and in the microtubes when the interface of color change from yellow to purple (relative to that of colors in control wells) was recorded in broth and agar (Figure 3c). A black and white scheme of a cross-sectional view of a microtube filled with broth and agar shows the effective flow of sample vapors in the closed testing system (Figure 4). The MIC values were determined as the lowest concentrations that inhibited bacterial growth compared with the compound-free control and are expressed in μg/mL. In the case of the vapor phase, the concentration was also expressed in $\mu g/cm^3$ as the weight of the volatile agent per volume unit of a microtube. DMSO, assayed as the negative control, did not inhibit any of the strains at the tested concentration ($\leq$1%). The respective susceptibilities of *H. influenzae*, *S. aureus*, *S. pneumoniae*, and *S. pyogenes* to ampicillin, oxacillin, amoxicillin, and tetracycline were checked as positive antibiotic controls [57]. All tests were performed as three independent experiments, each carried out in triplicate, and the results were presented as median/modal values. According to the widely accepted norm in MIC testing, the mode and median were used for the final value calculation when triplicate endpoints were within the two- and three-dilution ranges, respectively.

3.6. Cytotoxicity Assay

A modified method based on the metabolization of MTT to blue formazan by mitochondrial dehydrogenases in living lung cells previously described by Mosmann [58] was used. The lung fibroblast cells were treated for 72 h with the tested compounds dissolved in DMSO at a maximum concentration of 1% and diluted in the EMEM medium supplemented with 10% FBS. Twelve two-fold serially diluted concentrations of these agents ranging from 512 to 0.25 μg/mL were prepared. The microtiter plates were covered with EVA capmats^{TM} at 37 °C in a humidified atmosphere of 5% CO_2 in air and cultivated for 72 h. Thereafter, MTT reagent (1 mg/mL) in EMEM solution was added to each well and the plates were incubated for an additional 2 h at 37 °C in a humidified atmosphere of 5% CO_2 in air. The media were removed, and the intracellular formazan product was dissolved in 100 μL of DMSO. The solvent used did not affect the viability of the lung cells at the tested concentration ($\leq$1%). The absorbance was measured at 555 nm using a Tecan Infinite M200 spectrometer (Tecan Group, Mannedorf, Switzerland), and the viability was calculated in comparison to that of the untreated control. Three independent experiments (two replicates each) were performed for each test. The results of the cytotoxicity effect were calculated using GraphPad Prism software (GraphPad Software, La Jolla, CA, USA) and expressed as average IC_{50} with standard deviation in μg/mL. The levels of cytotoxic effects were classified according to the Special Programme for Research and Training in Tropical Diseases (WHO–Tropical Diseases) [46] as cytotoxic (IC_{50} < 2 μg/mL), moderately cytotoxic (IC_{50} 2–89 μg/mL), and non-toxic (IC_{50} > 90 μg/mL). Furthermore, IC_{80} was calculated as equivalent to the MIC endpoint [59] for comparison of microbiological and toxicological data. Therapeutic indices (TIs) were defined as the ratio of $\overline{x}$-IC_{80} and $\overline{x}$-MIC values with the aim of determining the amount of effective antibacterial agents with the quantity causing toxicity [60].

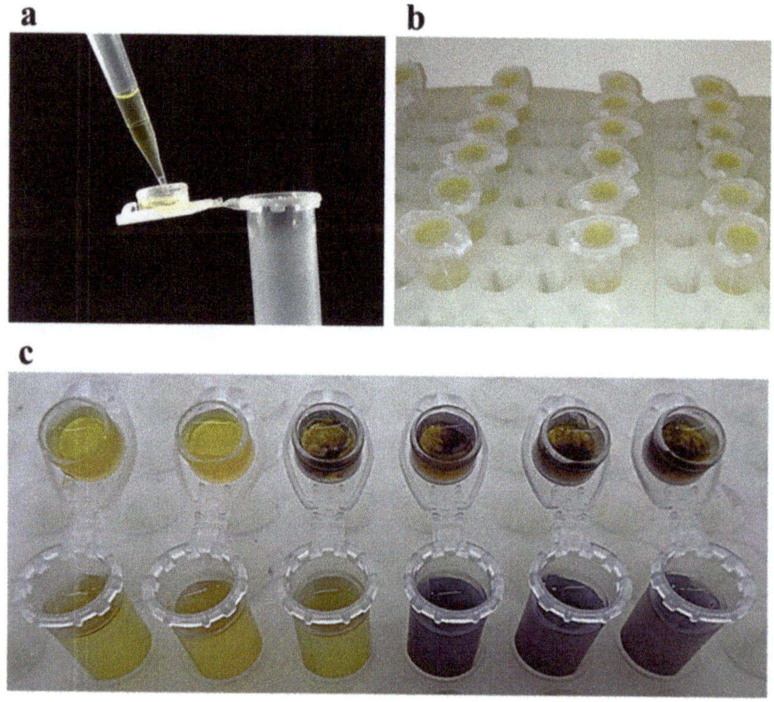

Figure 3. Broth macrodilution volatilization method (**a**) pipetting of agar in the microtube caps: 90 μL of agar is pipetted into rim of every cap; (**b**) incubation: after inoculation, microtubes containing liquid medium with serially diluted samples of tested volatiles and their caps containing solid medium are properly closed together to prevent the losses of active compounds; (**c**) MIC determination: the results are evaluated visually after coloring of living bacterial colonies with MTT dye.

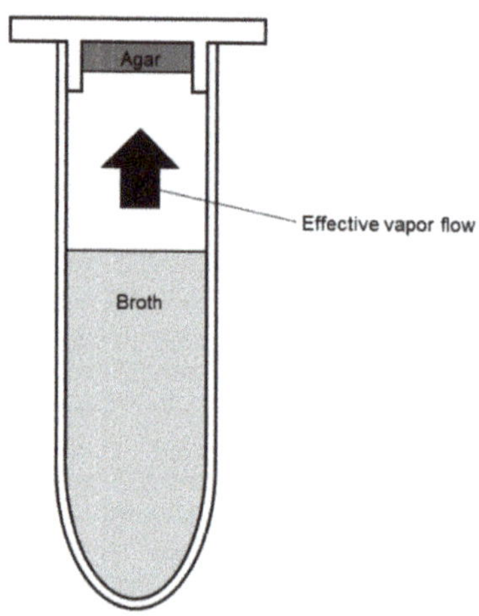

Figure 4. Detail of the cross-sectional view of the closed microtube with snap cap containing broth and agar media.

4. Conclusions

As a result of this study, a new broth macrodilution volatilization method was developed for the simultaneous determination of the antimicrobial effects of volatile agents in the liquid and the vapor phases at variable concentrations. This rapid, simple, cost- and labor-effective technique, which combines the principles of broth microdilution volatilization and standard broth macrodilution methods, is performed in commercially available microtubes and, therefore, does not require specialized equipment. It can also be a suitable option for the testing of slower growing organisms (e.g., fungi) that require longer incubation time to produce enough growth for MIC determination. Nevertheless, further research focusing on the optimization of the novel broth macrodilution volatilization method for susceptibility testing of a broader spectrum of microorganisms will be necessary to confirm this assumption. In addition, the validity of the method for the susceptibility testing of bacterial pathogens causing respiratory infection was evaluated using three antimicrobial phytochemicals (β-thujaplicin, thymohydroquinone, and thymoquinone). As a result of this research, thymohydroquinone was found to be a promising antibacterial agent for application in inhalation therapy that is safe to human lung cell lines. However, in vivo experiments are required to verify the therapeutic potential of this compound.

Author Contributions: Coordination and performance of the antimicrobial activity and cytotoxicity testing and its related data analysis, the manuscript drafting, M.H.; performance of the antimicrobial activity, A.C.; performance of the cytotoxicity testing, I.D.; performance of the NMR analysis, J.H.; conceptualization and coordination of the whole study and finalization of the manuscript, L.K. All authors have read and agreed to the published version of the manuscript.

Funding: This research was financially supported by the Czech University of Life Sciences Prague [project IGA 20213109] and METROFOOD-CZ Research Infrastructure Project [MEYS Grant No: LM2018100].

Conflicts of Interest: The authors declare no conflict of interest.

References

1. Brooks, W.A. Bacterial pneumonia. In *Hunter's Tropical Medicine and Emerging Infectious Diseases*, 10th ed.; Ryan, E.T., Hill, D.R., Solomon, T., Aronson, N.E., Endy, T.P., Eds.; Elsevier: Amsterdam, The Netherlands, 2020; pp. 446–453.
2. The Top 10 Causes of Death. Available online: https://www.who.int/news-room/fact-sheets/detail/the-top-10-causes-of-death (accessed on 19 March 2021).
3. Zafar, A.; Hasan, R.; Nizamuddin, S.; Mahmood, N.; Mukhtar, S.; Ali, F.; Morrissey, I.; Barker, K.; Torumkuney, D. Antibiotic susceptibility in *Streptococcus pneumoniae, Haemophilus influenzae* and *Streptococcus pyogenes* in Pakistan: A review of results from the Survey of Antibiotic Resistance (SOAR) 2002-15. *J. Antimicrob. Chemother.* **2016**, *71* (Suppl. 1), 103–109. [CrossRef] [PubMed]
4. Manohar, P.; Loh, B.; Nachimuthu, R.; Hua, X.T.; Welburn, S.C.; Leptihn, S. Secondary bacterial infections in patients with viral pneumonia. *Front. Med.* **2020**, *7*, 720. [CrossRef] [PubMed]
5. Borghardt, J.M.; Kloft, C.; Sharma, A. Inhaled therapy in respiratory disease: The complex interplay of pulmonary kinetic processes. *Can. Respir. J.* **2018**, *2018*, 2732017. [CrossRef] [PubMed]
6. Sorino, C.; Negri, S.; Spanevello, A.; Visca, D.; Scichilone, N. Inhalation therapy devices for the treatment of obstructive lung diseases: The history of inhalers towards the ideal inhaler. *Eur. J. Intern. Med.* **2020**, *75*, 15–18. [CrossRef]
7. Cock, I.E.; van Vuuren, S.F. The traditional use of southern African medicinal plants for the treatment of bacterial respiratory diseases: A review of the ethnobotany and scientific evaluations. *J. Ethnopharmacol.* **2020**, *263*, 113204. [CrossRef]
8. Jain, H.; Bairagi, A.; Srivastava, S.; Singh, S.B.; Mehra, N.K. Recent advances in the development of microparticles for pulmonary administration. *Drug Discov. Today* **2020**, *25*, 1865–1872. [CrossRef]
9. Ibrahim, M.; Verma, R.; Garcia-Contreras, L. Inhalation drug delivery devices: Technology update. *Med. Devices* **2015**, *8*, 131–139. [CrossRef]
10. Klepser, M.E. Role of nebulized infections antibiotics for the treatment of respiratory. *Curr. Opin. Infect. Dis.* **2004**, *17*, 109–112. [CrossRef]
11. Kokoska, L.; Kloucek, P.; Leuner, O.; Novy, P. Plant-derived products as antibacterial and antifungal agents in human health care. *Curr. Med. Chem.* **2019**, *26*, 1–38. [CrossRef]
12. Cos, P.; Vlietinck, A.J.; Vanden Berghe, D.; Maes, L. Anti-infective potential of natural products: How to develop a stronger in vitro 'proof-of-concept'. *J. Ethnopharmacol.* **2006**, *106*, 290–302. [CrossRef] [PubMed]
13. Mutlu-Ingok, A.; Devecioglu, D.; Dikmetas, D.N.; Karbancioglu-Guler, F.; Capanoglu, E. Antibacterial, antifungal, antimycotoxigenic, and antioxidant activities of essential oils: An updated review. *Molecules* **2020**, *25*, 4711. [CrossRef]
14. Taborsky, J.; Kunt, M.; Kloucek, P.; Lachman, J.; Zeleny, V.; Kokoska, L. Identification of potential sources of thymoquinone and related compounds in Asteraceae, Cupressaceae, Lamiaceae, and Ranunculaceae families. *Cent. Eur. J. Chem.* **2012**, *10*, 1899–1906. [CrossRef]

15. Fotopoulou, T.; Ciric, A.; Kritsi, E.; Calhelha, R.; Ferreira, I.C.F.R.; Sokovic, M.; Zoumpoulakis, P.; Koufaki, M. Antimicrobial/antibiofilm activity and cytotoxic studies of beta-thujaplicin derivatives. *Arch. Pharm.* **2016**, *349*, 698–709. [CrossRef]
16. Abdelazeem, A.H.; Mohamed, Y.M.A.; Gouda, A.M.; Omar, H.A.; Al Robaian, M.M. Novel thymohydroquinone derivatives as potential anticancer agents: Design, synthesis, and biological screening. *Aust. J. Chem.* **2016**, *69*, 1277–1284. [CrossRef]
17. Domon, H.; Hiyoshi, T.; Maekawa, T.; Yonezawa, D.; Tamura, H.; Kawabata, S.; Yanagihara, K.; Kimura, O.; Kunitomo, E.; Terao, Y. Antibacterial activity of hinokitiol against both antibiotic-resistant and -susceptible pathogenic bacteria that predominate in the oral cavity and upper airways. *Microbiol. Immunol.* **2019**, *63*, 213–222. [CrossRef] [PubMed]
18. Reyes-Jurado, F.; Navarro-Cruz, A.R.; Ochoa-Velasco, C.E.; Palou, E.; Lopez-Malo, A.; Avila-Sosa, R. Essential oils in vapor phase as alternative antimicrobials: A review. *Crit. Rev. Food Sci. Nutr.* **2020**, *60*, 1641–1650. [CrossRef] [PubMed]
19. Leigh-de Rapper, S.; van Vuuren, S.F. Odoriferous therapy: A review identifying essential oils against pathogens of the respiratory tract. *Chem. Biodivers.* **2020**, *17*, e2000062. [CrossRef] [PubMed]
20. Jaradat, N.A.; Al Zabadi, H.; Rahhal, B.; Hussein, A.M.; Mahmoud, J.S.; Mansour, B.; Khasati, A.I.; Issa, A. The effect of inhalation of *Citrus sinensis* flowers and *Mentha spicata* leave essential oils on lung function and exercise performance: A quasi-experimental uncontrolled before-and-after study. *J. Int. Soc. Sports Nutr.* **2016**, *13*, 36. [CrossRef] [PubMed]
21. Houdkova, M.; Rondevaldova, J.; Doskocil, I.; Kokoska, L. Evaluation of antibacterial potential and toxicity of plant volatile compounds using new broth microdilution volatilization method and modified MTT assay. *Fitoterapia* **2017**, *118*, 56–62. [CrossRef] [PubMed]
22. Kiani, S.; Minaei, S.; Ghasemi-Varnamkhasti, M. Application of electronic nose systems for assessing quality of medicinal and aromatic plant products: A review. *J. Appl. Res. Med. Aromat. Plants* **2016**, *3*, 1–9. [CrossRef]
23. BeruBe, K.; Aufderheide, M.; Breheny, D.; Clothier, R.; Combes, R.; Duffin, R.; Forbes, B.; Gaca, M.; Gray, A.; Hall, I.; et al. In vitro models of inhalation toxicity and disease. *Altern. Lab. Anim.* **2009**, *37*, 89–141.
24. Reyes-Jurado, F.; Franco-Vega, A.; Ramirez-Corona, N.; Palou, E.; Lopez-Malo, A. Essential oils: Antimicrobial activities, extraction methods and their modeling. *Food Eng. Rev.* **2015**, *7*, 275–297. [CrossRef]
25. Novy, P.; Kloucek, P.; Rondevaldova, J.; Havlik, J.; Kourimska, L.; Kokoska, L. Thymoquinone vapor significantly affects the results of *Staphylococcus aureus* sensitivity tests using the standard broth microdilution method. *Fitoterapia* **2014**, *94*, 102–107. [CrossRef]
26. Houdkova, M.; Albarico, G.; Doskocil, I.; Tauchen, J.; Urbanova, K.; Tulin, E.E.; Kokoska, L. Vapors of volatile plant-derived products significantly affect the results of antimicrobial, antioxidative and cytotoxicity microplate-based assays. *Molecules* **2020**, *25*, 6004. [CrossRef] [PubMed]
27. Clinical and Laboratory Standards Institute (CLSI). *Performance Standards for Antimicrobial Disk Susceptibility Tests*, 11th ed.; Approved Standard, CLSI Document M02-A11; CLSI: Wayne, PA, USA, 2012; p. 32, ISBN 1-56238-782-0.
28. European Committee on Antimicrobial Susceptibility Testing. Antimicrobial Susceptibility Testing EUCAST Disk Diffusion Method, Version 9. Available online: https://www.eucast.org/fileadmin/src/media/PDFs/EUCAST_files/Disk_test_documents/2021_manuals/Manual_v_9.0_EUCAST_Disk_Test_2021.pdf (accessed on 22 May 2021).
29. Clinical and Laboratory Standards Institute (CLSI). *Methods for Dilution Antimicrobial Susceptibility Tests for Bacteria That Grow Aerobically; Approves Standard*, 10th ed.; CLSI document M07-A10; CLSI: Wayne, PA, USA, 2015; p. 35, ISBN 1-56238-988-2.
30. National Committee for Clinical Laboratory Standards (NCCLS). *Methods for Dilution Antimicrobial Susceptibility Tests for Bacteria That Growth Aerobically; Approved Standard*, 6th ed.; NCCLS document M7-A6; NCCLS: Wayne, PA, USA, 2003.
31. Food and Drug Administration (FDA). *Guidance for industry and FDA. Class II Special Controls Guidance Document: Antimicrobial Susceptibility Test (AST) Systems*; Center for Devices and Radiological Health, FDA: Rockville, MD, USA, 2009.
32. International Organization for Standardization (ISO). *Susceptibility Testing of Infectious Agents and Evaluation of Performance of Antimicrobial Susceptibility Devices, Part 1. Broth Micro-Dilution Reference Method for Testing the In Vitro Activity of Antimicrobial Agents against Rapidly Growing Aerobic Bacteria Involved in Infectious Diseases*, 2nd ed.; ISO/DIS 20776-1; ISO: Geneva, Switzerland, 2019.
33. Houdkova, M.; Kokoska, L. Volatile antimicrobial agents and in vitro methods for evaluating their activity in the vapour phase: A review. *Planta Med.* **2020**, *86*, 822–857. [CrossRef]
34. Houdkova, M.; Urbanova, K.; Doskocil, I.; Rondevaldova, J.; Novy, P.; Nguon, S.; Chrun, R.; Kokoska, L. In vitro growth-inhibitory effect of Cambodian essential oils against pneumonia causing bacteria in liquid and vapour phase and their toxicity to lung fibroblasts. *S. Afr. J. Bot.* **2018**, *118*, 85–97. [CrossRef]
35. Houdkova, M.; Doskocil, I.; Urbanova, K.; Tulin, E.K.C.B.; Rondevaldova, J.; Tulin, A.B.; Kudera, T.; Tulin, E.E.; Zeleny, V.; Kokoska, L. Evaluation of antipneumonic effect of Philippine essential oils using broth microdilution volatilization method and their lung fibroblasts toxicity. *Nat. Prod. Commun.* **2018**, *13*, 1059–1066. [CrossRef]
36. Morita, Y.; Matsumura, E.; Tsujibo, H.; Yasuda, M.; Sakagami, Y.; Okabe, T.; Ishida, N.; Inamori, Y. Biological activity of alpha-thujaplicin, the minor component of *Thujopsis dolabrata* Sieb. et Zucc. var. *hondai Makino*. *Biol. Pharm. Bull.* **2001**, *24*, 607–611. [CrossRef] [PubMed]
37. Inoue, Y.; Suzuki, R.; Murata, I.; Nomura, H.; Isshiki, Y.; Kanamoto, I. Evaluation of antibacterial activity expression of the hinokitiol/cyclodextrin complex against bacteria. *Acs Omega* **2020**, *5*, 27180–27187. [CrossRef] [PubMed]
38. Chaieb, K.; Kouidhi, B.; Jrah, H.; Mahdouani, K.; Bakhrouf, A. Antibacterial activity of thymoquinone, an active principle of *Nigella sativa* and its potency to prevent bacterial biofilm formation. *BMC Complement Altern Med.* **2011**, *11*, 29. [CrossRef]
39. Halawani, E. Antibacterial activity of thymoquinone and thymohydroquinone of *Nigella sativa* L. and their interaction with some antibiotics. *Adv. Biol. Res.* **2009**, *3*, 148–152.

40. Muthaiyan, A.; Biswas, D.; Crandall, P.G.; Wilkinson, B.J.; Ricke, S.C. Application of orange essential oil as an antistaphylococcal agent in a dressing model. *BMC Complement. Altern. Med.* **2012**, *12*, 125. [CrossRef]
41. Valgas, C.; de Souza, S.M.; Smania, E.F.A.; Smania, A. Screening methods to determine antibacterial activity of natural products. *Braz. J. Microbiol.* **2007**, *38*, 369–380. [CrossRef]
42. Wang, T.H.; Hsia, S.M.; Wu, C.H.; Ko, S.Y.; Chen, M.Y.; Shih, Y.H.; Shieh, T.M.; Chuang, L.C.; Wu, C.Y. Evaluation of the antibacterial potential of liquid and vapor phase phenolic essential oil compounds against oral microorganisms. *PLoS ONE* **2016**, *11*, e0163147. [CrossRef] [PubMed]
43. Inouye, S.; Uchida, K.; Takizawa, T.; Yamaguchi, H.; Abe, S. Evaluation of the effect of terpenoid quinones on *Trichophyton mentagrophytes* by solution and vapor contact. *J. Infect. Chemother.* **2006**, *12*, 100–104. [CrossRef]
44. Espinel-Ingroff, A.; Canton, E. Antifungal susceptibility testing of yeasts. In *Antimicrobial Susceptibility Testing Protocols*; Schwalbe, R., Steele-Moore, L., Goodwin, A.C., Eds.; CRC Press: Boca Raton, FL, USA, 2007; pp. 173–208.
45. Vihanova, K.; Houdkova, M.; Promgool, T.; Urbanova, K.; Kanokmedhakul, S.; Kokoska, L. In vitro growth-inhibitory effect of essential oils and supercritical carbon dioxide extracts from *Cinnamomum* spp. barks and fruits against food bacterial pathogens in liquid and vapor phase. *J. Food Saf.* **2021**, e12900. [CrossRef]
46. Special Programme for Research and Training in Tropical Diseases. Available online: http://www.who.int/tdr/grants/workplans/en/cytotoxicity_invitro.pdf (accessed on 11 March 2021).
47. Gurung, R.L.; Lim, S.N.; Khaw, A.K.; Soon, J.F.F.; Shenoy, K.; Ali, S.M.; Jayapal, M.; Sethu, S.; Baskar, R.; Hande, M.P. Thymoquinone induces telomere shortening, DNA damage and apoptosis in human glioblastoma cells. *PLoS ONE* **2010**, *5*, e12124. [CrossRef]
48. Ivankovic, S.; Stojkovic, R.; Jukic, M.; Milos, M.; Milos, M.; Jurin, M. The antitumor activity of thymoquinone and thymohydroquinone in vitro and in vivo. *Exp. Oncol.* **2006**, *28*, 220–224.
49. Li, L.H.; Wu, P.; Lee, J.Y.; Li, P.R.; Hsieh, W.Y.; Ho, C.C.; Ho, C.L.; Chen, W.J.; Wang, C.C.; Yen, M.Y.; et al. Hinokitiol induces DNA damage and autophagy followed by cell cycle arrest and senescence in gefitinib-resistant lung adenocarcinoma cells. *PLoS ONE* **2014**, *9*, e104203. [CrossRef] [PubMed]
50. Lee, T.B.; Seo, E.J.; Lee, J.Y.; Jun, J.H. Synergistic anticancer effects of curcumin and hinokitiol on gefitinib resistant non-small cell lung cancer cells. *Nat. Prod. Commun.* **2018**, *13*, 1667–1671. [CrossRef]
51. Lee, T.B.; Jun, J.H. Can hinokitiol kill cancer cells? Alternative therapeutic anticancer agent via autophagy and apoptosis. *Korean J. Clin. Lab. Sci.* **2019**, *51*, 221–234. [CrossRef]
52. Allemailem, K.S.; Alnuqaydan, A.M.; Almatroudi, A.; Alrumaihi, F.; Aljaghwani, A.; Khalilullah, H.; Younus, H.; Khan, A.; Khan, M.A. Safety and therapeutic efficacy of thymoquinone-loaded liposomes against drug-sensitive and drug-resistant *Acinetobacter baumannii*. *Pharmaceutics* **2021**, *13*, 677. [CrossRef]
53. Tesarova, H.; Svobodova, B.; Kokoska, L.; Marsik, P.; Pribylova, M.; Landa, P.; Vadlejch, J. Determination of oxygen radical absorbance capacity of black cumin (*Nigella sativa*) seed quinone compounds. *Nat. Prod. Commun.* **2011**, *6*, 213–216. [CrossRef] [PubMed]
54. NIST WebBook Chemie. NIST Standard Reference Database Number 69. 2017. Available online: http://webbook.nist.gov/chemistry/ (accessed on 29 September 2020).
55. Castillo, A.M.; Patiny, L.; Wist, J. Fast and accurate algorithm for the simulation of NMR spectra of large spin systems. *J. Magn. Reson.* **2011**, *209*, 123–130. [CrossRef]
56. Steinbeck, C.; Krause, S.; Kuhn, S. NMRShiftDB-Constructing a free chemical information system with open-source components. *J. Chem. Inform. Comput. Sci.* **2003**, *43*, 1733–1739. [CrossRef]
57. Clinical and Laboratory Standards Institute (CLSI). *Performance Standards for Antimicrobial Susceptibility Testing*; 25th Informational Supplement M100-S25; CLSI: Wayne, PA, USA, 2015.
58. Mosmann, T. Rapid colorimetric assay for cellular growth and survival: Application to proliferation and cytotoxicity assays. *J. Immunol. Methods* **1983**, *65*, 56–63. [CrossRef]
59. Kokjohn, K.; Bradley, M.; Griffiths, B.; Ghannoum, M. Evaluation of in vitro activity of ciclopirox olamine, butenafine HCl and econazole nitrate against dermatophytes, yeasts and bacteria. *Int. J. Dermatol.* **2003**, *42*, 11–17. [CrossRef] [PubMed]
60. Trevor, A.J.; Katzung, B.G.; Kruidering-Hall, M. *Katzung and Trevor's Pharmacology Examination and Board Review*, 11th ed.; McGraw-Hill Education: New York, NY, USA, 2015; p. 20, ISBN 978-0-07-182639-6.

molecules

Article

In Vitro Antimicrobial Activity of Lavender, Mint, and Rosemary Essential Oils and the Effect of Their Vapours on Growth of *Penicillium* spp. in a Bread Model System

Veronika Valková [1,2,*], Hana Ďúranová [1], Lucia Galovičová [2], Nenad L. Vukovic [3], Milena Vukic [3] and Miroslava Kačániová [2,4,*]

1 AgroBioTech Research Centre, The Slovak University of Agriculture in Nitra, Tr. A. Hlinku 2, 94976 Nitra, Slovakia; hana.duranova@uniag.sk
2 Department of Fruit Sciences, Viticulture and Enology, Faculty of Horticulture and Landscape Engineering, Slovak University of Agriculture, Tr. A. Hlinku 2, 94976 Nitra, Slovakia; l.galovicova95@gmail.com
3 Department of Chemistry, Faculty of Science, University of Kragujevac, P.O. Box 12, 34000 Kragujevac, Serbia; nvukovic@kg.ac.rs (N.L.V.); milena.vukic@pmf.kg.ac.rs (M.V.)
4 Department of Bioenergy, Food Technology and Microbiology, Institute of Food Technology and Nutrition, University of Rzeszow, Zelwerowicza St. 4, 35601 Rzeszow, Poland
* Correspondence: veronika.valkova@uniag.sk (V.V.); miroslava.kacaniova@gmail.com (M.K.); Tel.: +421-37-641-4928 (V.V.); +421-37-641-4715 (M.K.)

Citation: Valková, V.; Ďúranová, H.; Galovičová, L.; Vukovic, N.L.; Vukic, M.; Kačániová, M. In Vitro Antimicrobial Activity of Lavender, Mint, and Rosemary Essential Oils and the Effect of Their Vapours on Growth of *Penicillium* spp. in a Bread Model System. *Molecules* **2021**, *26*, 3859. https://doi.org/10.3390/molecules26133859

Academic Editor: Ladislav Kokoska

Received: 29 May 2021
Accepted: 22 June 2021
Published: 24 June 2021

Publisher's Note: MDPI stays neutral with regard to jurisdictional claims in published maps and institutional affiliations.

Copyright: © 2021 by the authors. Licensee MDPI, Basel, Switzerland. This article is an open access article distributed under the terms and conditions of the Creative Commons Attribution (CC BY) license (https://creativecommons.org/licenses/by/4.0/).

Abstract: The chemical composition, antioxidant activity, and antimicrobial properties of three commercially available essential oils: rosemary (REO), lavender (LEO), and mint (MEO), were determined in the current study. Our data revealed that the major components of REO, MEO, and LEO were 1,8-cineole (40.4%), menthol (40.1%), and linalool acetate (35.0%), respectively. The highest DPPH radical-scavenging activity was identified in MEO (36.85 ± 0.49%) among the investigated EOs. Regarding antimicrobial activities, we found that LEO had the strongest inhibitory efficiencies against the growth of *Pseudomonas aeruginosa* and *Candida* (*C.*) *tropicalis*, MEO against *Salmonella* (*S.*) *enterica*, and REO against *Staphylococcus* (*S.*) *aureus*. The strongest antifungal activity was displayed by mint EO, which totally inhibited the growth of *Penicillium* (*P.*) *expansum* and *P. crustosum* in all concentrations; the growth of *P. citrinum* was completely suppressed only by the lowest MEO concentration. The lowest minimal inhibitory concentrations (MICs) against *S. enterica*, *S. aureus*, and *C. krusei* were assessed for MEO. In situ analysis on the bread model showed that 125 μL/L of REO exhibited the lowest mycelial growth inhibition (MGI) of *P. citrinum*, and 500 μL/L of MEO caused the highest MGI of *P. crustosum*. Our results allow us to make conclusion that the analysed EOs have promising potential for use as innovative agents in the storage of bakery products in order to extend their shelf-life.

Keywords: essential oils; volatile compounds; antioxidant activity; antifungal activity; antibacterial activity; bakery product; moisture content; water activity

1. Introduction

Bread is an important staple food worldwide. However, its fungal spoilage during storage is a serious problem that can result not only in economic losses, but also in human health hazards because of the presence of mycotoxins [1]. In general, bread rot is caused by microscopic fungi, such as *Penicillium* and *Aspergillus*, as well as *Mucor*, *Cladosporium*, *Fusarium*, and *Rhizopus* [2]. One of the potential alternatives to prevent the spoilage of bakery goods appears to be the application of essential oils (EOs) as natural preservatives [3].

EOs are volatile secondary metabolites derived from plants responsible for their typical smell and taste. They can be obtained from about 17,500 angiosperm plants (e.g., Rutaceae, Lamiaceae, Zingiberaceae, Myrtaceae, Asteraceae), and among them, only approximately 300 species of EOs are commercially available [4]. These highly concentrated aromatic materials can be extracted from various parts of the plant, including the leaves, stem, flowers, seeds, roots, fruit rind, resin, or bark [5]. The isolation of such oils is relatively simple, but their chemical composition depends on the extraction technique used. Hydrodistillation, solvent extraction, simultaneous distillation-extraction, supercritical carbon dioxide extraction, and the use of microwave ovens are the most frequently used extraction methods [6].

There are some properties, such as radical scavenging [7], as well as antiviral [8], antiprotozoal [9], antibacterial [10], and antifungal [11] properties, and many others that are well recognized regarding the biological activities of EOs. The wide range of activities is attributed to the diverse chemical composition of EOs. Generally, lipophilic and highly volatile components from many chemical classes are the most common substances found in EOs [12]. Terpenic and phenolic compounds [13], as well as alcohols and esters have shown significant biological effects [14].

Rosemary (*Rosmarinus officinalis* L.; Lamiaceae) is a rich source of phenolic compounds, such as carnosol, rosmanol, rosmaridiphenol, and rosmarquinone [15]. It is used in the treatment of various disorders and in food preservation as well [16]. Rosemary EO (REO) possesses a higher antioxidant activity than other EOs [15].

Lavender (*Lavandula* spp.) is a herb that belongs to the family Labiatae and is intensively cultivated for oil production [17]. The products derived from this popular garden herb are used as therapeutics as well as antibacterial agents, and lavender essential oil (LEO) is traditionally believed to have sedative, carminative, antidepressant, and anti-inflammatory properties [18].

Mint (*Mentha piperita* L.; Lamiaceae) EO (MEO) has been used in traditional medicine and its biological activity may be due to its major volatile components: carvone, menthol, and menthone [19].

The major purpose of our study was to evaluate the antifungal effects of selected EOs (REO, LEO, and MEO) against *Penicillium* spp. using the contact vapour method. In addition, the volatile compounds of the EOs, their antimicrobial properties and antioxidant activities, and basic technological properties of bread (as a model substrate for growth of fungi in situ) were determined. Summarily, the EO with the greatest potential and its effective concentration applied as a natural preservative used in the storage of bread in commercial practice were assessed.

2. Results

The vapour-phase antifungal activities of three selected EOs obtained from rosemary, lavender, and mint against *Penicillium* spp. inoculated on bread samples were evaluated in the current study. The data expand our findings related to bread preservation using natural alternatives, such as EOs [20,21].

2.1. Chemical Composition of EOs

The chemical composition of our EOs was determined by gas chromatography/mass spectrometry (GC/MS) analysis. Overall, there were 44, 43, and 40 components accounting for a total of 99.4%, 99.6%, and 99.5% of the EO identified in the LEO, MEO, and REO, respectively (Table 1). The main components of the LEO were linalool acetate (35.0%), linalool (32.7%), and 1,8-cineole (8.1%). The major chemical constituents of the MEO were represented by the menthol (40.1%), menthone (16.8%), and menthyl acetate (9.1%). However, 1,8-cineole (40.4%), camphor (11.9%), and α-pinene (8.7%) were detected as major compounds in the REO.

2.2. Antioxidant Activity of EOs

Table 2 presents the quenching of 2,2-diphenyl-1-picrylhydrazyl (DPPH) radicals from which it is evident that all analysed EOs displayed moderate antioxidant activity. Additionally, the results indicate the significantly ($p < 0.05$) strongest DPPH-scavenging activity of mint EO (36.85 ± 0.49%), whilst the EOs from lavender and rosemary displayed lower values for the activity (29.08 ± 0.99%, 28.76 ± 2.68%, respectively) without statistically significant differences.

Table 1. Chemical composition of the analysed EOs.

Components	LEO (%)	MEO (%)	REO (%)
1,8-cineole	8.1	5.2	40.4
menthol	-	40.1	-
linalool acetate	35.0	-	-
linalool	32.7	-	1.2
menthone	-	16.8	0.1
camphor	6.4	-	11.9
menthyl acetate	-	9.1	-
α-pinene	0.3	0.7	8.7
β-pinene	0.3	1.1	6.9
neo-menthol	-	4.7	-
(*E*)-caryophyllene	1.6	2.2	5.3
methofuran	-	4.6	-
borneol	2.4	-	3.9
camphene	0.2	tr	3.5
isomenthone	-	2.8	-
α-terpineol	1.3	-	2.7
α-limonene	0.9	1.8	2.4
ocimene	0.3	0.3	2.2
lavandulyl acetate	2.0	-	-
germacrene D	0.4	1.7	tr
pulegone	-	1.5	-
cis-sabinene hydrate	-	1.5	0.2
β-myrcene	0.6	0.2	1.5
bornyl acetate	-	-	1.4
4-terpineol	1.3	-	1.1
γ-terpinene	tr	0.5	1.2
(*Z*)-β-farnesene	1.0	-	-
(*E*)-β-ocimene	0.9	tr	0.1
hexyl butanoate	0.9	-	-
geranyl acetate	0.8	-	-
α-humulene	-	-	0.7
3-carvomenthenone	-	0.6	-
α-terpinene	-	0.3	0.6
caryophyllene oxide	-	tr	0.6
sabinene	tr	0.5	0.4
β-bourbonene	-	0.5	-
δ-cadinene	-	0.5	0.3
α-thujene	tr	tr	0.4
α-terpinolene	tr	0.4	0.4
α-copaene	-	tr	0.4
neryl acetate	0.4	-	-
iso-menthyl acetate	-	0.4	-
(*E*)-β-farnesene	-	0.4	-
α-amorphene	0.3	tr	0.2
hexyl tiglate	0.3	-	-
α-bisabolol	0.3	-	-
3-octanol	-	0.3	-
isomenthol	-	0.3	-
bicyclogermacrene	-	0.3	-
trans-linalool oxide	0.3	-	-
3-octanone	0.3	-	-
α-phellandrene	-	-	0.2
δ-3-carene	-	-	0.2
viridiflorol	-	0.2	-
n-amyl isovalerate	-	0.2	-

Table 1. *Cont.*

Components	LEO (%)	MEO (%)	REO (%)
n-hexanol	0.1	-	-
pinocarvone	-	-	0.1
tricyclene	tr	-	0.1
p-cimene	-	-	0.1
β-elemene	-	tr	-
carvone	-	tr	-
isopulegol	-	tr	-
cis-3-hexenol	tr	tr	-
β-thujone	-	-	tr
α-ylangene	-	-	tr
aromadendrene	-	-	tr
3-octanol	tr	-	-
ethyl hexanoate	tr	-	-
cis-linalool oxide	tr	-	-
capryl acetate	tr	-	-
nerol	tr	-	-
caryophyllene oxide	tr	-	-
epi-α-cadinol	tr	-	-
Total	**99.4**	**99.6**	**99.5**

Note: tr—compounds identified in amounts less than 0.1%; —not detected.

Table 2. Antioxidant activity of the analysed EOs.

	LEO	MEO	REO
Antioxidant Activity (%)	29.08 ± 0.99 ^{a}	36.85 ± 0.49 ^{b}	28.76 ± 2.68 ^{a}

Note: Mean ± standard deviation. MEO: Mint essential oil; LEO: Lavender essential oil; REO: Rosemary essential oil; Values with different superscripts within the same row are significantly different ($p < 0.05$).

2.3. *Antimicrobial Activity of EOs*

A disc diffusion method was used to evaluate the antimicrobial activities of selected EOs (LEO, MEO, REO) against Gram-positive (G^+) and Gram-negative (G^-) bacteria, yeasts, and microscopic fungi in the current study. As shown in Table 3, our results revealed that LEO had the strongest inhibitory efficiency against the growth of *Pseudomonas* (*P.*) *aeruginosa* and *Candida* (*C.*) *tropicalis*, with a zone of inhibition of 9.33 ± 0.58 mm and 9.66 ± 0.58 mm, respectively, which were significantly ($p < 0.05$) higher than those of MEO (7.33 ± 1.53, 6.00 ± 0.00 mm, respectively) and REO (7.00 ± 1.00, 8.33 ± 0.58 mm, respectively). On the other hand, LEO exhibited the least antimicrobial activity against the remaining bacteria and yeasts, with an inhibition zone ranging from 1.00 ± 0.00 (*Yersinia* (*Y.*) *enterocolitica*, *Staphylococcus* (*S.*) *aureus*) to 6.33 ± 0.58 mm (*C. krusei*). The antimicrobial activities were statistically ($p < 0.05$) different in comparison with MEO and REO. The EO from *M. piperita* showed the strongest antimicrobial activity against *Salmonella* (*S.*) *enterica*, with an inhibition zone of 9.00 ± 1.00 mm, which was demonstrably ($p < 0.05$) higher than those exhibited by LEO and REO. The inhibitory actions of MEO against the growth of *Y. enterocolitica* (7.00 ± 1.00 mm), *Enterococcus* (*E.*) *faecium* (8.00 ± 1.00 mm), *C. glabrata* (7.33 ± 0.58 mm), *C. albicans* (6.67 ± 0.58 mm), and *C. krusei* (9.67 ± 0.58 mm) were similar to those of REO (7.67 ± 1.53, 8.00 ± 1.00, 7.67 ± 0.58, 8.00 ± 1.00, 10.00 ± 1.00 mm) but considerably ($p < 0.05$) higher as compared to the actions of LEO (1.00 ± 0.00, 3.00 ± 0.00, 2.00 ± 0.00, 2.00 ± 0.00, 6.33 ± 0.58 mm, respectively).

The strongest antimicrobial activity of REO was shown to be against *S. aureus*, with an inhibition zone of 10.33 ± 0.58 mm, which significantly differed from those of MEO and LEO. Additionally, the REO activities against *S. enterica* and *C. tropicalis* were considerably ($p < 0.05$) higher as compared to LEO and MEO, respectively.

Table 3. Antimicrobial activity of EOs (inhibition zone in mm).

EOs	Gram-Negative Bacteria			Gram-Positive Bacteria		Yeasts			
	PA	SE	YE	EF	SA	CG	CA	CK	CT
	Inhibition Zone [mm]								
LEO	9.3 ± 0.6^{a}	1.3 ± 0.6^{a}	1.0 ± 0.0^{a}	3.0 ± 0.0^{a}	1.0 ± 0.0^{a}	2.0 ± 0.0^{a}	2.0 ± 0.0^{a}	6.3 ± 0.6^{a}	9.7 ± 0.6^{a}
MEO	7.3 ± 1.5^{b}	9.0 ± 1.0^{b}	7.0 ± 1.0^{b}	8.0 ± 1.0^{b}	5.3 ± 0.6^{b}	7.3 ± 0.6^{b}	6.7 ± 0.6^{b}	9.7 ± 0.6^{b}	6.0 ± 0.0^{b}
REO	7.0 ± 1.0^{b}	5.3 ± 0.6^{c}	7.7 ± 1.5^{b}	8.0 ± 1.0^{b}	10.3 ± 0.6^{c}	7.7 ± 0.6^{b}	8.0 ± 1.0^{b}	10.0 ± 1.0^{b}	8.3 ± 0.6^{c}
ATB	22.0 ± 1.0	23.0 ± 1.0	22.0 ± 1.0	25.0 ± 1.0	26.0 ± 1.0	23.0 ± 1.0	24.0 ± 1.0	25.0 ± 1.0	24.0 ± 1.0

Note: Means ± standard deviation. Values followed by superscript within the same column are significantly different ($p < 0.05$). MEO: Mint essential oil; LEO: Lavender essential oil; REO: Rosemary essential oil. *P. aeruginosa*—PA, *S. enterica*—SE, *Y. enterocolitica*—YE, *E. faecium*—EF, *S. aureus*—SA, *C. glabrata*—CG, *C. albicans*—CA, *C. krusei*—CK, *C. tropicalis*—CT. ATB—positive control (Cefoxitin for G^-, Gentamicin for G^+, Fluconazole for yeast).

Data from the inhibitory effects of the analysed EOs against three tested *Penicillium* (*P.*) spp. fungi (*P. crustosum, P. citrinum, P. expansum*) are shown in Table 4. Our results revealed that the growth inhibition of fungi strains depends on the type and concentration of the EO used. Remarkable antifungal activity was observed for the MEO among all investigated EOs, which completely inhibited the growth of *P. crustosum* and *P. expansum* in all used concentrations (125, 250, and 500 μL/L). The growth of *P. citrinum* was also totally inhibited by MEO in a concentration of 125 μL/L, whereas it showed significantly ($p < 0.05$) different zones of inhibition (6.67 ± 0.58 mm; 9.00 ± 1.00 mm, respectively) in the 250 and 500 μL/L concentrations. On the other hand, LEO (125 and 250 μL/L) and REO (in all concentrations) displayed no inhibitory impact on the growth of *P. crustosum*, and *P. citrinum* and *P. expansum* were significantly ($p < 0.05$) inhibited by the EOs in the highest concentrations.

Table 4. Antifungal activity of EOs (inhibition zone in mm).

EOs	*P. crustosum*			*P. citrinum*			*P. expansum*		
Conc.	125 (μL/L)	250 (μL/L)	500 (μL/L)	125 (μL/L)	250 (μL/L)	500 (μL/L)	125 (μL/L)	250 (μL/L)	500 (μL/L)
LEO	0.00 ± 0.00^{aA}	0.00 ± 0.00^{aA}	2.67 ± 0.58^{aB}	2.67 ± 0.58^{aA}	3.33 ± 0.58^{aAB}	4.00 ± 1.00^{aB}	2.33 ± 0.58^{aAB}	2.67 ± 0.58^{aBC}	4.00 ± 1.00^{aC}
MEO	N^{bA}	N^{bA}	N^{bA}	N^{bA}	6.67 ± 0.58^{bB}	9.00 ± 1.00^{bC}	N^{bA}	N^{bA}	N^{bA}
REO	0.00 ± 0.00^{aB}	0.00 ± 0.00^{aB}	0.00 ± 0.00^{cB}	1.67 ± 0.58^{aB}	2.33 ± 0.58^{aB}	4.00 ± 1.00^{aC}	3.33 ± 0.58^{aB}	5.67 ± 0.58^{cC}	6.00 ± 1.00^{aC}

Note: Means ± standard deviation. MEO: Mint essential oil; LEO: Lavender essential oil; REO: Rosemary essential oil. Values in the same column with different small letters, and those in the same row with different upper-case letters are significantly different ($p < 0.05$). Conc.—concentration; 0.00—total growth; N—without growth.

2.4. Minimum Inhibitory Concentrations of EOs against Gram-Negative and Gram-Positive Bacteria, and Yeasts

The MIC values of tested EOs against Gram-negative and Gram-positive bacteria and yeasts are represented in Tables 5 and 6. The EOs displayed a variable degree of inhibition activity against the different tested strains, with significant differences ($p < 0.05$) among the analysed EOs, as well as the microorganisms that were used. LEO had the lowest MICs against *C. albicans* regarding the effectiveness of selected EOs, whilst EO was the most effective against *S. enterica*. MEO exhibited the weakest action against the growth of *P. aeruginosa*, and also against *C. glabrata* and *C. albicans*. On the other hand, MEO was the most effective against *S. enterica* and *C. krusei*. Finally, REO had weak antimicrobial activity against *C. albicans* and a stronger effect against *S. enterica*.

2.5. Moisture Content and Water Activity of Bread Samples

The results from the moisture content (MC) and water activity (a_w) measurements showed that the parameters of bread analysed in our study had values of 41.65 ± 0.55% and 0.944 ± 0.001, respectively.

Table 5. Antimicrobial activity of EOs expressed as the minimum inhibitory concentration (MIC) in μL/mL against Gram-negative and Gram-positive bacteria.

EOs	Gram-Negative Bacteria								Gram-Positive Bacteria	
	PA		SE		YE		EF		SA	
	MIC 50 (μL/mL)	MIC90 (μL/mL)	MIC 50 (μL/mL)	MIC90 (μL/mL)	MIC 50 (μL/mL)	MIC90 (μL/mL)	MIC 50 (μL/mL)	MIC90 (μL/mL)	MIC 50 (μL/mL)	MIC90 (μL/mL)
LEO	232.15 ± 0.58 ^{aA}	288.41 ± 0.23 ^{aB}	94.26 ± 0.19 ^{aC}	115.15 ± 0.96 ^{aD}	243.11 ± 1.01 ^{aE}	391.10 ± 0.23 ^{aF}	134.18 ± 0.22 ^{aG}	255.21 ± 0.98 ^{aH}	205.88 ± 0.14 ^{aI}	286.99 ± 1.05 ^{aB}
MEO	2128.30 ± 0.41 ^{bA}	598.41 ± 0.74 ^{bB}	5.72 ± 0.44 ^{bC}	4.12 ± 0.15 ^{bD}	297.96 ± 0.17 ^{bE}	255.95 ± 0.07 ^{bF}	270.68 ± 0.81 ^{bG}	513.86 ± 0.69 ^{bH}	19.42 ± 0.62 ^{bI}	7.33 ± 0.46 ^{bJ}
REO	134.51 ± 0.19 ^{cA}	155.18 ± 0.09 ^{cB}	93.58 ± 0.25 ^{cC}	98.75 ± 0.11 ^{cD}	255.95 ± 0.65 c^{E}	299.76 ± 0.35 ^{cF}	270.68 ± 0.73 ^{bG}	313.86 ± 0.05 ^{cH}	198.58 ± 0.66 ^{cI}	331.18 ± 0.41 ^{cJ}

Note: Means ± standard deviation. MEO: Mint essential oil; LEO: Lavender essential oil; REO: Rosemary essential oil; *P. aeruginosa*—PA, *S. enterica*—SE, *Y. enterocolitica*—YE, *E. faecium*—EF, *S. aureus*—SA. Values in the same column with different small letters, and those in the same row with different upper-case letters are significantly different ($p < 0.05$).

Table 6. Antimicrobial activity of EOs expressed as the minimum inhibitory concentration (MIC) in μL/mL against yeasts.

EOs	Yeasts							
	CG		CA		CK		CT	
	MIC 50 (μL/mL)	MIC90 (μL/mL)	MIC 50 (μL/mL)	MIC90 (μL/mL)	MIC 50 (μL/mL)	MIC90 (μL/mL)	MIC 50 (μL/mL)	MIC90 (μL/mL)
LEO	179.61 ± 0.23 ^{aA}	241.63 ± 0.11 ^{aB}	432.40 ± 0.38 ^{aC}	724.99 ± 0.77 ^{aD}	121.35 ± 0.17 ^{aE}	226.40 ± 0.14 ^{aF}	144.25 ± 0.49 ^{aG}	191.35 ± 0.46 ^{aH}
MEO	562.30 ± 0.92 ^{bA}	944.85 ± 0.55 ^{bB}	459.91 ± 0.73 ^{bC}	644.58 ± 0.54 ^{bD}	5.50 ± 0.12 ^{bE}	8.60 ± 0.07 ^{bF}	432.40 ± 0.88 ^{bG}	139.81 ± 0.32 ^{bH}
REO	121.86 ± 0.47 ^{cA}	151.83 ± 0.67 ^{cB}	459.91 ± 0.56 ^{bC}	644.51 ± 0.33 ^{bD}	120.38 ± 0.64 ^{aE}	296.18 ± 0.09 ^{cF}	136.58 ± 0.76 ^{cG}	185.45 ± 0.82 ^{cH}

Note: Means ± standard deviation. *Candida glabrata*—CG, *Candida albicans*—CA, *Candida krusei*—CK, *Candida tropicalis*—CT. Values in the same column with different small letters, and those in the same row with different upper-case letters are significantly different ($p < 0.05$).

2.6. In Situ Antifungal Analysis on Bread

The antifungal properties of the analysed EOs on bread are presented in Table 7 and Figure 1. The results of the analysis revealed that LEO had the significantly ($p < 0.05$) highest inhibition against *P. crustosum* in all used concentrations (125, 250, and 500 μL/L) in a very slightly increasing manner (81.18 ± 2.78%, 85.88 ± 1.95%, 88.64 ± 2.74%, respectively), and against *P. citrinum* and *P. expansum* in the highest concentrations (89.38 ± 2.05%, 86.12 ± 3.04%, respectively). The strongest significant ($p < 0.05$) antifungal activity of MEO was observed against *P. crustosum* in concentrations of 125 and 500 μL/L (87.91 ± 1.06% and 90.19 ± 2.99%, respectively), and against *P. citrinum* and *P. expansum* in the lowest and the highest concentrations, respectively. Interestingly, REO exhibited ($p < 0.05$) the highest activity in different concentrations for individual fungal species: *P. crustosum* in 250 μL/L (92.48 ± 1.69%), *P. citrinum* at 500 μL/L (57.36 ± 2.63%), and *P. expansum* in 125 μL/L (86.48 ± 3.55%).

Table 7. Mycelial growth inhibition of the analysed EOs.

Fungi Strains	MGI (%)								
	LEO (μL/L)			MEO (μL/L)			REO (μL/L)		
	125	250	500	125	250	500	125	250	500
P. crustosum	81.18 ± 2.78 ^{aA}	85.88 ± 1.95 ^{aA}	88.64 ± 2.74 ^{aA}	87.91 ± 1.06 ^{aA}	77.65 ± 1.32 ^{aB}	90.19 ± 2.99 ^{aA}	73.73 ± 0.99 ^{aA}	92.48 ± 1.69 ^{aB}	−4.71 ± 2.18 ^{aC}
P. citrinum	61.12 ± 2.59 ^{bA}	70.49 ± 1.96 ^{bB}	89.38 ± 2.05 ^{aC}	42.54 ± 3.11 ^{bA}	18.30 ± 3.02 ^{bB}	23.03 ± 1.01 ^{bC}	14.03 ± 3.37 ^{bA}	39.02 ± 4.02 ^{bB}	57.36 ± 2.63 ^{bC}
P. expansum	63.45 ± 3.08 ^{bA}	77.62 ± 1.33 ^{cB}	86.12 ± 3.04 ^{aC}	62.68 ± 1.66 ^{cA}	67.05 ± 2.84 ^{cA}	82.07 ± 1.65 ^{cB}	86.48 ± 3.55 ^{cA}	36.05 ± 1.73 ^{bB}	41.10 ± 1.77 ^{cC}

Note: Means ± standard deviation. MEO: Mint essential oil; LEO: Lavender essential oil; REO: Rosemary essential oil. Values in the same column with different small letters, and those in the same row with different upper-case letters are different ($p < 0.05$).

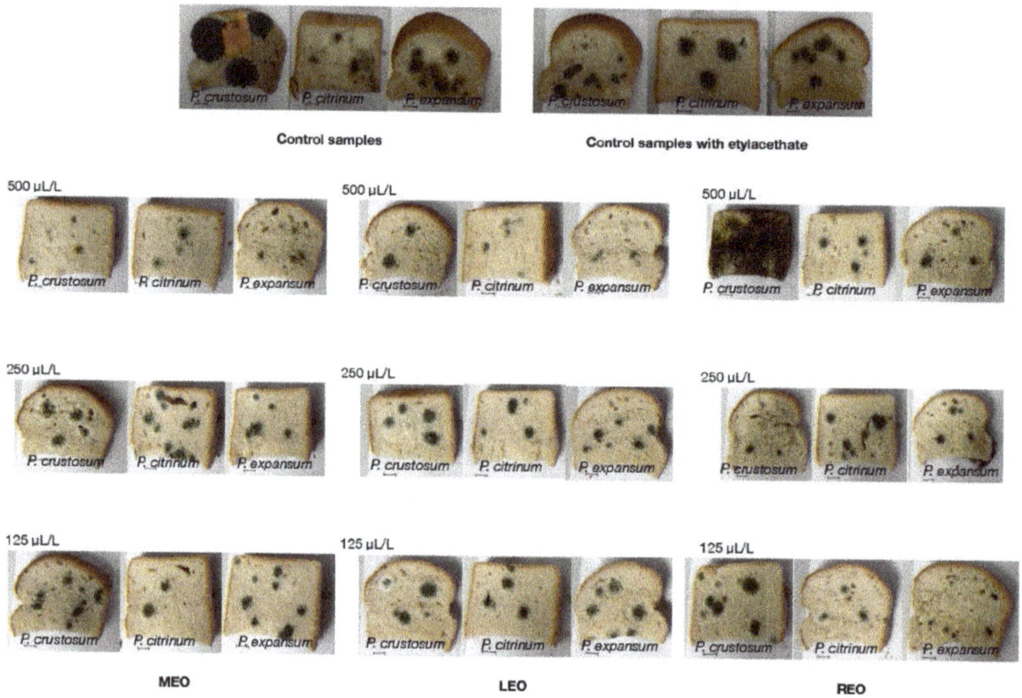

Figure 1. In situ analysis of antifungal activities of the EOs (MEO: Mint essential oil; LEO: Lavender essential oil; REO: Rosemary essential oil).

3. Discussion

It is generally known that the antibacterial effects of EOs depend on their chemical composition [22], which can be influenced by various factors, such as the plant developmental state, the plant part used for extraction, plant geographical location, and physical and chemical characteristics of the soil and climate in question [23].

According to ISO [24], the EOs obtained from *L. angustifolia* are mainly composed of linalool acetate (25.0–47.0%) and linalool (20.0–45.0%), which is in line with our study (35.0% and 32.7%, respectively). Similarly, the research by Zheljazkov et al. [25] and Baydar and Kineci [26] showed that linalool (23.3–43.4%; 34.0%) and linalool acetate (20.2–39.6%; 47.7%) were the main components of the EO from *L. angustifolia* Mill. and *L. x intermedia Emeric ex Loisel*, respectively. Other components in lavender EOs [24] are present in smaller quantities, such as E-β-ocimene (0.0–10.0%), Z-β-ocimene (0.0–6.0%), 4-terpineol (0–8.0%), lavandulyl acetate (0.0–8.0%), 3-octanone (0.0–5.0%), lavandulol (0.0–3.0%), α-terpineol (0.0–2.0%), β-phellandrene (0.0–1.0%), and limonene (0.0–1.0%), which is also in agreement with our results (0.9%, 0.0%, 1.3%, 2.0%, 0.3%, 0.0%, 8.1%, 1.3%, 0.0%, 0.9%, respectively). On the other hand, the concentrations of 1,8-cineole (8.1%) and camphor (6.4%) were higher in our LEO as compared to those (0.0–3.0% and 0.0–1.5%, respectively) reported by ISO [24]. The higher content of both components in the EO can not only significantly affect its aroma intensity [27], but the higher abundance of 1,8-cineole can also be associated with stronger antifungal properties [28].

Mentha EOs consist mainly of oxygenated monoterpenes as a major fraction [29]. Soković et al. [30] determined in the EO from *Mentha piperita*, that the most abundant substances were menthol (37.4%), menthyl acetate (17.4%), and menthone (12.7%). The chemical composition of the EO from *M. piperita* during two seasons (summer and winter) was analysed in the study by Hussain et al. [29]. The authors found that the main components of the EOs collected during summer and winter were menthone (28.13% and 25.54%), menthyl acetate (9.51% and 9.68%), limonene (7.58% and 7.73%), and isomenthone (4.04% and 7.63%), respectively. However, the content of the compounds (menthone 16.8%, menthyl acetate 9.1%, menthol 40.1%, limonene 1.8%, and isomenthone 2.8%) was different in our study, indicating that the aforementioned factors might contribute to the discrepancies in the chemical composition of the MEO analysed in the three studies. We assume that a lower content of methyl acetate in our MEO as compared to that by Soković et al. [30] may be connected with its higher antifungal activity since it was found that this compound causes a decrease in the antifungal properties of the EOs [28].

Regarding the REO chemical composition, our findings are confirmed by the many other studies [31–33] in which 1,8-cineole, camphor, and α-pinene were reported to be the major components of *R. officinalis* EOs. However, the percentages of the individual components (43.99%, 26.54%, and 37.6% and 47.2% for 1,8-cineole; 12.41%, 12.88%, and 7.1% and 13.3% for camphor; 10.09%, 20.14%, and 7.0% and 19.4% for α-pinene) of the rosemary EOs were different compared to our REO (40.4%, 8.7%, and 11.9%, respectively). Similarly, Elamrani et al. [34] indicated that the major compounds of *R. eriocalix* oil were 1,8-cineole (54.6%), camphor (8.6%), and β-pinene (6.8%). The commercially known EO from *R. officinalis* is characterized by the presence of 1,8-cineole (19.59%), camphor (18.35%), α-pinene (17.17%), camphene (10.10%), β-pinene (6.08%), and α-limonene (3.90%) [35], which is dissimilar in comparison with our findings (40.4%, 11.9%, 8.7%, 3.5%, 6.9%, and 2.4%, respectively).

As mentioned above, EOs are especially known for their variable range of biological functions, including an antioxidant purpose [36]. The antioxidant ability is dependent on compounds that protect the biological system against the deleterious influences of processes causing excessive oxidation exponentiation of reactive oxygen forms [37]. The DPPH, i.e., stable free radical, is a compound often used in methods for determining antioxidants' free radical scavenging activities [38].

The antioxidant activity of EOs may vary depending on their chemical composition [39]. Our results revealed that the effect of the MEO was stronger as compared to other analysed EOs. This fact may be related to the presence of individual volatile compounds, especially menthol and menthone, containing the hydroxyl radical (-OH), which improves the antioxidant activity strength [40]. We assume based on the data obtained from recent studies that other minor components in MEO, including 1,8-cineole, carvone, and γ-terpinene, could also increase the variable [41,42].

The antimicrobial properties of various plant EOs have been recognized since ancient times [43], and currently, the scientific community is increasingly focused on the evaluation of their capacity to inhibit the growth of diverse foodborne pathogens. Indeed, various studies showed antibacterial properties of many EOs against a wide range of bacterial strains (such as *Stenotrophomonas maltophilia, Bacillus subtilis, Y. enterocolitica, S. enterica* subs. *enterica, Bacillus cereus, S. aureus* subs. *aureus*), yeasts (*Candida albicans, C. kruseii, C. tropicalis*) [20,21,44,45], and fungal strains, including *Penicillium* spp. (*P. citrinum, P. crustosum, P. expansum, P. brevicompactum, P. funiculosum, P. glabrum, P. chrysogenum, P. oxalicum, P. polonicum*) [46,47].

The antimicrobial activity of *L. officinalis* EO (10 μg/disk) against *P. aeruginosa* was also evaluated in the study by Gavanji et al. [48]. However, the authors found that the zone of inhibition of their EO was 7.83 ± 0.03 mm, and *P. aeruginosa* proved to be more resistant toward a broad range of *L. officinalis* EO concentrations (0.08–100 μg/disk) as compared to *S. aureus*, which is inconsistent with our findings. Indeed, the LEO used in our study possessed a better inhibitory effect on the growth of *P. aeruginosa*, whilst the antibacterial activity against *S. aureus* was weak. This discrepancy between the two studies could be associated with the different chemical compositions of both lavender EOs employed. A particularity of *P. aeruginosa* is its high intrinsic resistance to antiseptics and antibiotics, which is partly caused by its low permeability of the outer membrane [49]. However, the study by Trombetta et al. [50] suggests that the antimicrobial effect of EO components, such as linalool acetate, may (at least partially) result from a perturbation of the lipid fraction of bacterial plasma membranes, thereby leading to alterations of membrane permeability and leakage of intracellular materials. In effect, the amount of linalool acetate was, in our LEO, quantified as a high content, whilst in the EO from *L. officinalis* used in the research by Gavanji et al. [48], it was completely absent. The hypothesis is also supported by the research of Hanamanthagouda et al. [51], in which EOs from dried leaves of *L. bipinnata* containing 3.37% of linalyl acetate exhibited low activity against *P. aeruginosa* (inhibition zone: 7 mm).

REO exhibited the strongest antibacterial activity against *S. aureus* in our research. Gomes Neto et al. [52], in line with this finding, reported significant inhibition of *S. aureus* viability and growth in meat broth induced by the effects of *R. officinalis* EO and by its majority compound 1,8-cineole itself, which was also quantified in our REO in the highest amount. The inhibitory and bactericidal efficiencies of *R. officinalis* EO, with the main component being 1,8-cineole (23.56%), against *S. aureus* were also reported by Jardak et al. [53].

Generally, MIC is a parameter that is often used for the measurement of Eos' antimicrobial activity, expressing the lowest concentration of the compound able to inhibit the growth of the analysed microorganisms [54].

Our results indicate that even antimicrobial highly resistant isolates, including *S. aureus* [55], *C. albicans* [56], and *E. faecium* [57], showed sensitivity to lavender, mint, and rosemary EOs, predicting their potential usage as promising detergents with the ability to inhibit the growth of a wide range of microorganisms. The differences in the susceptibility of the analysed bacteria and yeasts to the test EOs can be linked to variation in the rate of samples' penetration through the cell wall and cell membrane structures [58]. In addition to the EOs used, their concentration, and the type of microorganism tested, the differences in MIC values may be influenced by the cell size, cell damage, and EO oxidation [59].

EOs are known for their hydrophobicity through which they are capable of interacting with the fungal plasma membrane, leading to disruption of the membrane structures (leakage of some cellular components) or to alterations of the membrane permeability, reflecting their antifungal effects [60].

Many studies have shown strong antifungal activity of diverse plant EOs with a wide inhibition spectrum, pointing out their high potential as innovative preservative agents to replace synthetic fungicides [61]. Among other factors, the variations in the fungicidal activity of these aromatic compounds can be related to their active compounds, such as phenols, aldehydes, and ketones [62], which is consistent with the GC-MS analysis carried out in our study.

The results from MC and measurements showed that the parameters of bread analysed in our study had values of 41.65 $\pm$ 0.55% and 0.944 $\pm$ 0.001, respectively. Bread is considered as an intermediate moisture food, with MC typically ranging from 35–42%, and a_W above 0.95, which is consistent with our findings. Therefore, baked goods, including bread, are susceptible to microbial spoilage with high growth of various fungi strains [63], thereby offering its application as a suitable type of substrate in such experiments.

Generally, antimicrobial agents are applied in food products for two main reasons: (i) to control natural spoilage processes (preservation of food), and (ii) to prevent or control the growth of microorganisms (food safety) [43]. Our study was focused primarily on an evaluation of the antifungal effects of EOs on bread as a model substrate for fungal growth (in situ conditions) since fungal spoilage occurs more often than bacterial spoilage [64]. Vapour diffusion exposure was applied based on the fact that most of the antimicrobial activity of EOs is attributed to volatile compounds [65].

The EOs' antifungal activity upon solution contact (broth dilution and agar dilution methods) has been studied by many researchers. However, the activity by vapour-phase contact has been reported more rarely [66,67]. Different types of fungi, including *Penicillium* spp., are responsible for bread spoilage [68]. Although *P. expansum* is mainly associated with the degradation of apples, it was used in our study for its higher resistance than other species of the *Penicillium* strains [69]. Despite the high resistance, the antifungal effectiveness of all EOs tested against *P. expansum*, ranging from 36.05 ± 1.73% (250 μL/L of REO) to 86.48 ± 3.55% (125 μL/L of REO), was reported in the current research. Therefore, we assume that the EOs used can also be effective against other resistant species of microorganisms. Interestingly, *P. citrinum* was the most sensitive to the lowest concentration (125 μL/L) of MEO. We propose that the finding can be associated with the lower concentration of methyl acetate as compared to the MEO higher concentrations used as the chemical compound decreases the antifungal activity of MEO [28]. The results are in accordance with our previous studies, in which the antifungal effects of other EOs, such as coriander EO [20] or *Citrus aurantium* EO [21], against the same fungi species analysed (*P. citrinum, P. expansum, P. crustosum*) were confirmed.

4. Materials and Methods

4.1. Essential Oils

The following EOs were applied in this study: Lavender (LEO; *Lavandula angustifolia x latifolia*), mint (MEO; *Mentha x piperita* L.), and rosemary (REO; *Rosmarinus officinalis*). All essential oils were purchased from Hanus s.r.o. (Nitra, Slovakia) to complement our previous results [20,21] from such experiments, thus creating a comprehensive view of the biological actions of various commercially available EOs obtained from the same company.

4.2. Fungal Strains

Three *Penicillium* (*P.*) strains (*P. crustosum, P. citrinum, P. expansum*) were isolated from berry samples of *Vitis vinifera* and consequently classified using a reference-based MALDI-TOF MS Biotyper. The obtained results were also validated by comparison with the taxonomic identification obtained by 16S rDNA sequences analysis.

4.3. Microbial Strains

Three Gram-negative bacteria: *P. aeruginosa* CCM 1959, *S. enterica* subsp. *enterica* CCM 3807, *Y. enterocolitica* CCM 5671; two Gram-positive bacteria: *E. faecalis* CCM 4224, *S. aureus* subsp. *aureus* CCM 4223; and four yeasts: *C. glabrata* CCM 8270, *C. albicans* CCM 8186, *C. krusei* CCM 8271, and *C. tropicalis* CCM 8223 were used to evaluate the antimicrobial activities of the EOs. The microorganisms were obtained from the Czech Collection of Microorganisms (Brno, Czech Republic).

4.4. Evaluation of Antioxidant Activity of the EOs

The antioxidant activity of the three analysed EOs was assessed on the basis of the scavenging activity of the stable radicals 2,2-diphenyl-1-picrylhydrazyl (DPPH) according to the methodology used in the studies [20,21].

4.5. Chemical Characterization of EO Samples by Gas Chromatography/Mass Spectrometry (GC/MS) and Gas Chromatography (GC-FID)

Gas chromatography/mass spectrometry analyses of the selected EO samples were performed using an Agilent 6890N gas chromatograph (Agilent Technologies, Santa Clara, CA, USA) coupled to a quadrupole mass spectrometer 5975B (Agilent Technologies, Santa Clara, CA, USA). A HP-5MS capillary column (30 m × 0.25 mm × 0.25 m) was used. The temperature program was as follows: 60 °C to 150 °C (increasing rate 3 °C/min) and 150 °C to 280 °C (increasing rate 5 °C/min). The total run time was 60 min. Helium 5.0 was used as the carrier gas with a flow rate of 1 mL/min. The injection volume was 1 L (EO samples were diluted in pentane), while the split/splitless injector temperature was set at 280 °C. The investigated samples were injected in the split mode with a split ratio at 40.8:1. Electron-impact mass spectrometric data (EI-MS; 70 eV) were acquired in scan mode over the m/z range 35–550. The mass spectrometry ion source and MS quadrupole temperatures were 230 °C and 150 °C, respectively. Acquisition of data started after a solvent delay time of 3 min. Gas chromatography (GC-FID) analyses were performed on an Agilent 6890N gas chromatograph coupled to an FID detector. Column (HP-5MS) and chromatographic conditions were the same as for GC-MS. The FID detector temperature was set at 300 °C.

The individual volatile constituents of injected EO samples were identified based on their retention indices [69], and a comparison with reference spectra (Wiley and NIST databases). The retention indices were experimentally determined using the standard method [70], which included retention times of *n*-alkanes (C6–C34), injected under the same chromatographic conditions. The percentages of the identified compounds (amounts higher than 0.1%) were derived from their GC peak areas.

4.6. Evaluation of Antimicrobial Activity of the EOs

The evaluation of the antimicrobial activity of the EOs was performed using the agar disc diffusion method. For this purpose, there was an aliquot of 0.1 mL of fungal and bacterial suspension in Mueller Hinton Broth (MHB; Merck, Gernsheim, Germany) inoculated to Mueller Hinton Agar (MHA; Merck, Germany; 60 mm). Subsequently, the discs of filter paper (6 mm) were impregnated with 10 μL of the analysed EO samples and then applied on the MHA surface. Inoculated MHA plates were kept at 4 °C for 2 h and incubated aerobically at 37 °C for 24 h (bacteria). In the case of fungi, 10 μL of the analysed oils were applied at three concentrations (125, 250, and 500 μL/L, diluted in 0.1% dimethyl sulfoxide (DMSO)), and incubated at 25 °C for 5 days. Two antibiotics (Cefoxitin, Gentamicin) and one antifungal (Fluconazole) were used as positive controls for Gram-negative and Gram-positive bacteria and yeasts, respectively. Disks impregnated with ethanol served as negative controls.

The diameters of the inhibition zones were measured in mm after incubation. Each test was repeated three times (one repeat reflecting one separate plate).

4.7. Determination of Minimum Inhibitory Concentration

The minimum inhibitory concentration (MIC) was detected according to the National Committee for Clinical Laboratory Standards (NCCLS) as it was recently described by Kačániová et al. [20,21]. Chloramphenicol and nystatin, and DMSO served as positive and negative controls, respectively. MIC was detected at 570 nm with a spectrophotometer (Promega Inc., Madison, WI, USA).

4.8. Bread Preparation

Wheat bread used for analyses was baked in the Laboratory of Cereal technologies (AgroBioTech Research Centre) according to the methodology described by Kačaniova et al. [20,21].

4.9. Moisture Content and Water Activity of Bread

The moisture content (MC) and water activity (a_w) of bread were measured using the Lab Master a_w Standard (Novasina AG; Lachen, Switzerland) and the Kern DBS 60-3 moisture analyzer (Kern and Sohn GmbH, Balingen, Germany), respectively, after the bread cooling [20,21].

4.10. In Situ Antifungal Analyses on Bread Model

First, the bread samples were cut into slices with a 150 mm height and placed into 0.5 L sterile glass jars (Bormioli Rocco, Fidenza, Italy). A fungal spore suspension of each strain (in final concentration of 1×10^6 spores/mL) was diluted in 20 mL of sterile phosphate-buffered saline with 0.5% Tween 80 by adjusting the density to 1–1.2 McFarland; 5 μL of inoculum were used for bread inoculation. The EOs in concentrations of 125, 250, and 500 μL/L (EOs + ethyl acetate) were evenly distributed in a volume of 100 μL on a sterile paper–filter disc (6 cm), which was inserted into the cover of the jar, except for the treatment of the control group. The jars were hermetically closed and kept at 25 °C ± 1 °C for 14 days in the dark. The size of the microfungal colonies with visible mycelial growth and visible sporulation [20,21] was evaluated using stereological methods. In this concept, the volume density of the colonies was firstly assessed using ImageJ software (National Institutes of Health, Bethesda, MD, USA), counting the points of the stereological grid hitting the colonies and those falling to the reference space (growth substrate used, i.e., bread). The antifungal activities of the EOs were expressed as the percentage of mycelial growth inhibition (MGI), which was calculated using the formula: MGI = [(C − T)/C] × 100 [71], where C = volume density of the fungal colony in the control group and T = volume density of that in the treatment group.

4.11. Statistical Analysis

The data from the analyses was statistically evaluated using Prism 8.0.1 (GraphPad Software, San Diego, CA, USA). One-way analysis of variance (ANOVA) followed by Tukey's test was used to evaluate the statistical significance of differences between the analysed groups of samples. The level of significance was set at $p < 0.05$. MIC50 and MIC90 values (i.e., concentration causing 50% and 90% reduction of microbial growth) were estimated by the logit analysis.

5. Conclusions

The current study evaluated the chemical composition, antioxidant, antibacterial, and antifungal activities of rosemary, lavender, and mint EOs (125, 250, and 500 μL/L concentrations) against selected microorganisms. Our results revealed that MEO possessed the highest DPPH radical-scavenging activity, which was even significantly ($p < 0.05$) different from that of LEO and REO. Considering the antimicrobial activity, the EOs exhibited the strongest inhibitory efficiencies against the growth of *P. aeruginosa* and *C. tropicalis* (LEO), *S. enterica* (MEO), and *S. aureus* (REO). From the fungi strains, MEO (in all concentrations) was able to totally inhibit the growth of *P. expansum* and *P. crustosum*, whilst the growth of *P. citrinum* was completely suppressed only by its lowest concentration. From in situ analysis, REO (125 μL/L) exhibited the lowest MGI of *P. citrinum*, and 500 μL/L of MEO caused the highest MGI of *P. crustosum*. Our results suggest the analysed EOs have promising potential as innovative agents for use in the storage of bakery products to extend their shelf-life. Thus, their combination with other preservatives or modified atmosphere packaging could be a valuable alternative in the food industry. Moreover, our results complement our previous studies, thus creating a comprehensive view of the biological activities of various commercially available EOs obtained from the same company, Hanus s.r.o. (Nitra, Slovakia).

Author Contributions: Conceptualization, V.V., H.Ď. and M.K.; methodology, V.V., H.Ď. and M.K.; software, V.V.; validation, V.V., H.Ď. and M.K.; formal analysis, V.V.; investigation, V.V., H.Ď., L.G., N.L.V., M.V. and M.K.; resources, M.K.; data curation, M.K.; writing—original draft preparation, V.V., H.Ď. and M.K.; writing—review and editing, V.V., H.Ď. and M.K.; visualization, V.V.; supervision, M.K.; project administration, M.K.; funding acquisition, M.K. All authors have read and agreed to the published version of the manuscript.

Funding: This work was supported by the grant APVV SK-BY-RD-19-0014 "The formulation of novel compositions and properties study of the polysaccharides based edible films and coatings with antimicrobial and antioxidant plant additives".

Institutional Review Board Statement: Not applicable.

Informed Consent Statement: Not applicable.

Data Availability Statement: Data is contained within the article.

Acknowledgments: This work has been supported by the grants of the VEGA no. 1/0180/20.

Conflicts of Interest: The authors declare no conflict of interest.

Sample Availability: Essential oils samples are available from the authors.

References

1. Ju, J.; Xu, X.; Xie, Y.; Guo, Y.; Cheng, Y.; Qian, H.; Yao, W. Inhibitory effects of cinnamon and clove essential oils on mold growth on baked foods. *Food Chem.* **2018**, *240*, 850–855. [CrossRef]
2. Ju, J.; Xie, Y.; Yu, H.; Guo, Y.; Cheng, Y.; Zhang, R.; Yao, W. Synergistic inhibition effect of citral and eugenol against *Aspergillus niger* and their application in bread preservation. *Food Chem.* **2020**, *310*, 125974. [CrossRef]
3. Prakash, B.; Kedia, A.; Mishra, P.K.; Dubey, N.K. Plant essential oils as food preservatives to control moulds, mycotoxin contamination and oxidative deterioration of agri-food commodities–Potentials and challenges. *Food Control* **2015**, *47*, 381–391. [CrossRef]
4. Mérillon, J.M.; Riviere, C. *Natural Antimicrobial Agents*, 1st ed.; Springer: Cham, Switzerland, 2018.
5. Hanif, M.A.; Nisar, S.; Khan, G.S.; Mushtaq, Z.; Zubair, M. Essential oils. In *Essential Oil Research*; Springer: Cham, Switzerland, 2019; pp. 3–17.
6. Scheffer, J.J.C. The isolation of essential oils-factors influencing the oil composition. *Acta Hortic.* **1993**, *344*, 2–8. [CrossRef]
7. Mimica-Dukić, N.; Božin, B.; Soković, M.; Mihajlović, B.; Matavulj, M. Antimicrobial and antioxidant activities of three *Mentha* species essential oils. *Planta Med.* **2003**, *69*, 413–419.
8. Astani, A.; Reichling, J.; Schnitzler, P. Screening for antiviral activities of isolated compounds from essential oils. *Evid. Based Complementary Altern. Med.* **2011**. [CrossRef]
9. Leal, S.M.; Pino, N.; Stashenko, E.E.; Martínez, J.R.; Escobar, P. Antiprotozoal activity of essential oils derived from *Piper* spp. grown in Colombia. *J. Essent. Oil Res.* **2013**, *25*, 512–519. [CrossRef]
10. Kalemba, D.A.A.K.; Kunicka, A. Antibacterial and antifungal properties of essential oils. *Curr. Med. Chem.* **2003**, *10*, 813–829. [CrossRef]
11. Kumar, A.; Kudachikar, V.B. Antifungal properties of essential oils against anthracnose disease: A critical appraisal. *JPDP* **2018**, *125*, 133–144. [CrossRef]
12. Turek, C.; Stintzing, F.C. Stability of essential oils: A review. *Com. Rev. Food Sci. Food Saf.* **2013**, *12*, 40–53. [CrossRef]
13. Chrysargyris, A.; Mikallou, M.; Petropoulos, S.; Tzortzakis, N. Profiling of essential oils components and polyphenols for their antioxidant activity of medicinal and aromatic plants grown in different environmental conditions. *Agronomy* **2020**, *10*, 727. [CrossRef]
14. Satyavani, K.; Gurudeeban, S.; Manigandan, V.; Rajamanickam, E.; Ramanathan, T. Chemical compositions of medicinal mangrove species *Acanthus ilicifolius, Excoecaria agallocha, Rhizophora apiculata* and *Rhizophora mucronata. Curr. Res. Chem.* **2015**, *7*, 1–8. [CrossRef]
15. Oliveira Monteschio, J.; Souza, K.A.; Vital, A.C.P.; Guerrero, A.; Valero, M.V.; Kempinski, E.M.B.C.; Prado, I.N. Clove and rosemary essential oils and encapsuled active principles (eugenol, thymol and vanillin blend) on meat quality of feedlot-finished heifers. *Meat Sci.* **2017**, *130*, 50–57. [CrossRef]
16. Gao, M.; Feng, L.; Jiang, T.; Zhu, J.; Fu, L.; Yuan, D.; Li, J. The use of rosemary extract in combination with nisin to extend the shelf life of pompano (*Trachinotus ovatus*) fillet during chilled storage. *Food Control* **2014**, *37*, 1–8. [CrossRef]
17. Yang, J.; Gao, F.L. Study on antioxidant activities of total flavonoids from *Lavandula angustifolia* of Xinjiang. *China Food Addit* **2010**, *2*, 162–165.
18. Cavanagh, H.M.; Wilkinson, J.M. Lavender essential oil: A review. *Aust. Infect. Control.* **2005**, *10*, 35–37. [CrossRef]
19. Samber, N.; Khan, A.; Varma, A.; Manzoor, N. Synergistic anti-candidal activity and mode of action of *Mentha piperita* essential oil and its major components. *Pharm. Biol.* **2015**, *53*, 1496–1504. [CrossRef]
20. Kačániová, M.; Galovičová, L.; Ivanišová, E.; Vukovic, N.L.; Štefániková, J.; Valková, V.; Borotová, P.; Žiarovská, J.; Terentjeva, M.; Felšöciová, S.; et al. Antioxidant, antimicrobial and antibiofilm activity of coriander (*Coriandrum sativum* L.) essential oil for its application in foods. *Foods* **2020**, *9*, 282. [CrossRef] [PubMed]
21. Kačániová, M.; Terentjeva, M.; Galovičová, L.; Ivanišová, E.; Štefániková, J.; Valková, V.; Borotová, P.; Kowalczewski, P.L.; Kunová, S.; Felšöciová, S.; et al. Biological activity and antibiofilm molecular profile of *Citrus aurantium* essential oil and its application in a food model. *Molecules* **2020**, *25*, 3956. [CrossRef] [PubMed]
22. Jianu, C.; Pop, G.; TGruia, A.; Horhat, F.G. Chemical composition and antimicrobial activity of essential oils of lavender (*Lavandula angustifolia*) and lavandin (*Lavandula x intermedia*) grown in Western Romania. *Int. J. Agric. Biol.* **2013**, *15*, 772–776.
23. Gende, L.B.; Maggi, M.; Van Baren, C.; Lira, A.D.L.; Bandoni, A.; Fritz, R.; Eguaras, M. Antimicrobial and miticide activities of *Eucalyptus globulus* essential oils obtained from different Argentine regions. *Span. J. Agric. Res.* **2010**, 642–650. [CrossRef]
24. ISO (International Standards Organisation). *Oil of Lavender (Lavandula angustifolia Mill.)*, 3rd ed.; ISO 3515:2002; ISO/TC 54; ISO: Genewa, Switherland, 2002; pp. 1–16.
25. Zheljazkov, V.D.; Astatkie, T.K.; Hristov, A.N. Lavender and hyssop productivity, oil content, and bioactivity as a function of harvest time and drying. *Ind. Crop. Prod.* **2012**, *36*, 222–228. [CrossRef]
26. Baydar, H.; Kineci, S. Scent composition of essential oil, concrete, absolute and hydrosol from lavandin (*Lavandula x intermedia Emeric ex Loisel.*). *J. Essent. Oil Bear. Plants* **2009**, *12*, 131–136. [CrossRef]
27. Wińska, K.; Mączka, W.; Łyczko, J.; Grabarczyk, M.; Czubaszek, A.; Szumny, A. Essential oils as antimicrobial agents—Myth or real alternative? *Molecules* **2019**, *24*, 2130. [CrossRef]
28. Griffin, S.G.; Markham, J.L.; Leach, D.N. An agar dilution method for the determination of the minimum inhibitory concentration of essential oils. *J. Essent. Oil Res.* **2000**, *12*, 249–255. [CrossRef]
29. Hussain, A.I.; Anwar, F.; Nigam, P.S.; Ashraf, M.; Gilani, A.H. Seasonal variation in content, chemical composition and antimicrobial and cytotoxic activities of essential oils from four *Mentha* species. *J. Sci. Food Agric.* **2010**, *90*, 1827–1836. [CrossRef]
30. Soković, M.D.; Vukojević, J.; Marin, P.D.; Brkić, D.D.; Vajs, V.; Van Griensven, L.J. Chemical composition of essential oils of thymus and mentha species and their antifungal activities. *Molecules* **2009**, *14*, 238–249. [CrossRef]

31. Ait-Ouazzou, A.; Lorán, S.; Bakkali, M.; Laglaoui, A.; Rota, C.; Herrera, A.; Conchello, P. Chemical composition and antimicrobial activity of essential oils of *Thymus algeriensis*, *Eucalyptus globulus* and *Rosmarinus officinalis* from Morocco. *J. Sci. Food Agric.* **2011**, *91*, 2643–2651. [CrossRef]
32. Jiang, Y.; Wu, N.; Fu, Y.J.; Wang, W.; Luo, M.; Zhao, C.J.; Liu, X.L. Chemical composition and antimicrobial activity of the essential oil of *Rosemary*. *Environ. Toxicol. Pharmacol.* **2011**, *32*, 63–68. [CrossRef]
33. Moumni, S.; Elaissi, A.; Trabelsi, A.; Merghni, A.; Chraief, I.; Jelassi, B.; Ferchichi, S. Correlation between chemical composition and antibacterial activity of some Lamiaceae species essential oils from Tunisia. *BMC Complementary Med. Ther.* **2020**, *20*, 1–15. [CrossRef]
34. Elamrani, A.; Zrira, S.; Benaissa, M. Isolation of Moroccan Rosmarinus eriocalix oil: Comparison between hydrodistillation and microwave extraction. *J. Essent. Oil Bear. Plants* **2003**, *6*, 1–8. [CrossRef]
35. Chalchat, J.C.; Garry, R.P.; Michet, A.; Benjilali, B.; Chabart, J.L. Essential oils of rosemary (*Rosmarinus officinalis* L.). The chemical composition of oils of various origins (Morocco, Spain, France). *J. Essent. Oil Res.* **1993**, *5*, 613–618. [CrossRef]
36. Benedet, J.A.; Umeda, H.; Shibamoto, T. Antioxidant activity of flavonoids isolated from young green barley leaves toward biological lipid samples. *J. Agric. Food Chem.* **2007**, *55*, 5499–5504. [CrossRef] [PubMed]
37. López-Alarcón, C.; Denicola, A. Evaluating the antioxidant capacity of natural products: A review on chemical and cellular-based assays. *Anal. Chim. Acta* **2013**, *763*, 1–10. [CrossRef] [PubMed]
38. Hussain, A.I.; Anwar, F.; Sherazi, S.T.H.; Przybylski, R. Chemical composition, antioxidant and antimicrobial activities of basil (*Ocimum basilicum*) essential oils depends on seasonal variations. *Food Chem.* **2008**, *108*, 986–995. [CrossRef] [PubMed]
39. Diniz do Nascimento, L.; Moraes, A.A.B.D.; Costa, K.S.D.; Pereira Galúcio, J.M.; Taube, P.S.; Costa, C.M.L.; Faria, L.J.G.D. Bioactive natural compounds and antioxidant activity of essential oils from spice plants: New findings and potential applications. *Biomolecules* **2020**, *10*, 988. [CrossRef]
40. Sun, Z.; Wang, H.; Wang, J.; Zhou, L.; Yang, P. Chemical composition and anti-inflammatory, cytotoxic and antioxidant activities of essential oil from leaves of *Mentha piperita* grown in China. *PLoS ONE* **2014**, *9*, e114767. [CrossRef]
41. Ladan Moghadam, A.R. Antioxidant activity and chemical composition of *Rosmarinus officinalis* L. essential oil from Iran. *J. Essent. Oil Bear. Plants* **2015**, *18*, 1490–1494. [CrossRef]
42. Ruberto, G.; Baratta, M.T. Antioxidant activity of selected essential oil components in two lipid model systems. *Food Chem.* **2000**, *69*, 167–174. [CrossRef]
43. Shaaban, H.A. Essential Oil as Antimicrobial Agents: Efficacy, Stability, and Safety Issues for Food Application. In *Essential Oils-Bioactive Compounds. New Perspectives and Applications*; IntechOpen: London, UK, 2020. [CrossRef]
44. Juricová, V.; Ivanišová, E.; Árvay, J.; Godočíková, L.; Kačániová, M. Technological, antioxidant, antimicrobial and sensory profiles of selected kinds of grape oils. *JMBFS* **2021**, 500–504. [CrossRef]
45. Semerdjieva, I.B.; Zheljazkov, V.D.; Dincheva, I.; Astatkie, T.; Kačániová, M. Chemotypes of Juniperus oxycedrus in Bulgaria and the antimicrobial activity of galbuli essential oils. *Ind. Crops Prod.* **2020**, *158*, 113005. [CrossRef]
46. Felšöciová, S.; Vukovic, N.; Jeżowski, P.; Kačániová, M. Antifungal activity of selected volatile essential oils against *Penicillium* sp. *Open Life Sci.* **2020**, *15*, 511–521. [CrossRef] [PubMed]
47. Kačániová, M.; Galovičová, L.; Valková, V.; Tvrdá, E.; Terentjeva, M.; Žiarovská, J.; Kunová, S.; Savitskaya, P.; Grinshpan, D.; Štefániková, J.; et al. Antimicrobial and antioxidant activities of *Cinnamomum cassia* essential oil and its application in food preservation. *Open Chem.* **2021**, *19*, 214–227. [CrossRef]
48. Gavanji, S.; Mohammadi, E.; Larki, B.; Bakhtari, A. Antimicrobial and cytotoxic evaluation of some herbal essential oils in comparison with common antibiotics in bioassay condition. *Integr. Med. Res.* **2014**, *3*, 142–152. [CrossRef]
49. Chevalier, S.; Bouffartigues, E.; Bodilis, J.; Maillot, O.; Lesouhaitier, O.; Feuilloley, M.G.; Cornelis, P. Structure, function and regulation of Pseudomonas aeruginosa porins. *FEMS Microb. Rev.* **2017**, *41*, 698–722. [CrossRef] [PubMed]
50. Trombetta, D.; Castelli, F.; Sarpietro, M.G.; Venuti, V.; Cristani, M.; Daniele, C.; Bisignano, G. Mechanisms of antibacterial action of three monoterpenes. *Antimicrob. Agents Chemother.* **2005**, *49*, 2474–2478. [CrossRef]
51. Hanamanthagouda, M.S.; Kakkalameli, S.B.; Naik, P.M.; Nagella, P.; Seetharamareddy, H.R.; Murthy, H.N. Essential oils of Lavandula bipinnata and their antimicrobial activities. *Food Chem.* **2010**, *118*, 836–839. [CrossRef]
52. Gomes Neto, N.J.; Luz, I.D.S.; Tavares, A.G.; Honório, V.G.; Magnani, M.; de Souza, E.L. *Rosmarinus officinalis* L. essential oil and its majority compound 1, 8-cineole at sublethal amounts induce no direct and cross protection in *Staphylococcus aureus* ATCC 6538. *Foodborne Pathog. Dis.* **2012**, *9*, 1071–1076. [CrossRef]
53. Jardak, M.; Elloumi-Mseddi, J.; Aifa, S.; Mnif, S. Chemical composition, anti-biofilm activity and potential cytotoxic effect on cancer cells of *Rosmarinus officinalis* L. essential oil from Tunisia. *Lipids Health Dis.* **2017**, *16*, 1–10. [CrossRef]
54. Southwell, I.A.; Hayes, A.J.; Markham, J.; Leach, D.N. The search for optimally bioactive Australian tea tree oil. *Acta Hotic.* **1993**, 256–265. [CrossRef]
55. Elsom, G.K.; Hide, D. Susceptibility of methicillin-resistant Staphylococcus aureus to tea tree oil and mupirocin. *J. Antimicrob. Chemother.* **1999**, *43*, 427–428. [CrossRef]
56. Vazquez, J.A.; Arganoza, M.T.; Boikov, D.; Akins, R.A.; Vaishampayan, J.K. In vitro susceptibilities of *Candida* and *Aspergillus* species to *Melaleuca alternafolia* (tea tree) oil. *Rev. Iberoam Micolgía* **2000**, *17*, 60–63.
57. May, J.; Chan, C.H.; King, A.; Williams, L.; French, G.L. Time-kill studies of tea tree oils on clinical isolates. *J. Antimicrob. Chemother.* **2000**, *45*, 639–643. [CrossRef] [PubMed]

58. Cox, S.D.; Mann, C.M.; Markham, J.L.; Bell, H.C.; Gustafson, J.E.; Warmington, J.R.; Wyllie, S.G. The mode of antimicrobial action of the essential oil of *Melaleuca alternifolia* (tea tree oil). *J. Appl. Microb.* **2000**, *88*, 170–175. [CrossRef] [PubMed]
59. Skandamis, P.; Koutsoumanis, K.; Nychas, G.J.E.; Fasseas, K. Inhibition of oregano essential oil and EDTA on *Escherichia coli* O157: H7. *Ital. J. Food Sci.* **2001**, *13*, 65–75.
60. Sivakumar, D.; Bautista-Baños, S. A review on the use of essential oils for postharvest decay control and maintenance of fruit quality during storage. *Crop Prot.* **2014**, *64*, 27–37. [CrossRef]
61. Tomazoni, E.Z.; Griggio, G.S.; Broilo, E.P.; da Silva Ribeiro, R.T.; Soares, G.L.G.; Schwambach, J. Screening for inhibitory activity of essential oils on fungal tomato pathogen *Stemphylium solani* Weber. *Biocatal. Agric. Biotechnol.* **2018**, *16*, 364–372. [CrossRef]
62. Oussalah, M.; Caillet, S.; Saucier, L.; Lacroix, M. Inhibitory effects of selected plant essential oils on the growth of four pathogenic bacteria: E. coli O157: H7, *Salmonella typhimurium*, *Staphylococcus aureus* and *Listeria monocytogenes*. *Food Control* **2007**, *18*, 414–420. [CrossRef]
63. Day, L. Cereal Food Production with Low Salt. In *Reference Module in Food Science*; Elsevier: Amsterdam, The Netherlands, 2016. [CrossRef]
64. Leuschner, R.G.K.; O'Callaghan, M.J.A.; Arendt, E.K. Moisture distribution and microbial quality of part baked breads as related to storage and rebaking conditions. *J. Food Sci.* **1999**, *64*, 543–546. [CrossRef]
65. Becerril, R.; Gómez-Lus, R.; Goni, P.; López, P.; Nerín, C. Combination of analytical and microbiological techniques to study the antimicrobial activity of a new active food packaging containing cinnamon or oregano against *E. coli* and *S. aureus*. *ABC* **2007**, *388*, 1003–1011. [CrossRef] [PubMed]
66. Matsuoka, H.; Ii, Y.; Takekawa, Y.; Teraoka, T. Evaluation of antifungal volatile compounds on the basis of the elongation rate of a single hypha. *Appl. Environ. Microbiol.* **1990**, *56*, 3779–3784. [CrossRef] [PubMed]
67. Inouye, S.; Tsuruoka, T.; Watanabe, M.; Takeo, K.; Akao, M.; Nishiyama, Y.; Yamaguchi, H. Inhibitory effect of essential oils on apical growth of Aspergillus fumigatus by vapour contact. *Mycoses* **2000**, *43*, 17–23. [CrossRef]
68. Passarinho, A.T.P.; Dias, N.F.; Camilloto, G.P.; Cruz, R.S.; Otoni, C.G.; Moraes, A.R.F.; Soares, N.D.F.F. Sliced bread preservation through oregano essential oil-containing sachet. *J. Food Proc. Eng.* **2014**, *37*, 53–62. [CrossRef]
69. Adams, R.P. *Identification of Essential Oil Components by Gas Chromatography/Mass Spectrometry*; Allured Publishing Corporation: Carol Stream, IL, USA, 2007; Volume 456.
70. Van Den Dool, H.; Kratz, P.D. A Generalization of the Retention Index System Including Linear Temperature Programmed Gas-Liquid Partition Chromatography. *J. Chromatogr. A* **1963**, *11*, 463–471. [CrossRef]
71. Sempere-Ferre, F.; Asamar, J.; Castell, V.; Roselló, J.; Santamarina, M.P. Evaluating the Antifungal Potential of Botanical Compounds to Control *Botryotinia fuckeliana* and *Rhizoctonia solani*. *Molecules* **2021**, *26*, 2472. [CrossRef] [PubMed]

molecules

Article

Inhibition of Fungal Strains Isolated from Cereal Grains via Vapor Phase of Essential Oils

Tereza Střelková [1], Bence Nemes [1], Anett Kovács [1], David Novotný [2], Matěj Božik [1] and Pavel Klouček [1,*]

1 Department of Food Science, Faculty of Agrobiology, Food and Natural Resources, Czech University of Life Sciences Prague, Kamýcká 129, 16500 Prague, Czech Republic; strelkova@af.czu.cz (T.S.); kovikussz@gmail.com (B.N.); kovacs.anett007@gmail.com (A.K.); bozik@af.czu.cz (M.B.)
2 Department of Ecology and Diagnostics of Fungal Pathogens, Crop Research Institute, Drnovská 507/73, 16106 Prague, Czech Republic; novotny@vurv.cz
* Correspondence: kloucek@af.czu.cz

Abstract: Fungal contamination in stored food grains is a global concern and affects food economics and human and animal health. It is clear that there is a need to develop new technologies with improved performances that are also eco-friendly in nature. Due to the bioactivity of essential oils (EOs) in the vapor phase, their low toxicity for humans, and their biodegradability and antifungal properties, EOs could be a suitable solution. In this study, we explored the potential of thyme, oregano, lemongrass, clove, and cajeput EOs in the vapor phase. For 17 days, inhibitory activity was assessed against five strains of postharvest pathogens—*Aspergillus* spp., *Fusarium* s. l. spp., and *Penicillium ochrochloron*—isolated from cereal grains. A modified disc volatilization method was used, which is more effective in comparison to traditional screening methods. Three concentrations were tested (250, 125, and 62.5 μL/L). The two highest concentrations resulted in complete inhibition of fungal growth; however, even 62.5 μL/L showed a significant antifungal effect. The efficiency of EOs followed this order: thyme > oregano > lemongrass > clove > cajeput. From our findings, it appears that the use of EOs vapors is a better option not only for laboratory experiments, but for subsequent practice.

Keywords: antifungal; cereal; essential oil; fast screening; fungi; inhibition; vapour phase

Citation: Střelková, T.; Nemes, B.; Kovács, A.; Novotný, D.; Božik, M.; Klouček, P. Inhibition of Fungal Strains Isolated from Cereal Grains via Vapor Phase of Essential Oils. *Molecules* **2021**, *26*, 1313. https://doi.org/10.3390/molecules26051313

Academic Editor: Gianni Sacchetti

Received: 22 January 2021
Accepted: 24 February 2021
Published: 1 March 2021

Publisher's Note: MDPI stays neutral with regard to jurisdictional claims in published maps and institutional affiliations.

Copyright: © 2021 by the authors. Licensee MDPI, Basel, Switzerland. This article is an open access article distributed under the terms and conditions of the Creative Commons Attribution (CC BY) license (https://creativecommons.org/licenses/by/4.0/).

1. Introduction

Food grains constitute a vital part of the daily diet of the population worldwide [1], and the most cultivated crop is wheat [2]. However, cereal grains contain a large number of microorganisms that deteriorate the products' nutritive value and are dangerous to human and animal health [3]. Fungal contamination in stored food grains is a global concern and affects food economics both directly and indirectly [1]. Approximately 20% of wheat that would otherwise be available each year is lost due to diseases [2,4]. In addition, the mycotoxins secreted by different seed-borne fungi cause qualitative losses of commodities, and potentially induce various health problems in consumers [5]. Approximately 25–40% of cereals consumed all over the world are contaminated by mycotoxins [5,6], and the European Food Safety Authority, in their Panel on Contaminations in the Food Chain [7], stated grains and grain-based products are one of three main chronic dietary sources of ochratoxin A, a mycotoxin of *Aspergillus ochraceus*. Among the mycotoxins, aflatoxins chiefly produced by *Aspergillus flavus* are the most dangerous, and approximately 4.5 billion people in underdeveloped countries are exposed to aflatoxicoses [6].

However, the major factors responsible for fungal growth depend on various intrinsic and extrinsic factors, such as the meteorological conditions during vegetation and harvesting, the duration of storage, the water content of grains, the storage temperature, the humidity during storage, and the type of storage technology [1,3]. *Fusarium* spp., *Aspergillus* spp., and *Penicillium* spp. were found to be dominant [1,3,8,9], but the higher incidence of *Aspergilli* than other fungi may be due to their saprophytic nature and ability to colonize diverse substrates because of secretions of various hydrolytic enzymes [5]. Although *Fusarium* species are predominantly considered to be field fungi, it has been reported that the production of fumonisins (*Fusarium* mycotoxins) can occur post-harvest when storage conditions are inadequate [10]. The removal of mycotoxins from the food chain is one of the major challenges for food scientists. Therefore, efforts should be made to prevent sources of mycotoxins from being present, i.e., the mycotoxin-producing fungi in the stored grains [1].

Today, synthetic pesticides play a major role in crop protection, but the widespread use of pesticides has resulted in the development of pest resistance, outbreaks of new pests, toxicity to non-target organisms, and harmful effects on the environment [11]. During storage, phosphine (PH_3) and methyl bromide (CH_3Br) are used for grain disinfection, but limitations regarding the application of the latter in Europe are increasing, and in America, their use is totally banned [3]. Hence, there is a need to develop new fungicides/preservatives with improved performances that are also eco-friendly in nature [6].

Many studies have explored various nonchemical fungi management practices, including traditional methods such as drying to a safe moisture level, aeration, and dry heating and novel control measures such as hermetic storage, microwave heating, and applications of gaseous ozone, cold plasma, ionizing radiation, pulsed light, or supercritical carbon dioxide ($SCeCO_2$) [1]. In addition, the use of plant extracts and essential oils (EOs) for the control of seed-associated fungi could be an eco-friendly solution, resulting in a lower chance of pathogens developing resistance [4,8].

EOs are volatile, oily liquids extracted from various plant materials; they are complex mixtures of chemical compounds with predominant terpenes associated with alcohols, aldehydes, and ketones [4]. Due to their bioactivity in the vapor phase, their low toxicity for humans, and their biodegradability and antifungal properties, EOs could find applications as fumigants for the protection of cereals and cereal-based products [12–14]. The antimicrobial effect could possibly be attributed to the presence of various antifungal substances, mainly phenolic compounds such as the monoterpenes thymol and carvacrol [15]. It is postulated that EOs, through their lipophilicity, have the ability to penetrate the plasma membrane, causing morphological changes in the hyphae, damaging the enzymatic cell systems, disrupting in the plasma membrane, and eventually destroying the mitochondria, thereby killing the fungi [1,8]. Many tested EOs show inhibitory effects on fungal postharvest pathogens, always in a dose-dependent manner [5,9–11,16,17].

The determination of the minimum inhibitory concentration (MIC) is important for setting a minimum dose for controlling fungal populations while using the lowest possible amount of pesticide [5]. The most common methodology for testing antifungal properties is the poisoned food technique—the use of culture media mixed with different amounts of an EO [2,18,19]. However, the less explored use of EOs in the vapor phase [13,20,21] seems to be a better option, as it should be more practical and more realistic for applications during storage. The main aims of the present study were to determine the antifungal activities of thyme, oregano, clove, lemongrass, and cajeput EOs in the vapor phase, and their suitability for use as disinfectants against *Aspergillus* spp., *Fusarium* spp., and *Penicillium ochrochloron* isolated from cereal grains.

2. Results

2.1. Compositions of Essential Oils

The major components of the EOs tested were identified and assessed by GC–MS. The main compounds found in thyme oil were thymol (58%), p-cymene (22%), and linalool (3%). Carvacrol was identified and determined as the major component of oregano oil (70%), followed by p-cymene (11%) and thymol (3%). Lemongrass oil contained, above all, geranial (42%) and neral (28%), and in smaller quantities geraniol (5%) and geranyl acetate (4%). The major components of clove oil were eugenol (80%), eugenol acetate (7%), and caryophyllene (7%). The minor components are listed in an earlier related publication [13].

The main components of cajeput oil were terpinen-4-ol (44%), γ-terpinene (20%), and p-cymene (14%); α-terpineol (4%) and 1,8-cineole (3%) were present in smaller quantities.

2.2. Antifungal Effect of the Essential Oils

After testing the highest concentration, namely, 250 μL/L, it was found that all of the EOs completely inhibited all strains. Therefore, 250 μL/L of any EO demonstrated fungicidal effect against the fungal strains we selected. Due to this finding, it was possible to start testing a lower concentration.

A lower concentration—125 μL/L—of all of the EOs except cajeput fully inhibited the mycelial growth of all strains. The cajeput EO was the least effective, so testing with a lower concentration of this EO was not appropriate.

Using the lowest concentration, namely, 62.5 μL/L, all of the EOs showed some inhibitory effect. A heat map of these results is available in Supplementary Table S2. The most effective was the thyme EO (Figure 1a), followed by the oregano EO (Figure 1b). These two EOs had the strongest inhibitory effect on the mycelial growth of every fungal strain. Among the strains tested, *A. niger* and *A. flavus* were the most resistant (significantly) and had almost the same mycelial growth development. On the sixth and seventh days, for the first time, oregano treatment on both *Aspergilli* did not statistically differ from the treatments of clove and lemongrass, respectively. Two days later, there was no statistical difference between the clove, lemongrass, oregano, or thyme treatments. However, the thyme EO as the first and the oregano EO as the second were still the most effective EOs. Control-equivalent full growth (4.75 cm diameter) for the thyme and oregano EOs was reached after 13 and 10 days of incubation, respectively. In the case of the thyme EO against *Fusarium sporotrichioides* and *F. solani*, full growth was not observed, even after 17 days. In general, the mycelia of the *Fusarium* strains grew very slowly due to the treatments of the thyme and oregano EOs. Until the tenth day, they were the most susceptible strains; after the tenth day, faster growth was observed. This mycelial growth development was significantly different compared to the controls (Figure 1e). However, no statistical difference from the control was observed at the end of the experiment (day 17). A table with statistical comparisons of the EOs' efficiencies is available in the Supplementary Materials.

The lemongrass EO showed a lower inhibitory effect, with similar activity against *F. sporotrichioides*, *F. solani*, and *A. niger* (Figure 1c). A very similar growth pattern was observed—intensive growth from the third day to the maximum on the sixth day—which means it delayed the fungal growth on average for two days compared to the control. The most resistant strain was *A. flavus*.

The clove EO (Figure 1d) at this concentration was the least effective EO against all of the tested strains, except *A. flavus*, against which the least effective was lemongrass EO. After 24 h, only the clove treatment was statistically equal to the growth control. Among the EOs tested, the clove EO was significantly the least effective against *F. solani*, whose mycelial growth was comparable to that of the controls (entirely grown on the fourth day). From the fifth day, the clove and lemongrass treatments did not significantly differ from the control for all fungi except *P. ochrochloron*. Of all the strains tested, *P. ochrochloron* was the least invasive and the most susceptible. The development of this fungus after treatment of EOs was significantly lower; even at the end of the assay, the mycelial growth did not reach the maximum. Additionally, the controls indicated that this strain is less invasive than others. Only when treated with the thyme or oregano EO was the mycelial growth of *P. ochrochloron* less inhibited than the growth of *F. sporotrichioides* and *F. solani*, but after the tenth day, in contrast to *Fusaria*, it did not increase pronouncedly. To briefly summarize the effectiveness of the EOs, they can be sorted in this order: thyme > oregano > lemongrass > clove.

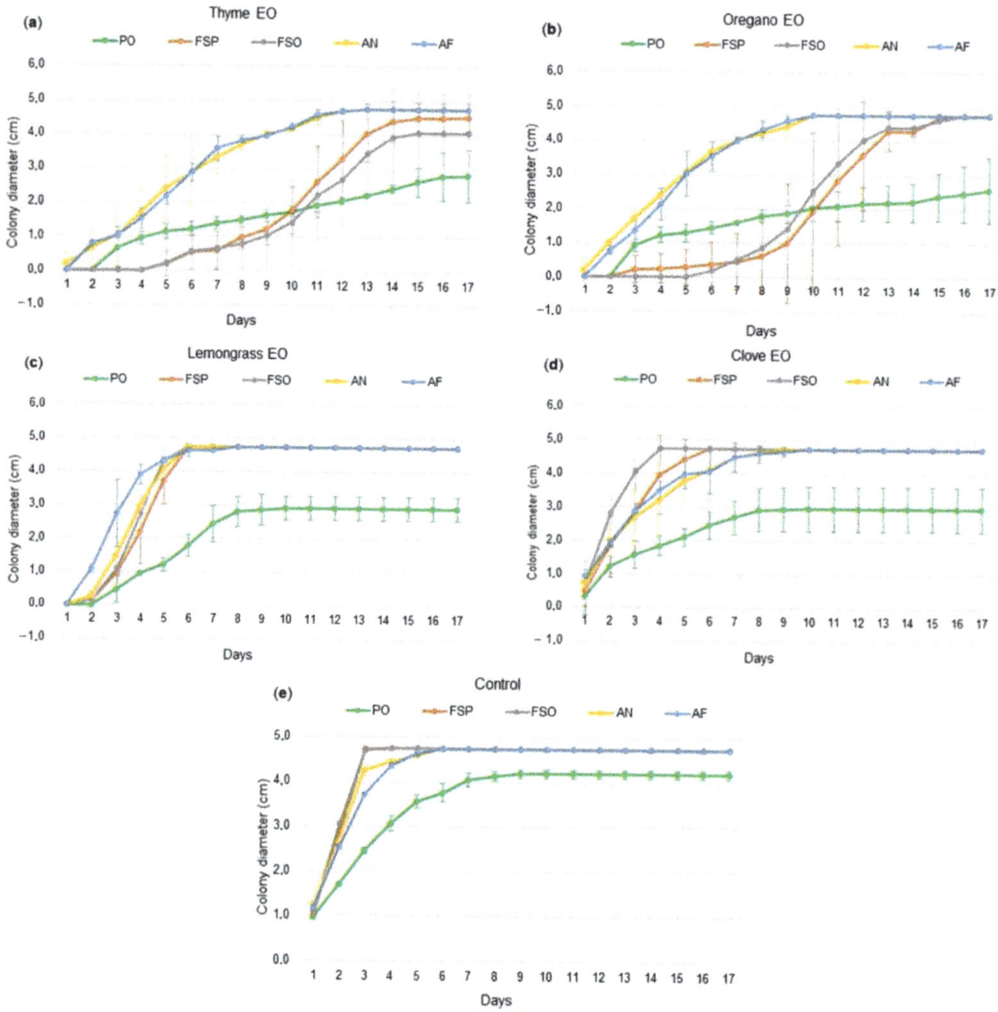

Figure 1. Each graph shows the efficiency of one essential oil (EO) at 62.5 μL/L against particular fungal strains (PO = *Penicillium ochrochloron*, FSP = *Fusarium sporotrichioides*, FSO = *F. solani*, AN = *Aspegillus niger*, and AF = *A. flavus*) over 17 days. The data are the average of three repetitions. (**a**) Efficiency of thyme EO; (**b**) efficiency of oregano EO; (**c**) efficiency of lemongrass EO; (**d**) efficiency of clove EO; (**e**) control sample without any EO.

3. Discussion

In grain cereals, mycotoxins are produced by fungal species such as *Aspergillus*, *Penicillium*, and *Fusarium* that colonize the plants in a field and can spread during the post-harvest period [22]. Spoilage of stored food commodities is a chronic problem and can produce qualitative and quantitative losses throughout the world [23]. Most EOs are considered to be "generally recognized as safe" (GRAS) food additives by the Food and Drug Administration (FDA), which makes them potential bio-resources of eco-friendly antifungal agents [24].

However, the dissimilar methods used for monitoring of antifungal efficiency constitute a problem. No standardized test has been developed and adopted for evaluating the possible antifungal activity of EOs against seed-borne fungi [25]. Not many studies have been performed that have applied EOs via gaseous contact, or have used similar EOs and similar strains as in this study. Until now, studies have been carried out using a combination of the same strains and different EOs or vice versa, and of varying concentrations. Herein, we used a method by Kloucek et al. [21]—a modified version of the commonly used disc volatilization method that uses a four-section Petri dish, a large filter paper disc evenly impregnated with EO, and a medium-containing lid. In comparison to the normal disc volatilization method, the labor and materials needed are reduced by several fold, and the composition of headspace is more uniform than in the case of a 6 mm disc, wherein different volatilities of particular compounds could influence the results. On the contrary, a number of studies have been performed using contact assays, such as the poisoned food technique. However, several researchers have concurred that the best antifungal activity of volatile compounds is achieved by gaseous contact, as opposed to aqueous solutions or agar contact [25–30].

Tullio et al. [27] reported that the inhibition effect of certain EOs (thyme red, clove, etc.) in the gaseous phase is generally higher than that in liquid state. They tested a concentration ranging from 10 mL/L to 19 μL/L against some fungi of the species *Mucor*, *Rhizopus*, *Penicillium*, *Alternaria*, and *Cladosporium*. In the disc volatilization method study of *Mentha piperita* EO, Tyagi and Malik [31] found that the minimum inhibitory concentrations (MICs) for *A. flavus* and *A. niger* varied from 1130 to 2250 μL/L, and the minimum fungicidal concentrations (MFCs) were 1130–2250 and 2250–4500 μL/L, respectively. In all strains, the zone of inhibition resulting from the exposure to EO vapors was again significantly larger than that due to same concentration of EO in the liquid phase, which supports our experiment. Yahyazadeh et al. [32] reported the fungicidal effect of thyme and clove EOs applied by gaseous contact against *Penicillium digitatum*. These EOs completely inhibited fungal growth by their volatiles at 340 μL/L, but when they were added to medium (direct contact), even at 600 μL/L, the effect was just fungistatic. However, even at concentrations that caused less than 100% mycelial growth inhibition, which is also our case, conidia lost their pigmentation (became hyaline). According to Yigit et al. [33], this effect might decrease the virulence of pathogens. In the case of clove EO, Bluma et al. [34] found that this oil did not show homogeneous antifungal activity against *Aspergillus Flavi*, and its efficacy depended on the water activity. In our study, evaluation of water activity was not included. However, the efficiency of the clove EO could have been affected due to this.

Additionally, a few contact assay studies and their results can be mentioned here. Linde et al. [35] used a modified microdilution technique and found that EO from *Petroselinum crispum* (parsley) exhibited fungistatic activity against all tested fungi, mainly *P. ochrochloron*. The in vitro results of Jahani et al. [36] showed that the growth of *A. niger* was completely inhibited by the contact assay with clove EO at concentrations of 200, 400, 600, and 800 μL/L on the first and tenth days, and thyme EO application at 800 μL/L on the tenth day. Morphological evaluation performed by both light microscopy and scanning electron microscopy conducted by Kohiyama et al. [37] showed that the antifungal activity of thyme EO against *A. flavus* could be detected at a concentration of 50 μL/L and the fungicidal effect at 250 μL/L. Oliveira et al. [24] also tested thyme EO, but at a higher concentration of 500 μL/L, and observed the complete inhibition of *A. flavus* growth. However, this is a very high concentration that cannot be used in practice. Krzyśko-Łupicka et al. [38] compared lemongrass, thyme, and cajeput EOs against *Fusarium* phytopathogens by a poisoned substrate assay and found out that thyme oil fully inhibited the growth at the lowest concentration (250 μL/L), lemongrass oil caused the same result at a slightly higher concentration (500 μL/L), and cajeput had a weaker effect even at 5 mL/L, which is consistent with its exclusion from our testing.

The determination of the antifungal activity of EOs appears to be influenced by the method used, as evidenced by the differences between the results just cited. It is also possible to assume that different chemical composition of specific tested EOs and different sensitivity or resistance of individual fungal strains are influential factors. Tullio et al. [27] described the content of thymol, a monoterpene phenol occurring in thyme EO, as a major contributor of bioactivity, and its connection with the efficacy of thyme EO, which corresponds to our findings. When multiple species of thyme EO were tested [21], a consistently significant inhibitory effect was obtained, although the composition differed. In addition, thymol was the most abundant compound in our EO. However, even EO rich in components such as carvacrol, p-cymene, or geranial has good preconditions for antifungal action. A eugenol-rich EO should act in the same way, in our case of clove EO; however, the factors mentioned above play important roles.

The advantages of using the volatile gas phase of EOs for agricultural and food products are that they may have less of an influence on the final taste and aroma, and their release may be regulated [34]. In addition, our results suggest that, thanks to the use of the gas phase, it is possible to achieve the inhibition of fungal growth using significantly lower concentrations than have been tested thus far. The above-mentioned studies have tested substantially higher concentrations by applying a gas phase or by direct contact, the efficiency of which could have been higher, but the need remains to find the lowest possible usable concentration. Higher concentrations could adversely affect sensory properties and would require more EOs to be used, which would be reflected in costs. In addition, our method makes it possible to monitor the inhibition of fungal growth over time, from which the need for possible repeated applications of EOs to different strains can be deduced. This could then be carried out more efficiently in practice, thanks to the use of the gas phase.

4. Materials and Methods

4.1. Essential Oils

Six essential oils were tested for their antifungal potential. The EOs from thyme (*Thymus vulgaris* L.) and lemongrass (*Cymbopogon citratus* L.) were purchased from the commercial supplier Sigma-Aldrich (Hamburg, Germany); oregano (*Origanum vulgare* L.), clove (*Syzygium aromaticum* L.), and cajeput (*Melaleuca alternifolia* Cheel) were purchased from Biomedica (Prague, Czech Rep.). These EOs were chosen according to previous results [21]. All of the used EOs were stored in glass bottles at 4 °C until used.

4.2. Essential Oil Analysis

Relative proportions of essential oils constituents were assessed by gas chromatography with flame ionization detector (GC–FID), Agilent 7890A (Agilent Technologies, Palo Alto, CA, USA) with HP-5MS, 30 m × 0.25 mm, and a 0.25 µm film thickness. The oven temperature was increased from 60 °C at a rate of 3 °C/min to a maximum of 231 °C, where it was kept constant for 10 min. Prior to the analysis, the EOs were diluted in hexane to a concentration of 1 µL/L. One microliter of the sample was injected in the split mode 1:12. The carrier gas was nitrogen (constant flow of 1 mL/min, 99.999% purity); the injector and detector temperatures were 250 °C. The relative proportions were calculated by dividing the individual peak area by the total area of all peaks; the response factor was not taken into account. Only compounds over 3% were included. The EO constituents were identified by mass spectrometry (GC–MS), the results of which were previously reported [13].

4.3. Microorganisms

The fungal strains used in present study were isolated in our laboratory from grain samples of wheat (*Triticum aestivum*), barley (*Hordeum vulgare*), and triticale (*Triticosecale*) of organic quality collected from experimental station of the Czech University of Life Sciences, Prague—Uhříněves. These included *Aspergillus flavus* Link (strain number VURV F-778), *Aspergillus niger* Tiegh. (strain number VURV F-779), *Penicillium ochrochloron* Biourge (strain number VURV F-780), *Fusarium sporotrichioides* Sherb. (strain number VURV F-804), and *Fusarium solani* (according to the new taxonomy—*Neocosmospora solani* (Mart.) L. Lombart and Crous). The fungal isolates were grown on Sabouraud Dextrose Agar (SDA) (Oxoid CZ s.r.o., Brno, Czech Rep.) at 25 °C. Pure cultures were obtained after repeated sub-culturing of isolated fungi. After the determination, the strains were preserved in liquid nitrogen in the form of freeze-dried conserves, and under paraffin oil on an agar slant in test tubes.

4.4. Determination of Microorganisms

A Phire Plant Direct PCR Kit was used to obtain DNA for preparation and PCR from five-day-old cultures grown on 2% malt extract agar. Translation elongation factor 1-alpha fragment strains belonging to genus *Fusarium* were amplified with primers EF1 and EF2 under conditions according to O'Donnell et al. [39]. Beta-tubulin fragment was amplified with primers Ben2f [40] and Bt2b [41] in strains belong to genera *Aspergillus* and *Penicillium*. PCR amplification of the section was performed in touch-down PCR mode: denaturation at 95 °C for 2 min; five cycles of 95 °C for 30 s, annealing for 30 s with temperature starting at 65 °C and decreasing by 1 °C each cycle and extension for 30 s; followed by 30 cycles of 95 °C for 15 s, 60 °C for 30 s, and 72 °C for 30 s; and final extension at 72 °C for 1 min. The obtained sequences were checked and compared using the Chromas and BioEdit programs. The species identity was determined by comparing the DNA sequence to the NCBI database using BLAST [42].

4.5. Antimicrobial Assay

The antifungal tests were carried out according to the method reported by Kloucek et al. [21] with several modifications. In the first phase of the study, all EOs were tested at the highest concentration (250 µL/L of air); then, 125 and 62.5 µL/L were tested. The tests were performed in 90 mm Petri dishes (PDs) divided into four sections. Into each section, 5 mL of SDA medium was poured, and SDA medium was poured onto the lid as well. After solidification, different mycelia were inoculated onto the middle of the three compartments by sterile loops. The fourth compartment remained empty as a purity control. The EOs were diluted in ethyl acetate to obtain the final volumes required. Each solution was equally distributed on 85 mm round sterile filter paper using a micropipette, and the paper was left to dry for 1 min for the evaporation of ethyl acetate. Finally, the filter paper was laid onto the walls of the compartments, so there was no direct contact with the medium in the Petri dish or lid containing solidified medium; then, the PD was hermetically closed and sealed with parafilm. The PDs were incubated at 25 °C in reverse position for 17 days, and every day the fungal growth was evaluated by measuring two perpendicular diameters of the colony using a ruler. The radial growth inhibition was observed and compared with blank filter paper with and without ethyl acetate, which served as negative controls. All the assays were carried out in triplicate and under aseptic conditions.

4.6. Statistical Analysis

The data were homogeneous and normal according to Bartlett and Shapiro tests. The results were tested by one-way ANOVA and Scheffe's method of homogeneous subsets (Statistica12, StatSoft CR s.r.o., Prague, Czech Rep.). The data were normalized to a percentage of fungal growth each day, with the growth of the control that day equaling 100%. A table containing the statistical data is available in the Supplementary Materials (Table S1).

5. Conclusions

In this study, we demonstrated that antifungal effects of the EOs from thyme, oregano, clove, and lemongrass can be achieved with lower doses and for longer time than those tested in previous studies. Complete inhibition was achieved at concentrations equal to or higher than 125 µL/L; at 62.5 µL/L, the fungal growth was significantly slowed down. The above discussion supports the suitability of the selected method, as studies have demonstrated a more efficient action of EOs in the gaseous phase. With this method, it is also possible to reduce the required amount of applied EO. However, this depends on the combination of the type of EO and the fungal strain. In general, the EOs tested can be ranked according to effectiveness as follows: thyme > oregano > lemongrass > clove. The most suitable uses would be EOs that have the broadest antifungal effect due to the real contamination of grains being by several strains. Future research needs to focus on testing these low-dose EOs in agricultural and food matrices, as they could become a part of organic production, could extend shelf life, and would not compromise sensory quality. In addition, our method is suitable for the screening of large quantities of samples in a shorter time and could be standardized for testing antimicrobial activity in the gas phase due to the uniformity of the headspace.

Supplementary Materials: The following are available online. Table S1: Relative fungal growth after EOs treatment at 62.5 μL/L; Table S2: A heat map of EOs efficiency at 62.5 μL/L against fungal strains.

Author Contributions: Conceptualization: P.K. and T.S.; methodology: P.K.; writing—original draft preparation: T.S.; writing—review and editing: P.K.; supervision: P.K. and M.B.; investigation: T.S., B.N., A.K., and D.N.; project administration: P.K.; funding acquisition: P.K. All authors read and agreed to the published version of the manuscript.

Funding: This research was funded by the European Regional Development Fund—Project "Centre for the investigation of synthesis and transformation of nutritional substances in the food chain in interaction with potentially harmful substances of anthropogenic origin: comprehensive assessment of soil contamination risks for the quality of agricultural products" (number CZ.02.1.01/0.0/0.0/16_019/0000845); by the National Agency for Agricultural Research of the Ministry of Agriculture of the Czech Republic under project number QK21010064; and by the Research Infrastructure METROFOOD-CZ supported by the Ministry of Education, Youth, and Sports of the Czech Republic under project number LM2018100.

Informed Consent Statement: Not applicable.

Data Availability Statement: The data presented in this study are available in Figure 1 and Tables S1 and S2.

Acknowledgments: We would like to acknowledge Libor Mičák from the experimental station of the Czech University of Life Sciences Prague—Uhříněves for the donation of grains.

Conflicts of Interest: The authors declare no conflict of interest. The funders had no role in the design of the study; in the collection, analyses, or interpretation of data; in the writing of the manuscript; or in the decision to publish the results.

Sample Availability: Samples of the compounds are available from the authors.

References

1. Mohapatra, D.; Kumar, S.; Kotwaliwale, N.; Singh, K.K. Critical factors responsible for fungi growth in stored food grains and non-chemical approaches for their control. *Ind. Crops Prod.* **2017**, *108*, 162–182. [CrossRef]
2. Roselló, J.; Sempere, F.; Sanz-Berzosa, I.; Chiralt, A.; Santamarina, M.P. Antifungal activity and potential use of essential oils against Fusarium culmorum and Fusarium verticillioides. *J. Essent. Oil Bearing Plants* **2015**, *18*, 359–367. [CrossRef]
3. Raila, A.; Lugauskas, A.; Steponavičius, D.; Railienė, M.; Steponavičienė, A.; Zvicevicius, E. E. Application of Ozone for Reduction of Mycological Infection. *Ann. Agric. Environ. Med.* **2006**, *13*, 287–294.
4. Ishaq, M.; Tahira, R.; Javed, A.; Jamal, A.; Raja, M.U.; Munir, A.; Ur-Rehman, A. Lemongrass essential oil as an alternate approach to manage seed associated fungi of wheat and rice. *Int. J. Agric. Biol.* **2017**, *19*, 1301–1306. [CrossRef]
5. Kedia, A.; Prakash, B.; Mishra, P.K.; Dubey, N.K. Antifungal and antiaflatoxigenic properties of Cuminum cyminum (L.) seed essential oil and its efficacy as a preservative in stored commodities. *Int. J. Food Microbiol.* **2014**, *168–169*, 1–7. [CrossRef]
6. Kumar, S.A.; Chandrabhan, S.; Shriram, P. Isolation of Aspergillus flavus from stored food commodities and Thymus vulgaris (L.) essential oil used as a safe plant based preservative. *Pharmacogn. Mag.* **2009**, *5*, 343–349. [CrossRef]
7. Schrenk, D.; Bodin, L.; Chipman, J.K.; del Mazo, J.; Grasl-Kraupp, B.; Hogstrand, C.; Hoogenboom, L.; Leblanc, J.C.; Nebbia, C.S.; Nielsen, E.; et al. Risk assessment of ochratoxin A in food. *EFSA J.* **2020**, *18*. [CrossRef]
8. Sreenivasa, M.Y.; Dass, R.S.; Charith Raj, A.P.; Nagendra Prasad, M.N.; Achar, P.N.; Janardhana, G.R. Assessment of the growth inhibiting effect of some plant essential oils on different Fusarium species isolated from sorghum and maize grains. *J. Plant Dis. Prot.* **2011**, *118*, 208–213. [CrossRef]
9. Anžlovar, S.; Likar, M.; Koce, J.D. Antifungal potential of thyme essential oil as a preservative for storage of wheat seeds. *Acta Bot. Croat.* **2017**, *76*, 64–71. [CrossRef]
10. Sumalan, R.M.; Alexa, E.; Poiana, M.A. Assessment of inhibitory potential of essential oils on natural mycoflora and Fusarium mycotoxins production in wheat. *Chem. Cent. J.* **2013**, *7*, 1–12. [CrossRef]
11. Matusinsky, P.; Zouhar, M.; Pavela, R.; Novy, P. Antifungal effect of five essential oils against important pathogenic fungi of cereals. *Ind. Crops Prod.* **2015**, *67*, 208–215. [CrossRef]

12. Gutiérrez, L.; Sánchez, C.; Batlle, R.; Nerín, C. New antimicrobial active package for bakery products. *Trends Food Sci. Technol.* **2009**, *20*, 92–99. [CrossRef]
13. Božik, M.; Císarová, M.; Tančinová, D.; Kouřimská, L.; Hleba, L.; Klouček, P. Selected essential oil vapours inhibit growth of Aspergillus spp. in oats with improved consumer acceptability. *Ind. Crops Prod.* **2017**, *98*, 146–152. [CrossRef]
14. Císarová, M.; Hleba, L.; Medo, J.; Tančinová, D.; Mašková, Z.; Čuboň, J.; Kováčik, A.; Foltinová, D.; Božik, M.; Klouček, P. The in vitro and in situ effect of selected essential oils in vapour phase against bread spoilage toxicogenic aspergilli. *Food Control* **2020**, *110*. [CrossRef]
15. Garcia, D.; Ramos, A.J.; Sanchis, V.; Marín, S. Equisetum arvense hydro-alcoholic extract: Phenolic composition and antifungal and antimycotoxigenic effect against Aspergillus flavus and Fusarium verticillioides in stored maize. *J. Sci. Food Agric.* **2013**, *93*, 2248–2253. [CrossRef] [PubMed]
16. Tzortzakis, N.G.; Economakis, C.D. Antifungal activity of lemongrass (Cympopogon citratus L.) essential oil against key postharvest pathogens. *Innov. Food Sci. Emerg. Technol.* **2007**, *8*, 253–258. [CrossRef]
17. Kumar, P.; Mishra, S.; Kumar, A.; Sharma, A.K. Antifungal efficacy of plant essential oils against stored grain fungi of Fusarium spp. *J. Food Sci. Technol.* **2016**, *53*, 3725–3734. [CrossRef]
18. Maxia, A.; Falconieri, D.; Piras, A.; Porcedda, S.; Marongiu, B.; Frau, M.A.; Gonçalves, M.J.; Cabral, C.; Cavaleiro, C.; Salgueiro, L. Chemical Composition and Antifungal Activity of Essential Oils and Supercritical CO2Extracts of Apium nodiflorum (L.) Lag. *Mycopathologia* **2012**, *174*, 61–67. [CrossRef]
19. Alexa, E.; Sumalan, R.M.; Danciu, C.; Obistioiu, D.; Negrea, M.; Poiana, M.A.; Rus, C.; Radulov, I.; Pop, G.; Dehelean, C. Synergistic antifungal, allelopatic and anti-proliferative potential of salvia officinalis L., and thymus vulgaris L. Essential oils. *Molecules* **2018**, *23*, 185. [CrossRef] [PubMed]
20. Tang, X.; Shao, Y.L.; Tang, Y.J.; Zhou, W.W. Antifungal activity of essential oil compounds (geraniol and citral) and inhibitory mechanisms on grain pathogens (aspergillus flavus and aspergillus ochraceus). *Molecules* **2018**, *23*, 2108. [CrossRef]
21. Kloucek, P.; Smid, J.; Frankova, A.; Kokoska, L.; Valterova, I.; Pavela, R. Fast screening method for assessment of antimicrobial activity of essential oils in vapor phase. *Food Res. Int.* **2012**, *47*, 161–165. [CrossRef]
22. Morcia, C.; Malnati, M.; Terzi, V. In vitro antifungal activity of terpinen-4-ol, eugenol, carvone, 1,8-cineole (eucalyptol) and thymol against mycotoxigenic plant pathogens. *Food Addit. Contam. Part A Chem. Anal. Control. Expo. Risk Assess.* **2012**, *29*, 415–422. [CrossRef] [PubMed]
23. Pandey, A.K.; Kumar, P.; Singh, P.; Tripathi, N.N.; Bajpai, V.K. Essential oils: Sources of antimicrobials and food preservatives. *Front. Microbiol.* **2017**, *7*, 1–14. [CrossRef] [PubMed]
24. Oliveira, R.C.; Carvajal-Moreno, M.; Correa, B.; Rojo-Callejas, F. Cellular, physiological and molecular approaches to investigate the antifungal and anti-aflatoxigenic effects of thyme essential oil on Aspergillus flavus. *Food Chem.* **2020**, *315*. [CrossRef]
25. Mateo, E.M.; Gómez, J.V.; Domínguez, I.; Gimeno-Adelantado, J.V.; Mateo-Castro, R.; Gavara, R.; Jiménez, M. Impact of bioactive packaging systems based on EVOH films and essential oils in the control of aflatoxigenic fungi and aflatoxin production in maize. *Int. J. Food Microbiol.* **2017**, *254*, 36–46. [CrossRef] [PubMed]
26. Singh, G.; Maurya, S.; de Lampasona, M.P.; Catalan, C. Chemical constituents, antifungal and antioxidative potential of Foeniculum vulgare volatile oil and its acetone extract. *Food Control* **2006**, *17*, 745–752. [CrossRef]
27. Tullio, V.; Nostro, A.; Mandras, N.; Dugo, P.; Banche, G.; Cannatelli, M.A.; Cuffini, A.M.; Alonzo, V.; Carlone, N.A. Antifungal activity of essential oils against filamentous fungi determined by broth microdilution and vapour contact methods. *J. Appl. Microbiol.* **2007**, *102*, 1544–1550. [CrossRef]
28. Vilela, G.R.; de Almeida, G.S.; D'Arce, M.A.B.R.; Moraes, M.H.D.; Brito, J.O.; da Silva, M.F.d.G.F.; Silva, S.C.; de Stefano Piedade, S.M.; Calori-Domingues, M.A.; da Gloria, E.M. Activity of essential oil and its major compound, 1,8-cineole, from Eucalyptus globulus Labill., against the storage fungi Aspergillus flavus Link and Aspergillus parasiticus Speare. *J. Stored Prod. Res.* **2009**, *45*, 108–111. [CrossRef]
29. Gómez, J.V.; Tarazona, A.; Mateo-Castro, R.; Gimeno-Adelantado, J.V.; Jiménez, M.; Mateo, E.M. Selected plant essential oils and their main active components, a promising approach to inhibit aflatoxigenic fungi and aflatoxin production in food. *Food Addit. Contam. Part A Chem. Anal. Control. Expo. Risk Assess.* **2018**, *35*, 1581–1595. [CrossRef]
30. Tyagi, A.K.; Malik, A. Antimicrobial potential and chemical composition of Eucalyptus globulus oil in liquid and vapour phase against food spoilage microorganisms. *Food Chem.* **2011**, *126*, 228–235. [CrossRef]
31. Tyagi, A.K.; Malik, A. Antimicrobial potential and chemical composition of Mentha piperita oil in liquid and vapour phase against food spoiling microorganisms. *Food Control* **2011**, *22*, 1707–1714. [CrossRef]
32. Yahyazadeh, M.; Omidbaigi, R.; Zare, R.; Taheri, H. Effect of some essential oils on mycelial growth of Penicillium digitatum Sacc. *World J. Microbiol. Biotechnol.* **2008**, *24*, 1445–1450. [CrossRef]
33. Yigit, F.; Özcan, M.; Akgül, A. Inhibitory effect of some spice essential oils on Penicillium digitatum causing postharvest rot in citrus. *Grasas Aceites* **2000**, *51*, 237–240. [CrossRef]
34. Bluma, R.; Landa, M.F.; Etcheverry, M. Impact of volatile compounds generated by essential oils on Aspergillus section Flavi growth parameters and aflatoxin accumulation. *J. Sci. Food Agric.* **2009**, *89*, 1473–1480. [CrossRef]
35. Linde, G.A.; Gazim, Z.C.; Cardoso, B.K.; Jorge, L.F.; Tešević, V.; Glamoćlija, J.; Soković, M.; Colauto, N.B. Antifungal and antibacterial activities of Petroselinum crispum essential oil. *Genet. Mol. Res.* **2016**, *15*, 1–11. [CrossRef]

36. Jahani, M.; Pira, M.; Aminifard, M.H. Antifungal effects of essential oils against Aspergillus niger in vitro and in vivo on pomegranate (Punica granatum) fruits. *Sci. Hortic.* **2020**, *264*, 109188. [CrossRef]
37. Kohiyama, C.Y.; Yamamoto Ribeiro, M.M.; Mossini, S.A.G.; Bando, E.; Bomfim, N.D.S.; Nerilo, S.B.; Rocha, G.H.O.; Grespan, R.; Mikcha, J.M.G.; Machinski, M. Antifungal properties and inhibitory effects upon aflatoxin production of Thymus vulgaris L. by Aspergillus flavus Link. *Food Chem.* **2015**, *173*, 1006–1010. [CrossRef]
38. Krzyśko-Łupicka, T.; Sokół, S.; Piekarska-Stachowiak, A. Evaluation of fungistatic activity of eight selected essential oils on four heterogeneous Fusarium isolates obtained from cereal grains in southern Poland. *Molecules* **2020**, *25*, 292. [CrossRef]
39. O'Donnell, K.; Kistlerr, H.C.; Cigelnik, E.; Ploetz, R.C. Multiple evolutionary origins of the fungus causing panama disease of banana: Concordant evidence from nuclear and mitochondrial gene genealogies. *Proc. Natl. Acad. Sci. USA* **1998**, *95*, 2044–2049. [CrossRef]
40. Hubka, V.; Kolarik, M. β-tubulin paralogue tubC is frequently misidentified as the benA gene in Aspergillus section Nigri taxonomy: Primer specificity testing and taxonomic consequences. *Persoonia Mol. Phylogeny Evol. Fungi* **2012**, *29*, 1–10. [CrossRef]
41. Glass, N.L.; Donaldson, G.C. Development of primer sets designed for use with the PCR to amplify conserved genes from filamentous ascomycetes. *Appl. Environ. Microbiol.* **1995**, *61*, 1323–1330. [CrossRef] [PubMed]
42. Madden, T. NCBI_blast Information. In *The NCBI Handbook*; National Center for Biotechnology Information: Bethesda, MD, USA, 2013; pp. 1–15.

molecules

MDPI

Article

Chemical Composition and Determination of the Antibacterial Activity of Essential Oils in Liquid and Vapor Phases Extracted from Two Different Southeast Asian Herbs—*Houttuynia cordata* (Saururaceae) and *Persicaria odorata* (Polygonaceae)

Kristýna Řebíčková [1], Tomáš Bajer [1], David Šilha [2], Markéta Houdková [3], Karel Ventura [1] and Petra Bajerová [1,*]

1 Department of Analytical Chemistry, Faculty of Chemical Technology, University of Pardubice, Studentská 573, 532 10 Pardubice, Czech Republic; kristyna.rebickova@student.upce.cz (K.Ř.); tomas.bajer@upce.cz (T.B.); karel.ventura@upce.cz (K.V.)

2 Department of Biological and Biochemical Sciences, Faculty of Chemical Technology, University of Pardubice, Studentská 573, 532 10 Pardubice, Czech Republic; david.silha@upce.cz

3 Department of Crop Sciences and Agroforestry, Faculty of Tropical AgriSciences, Kamýcká 129, Czech University of Life Sciences Prague, 165 00 Prague, Czech Republic; houdkovam@ftz.czu.cz

* Correspondence: petra.bajerova@upce.cz; Tel.: +420-466-037-078

Academic Editor: Derek J. McPhee
Received: 30 April 2020; Accepted: 20 May 2020; Published: 22 May 2020

Abstract: Essential oils obtained via the hydrodistillation of two Asian herbs (*Houttuynia cordata* and *Persicaria odorata)* were analyzed by gas chromatography coupled to mass spectrometry (GC–MS) and gas chromatography with flame ionization detector (GC–FID). Additionally, both the liquid and vapor phase of essential oil were tested on antimicrobial activity using the broth microdilution volatilization method. Antimicrobial activity was tested on Gram-negative and Gram-positive bacteria—*Escherichia coli, Staphylococcus aureus, Pseudomonas aeruginosa, Enterococcus faecalis, Streptococcus pyogenes, Klebsiella pneumoniae, Seratia marcescense* and *Bacillus subtilis*. Hydrodistillation produced a yield of 0.34% (*Houttuynia cordata*) and 0.40% (*Persicaria odorata*). 41 compounds were identified in both essential oils. Essential oils contained monoterpenes and their oxidized forms, sesquiterpenes and their oxidized forms, oxidized diterpenes, derivates of phenylpropene and other groups, such as, for example, aldehydes, alcohols or fatty acids. Both essential oils were antimicrobial active in both vapor and liquid phases at least in case of one bacterium. They expressed various antimicrobial activity in the range of 128–1024 $\mu g \cdot mL^{-1}$, 512–1024 $\mu g \cdot mL^{-1}$ in broth and 1024 $\mu g \cdot mL^{-1}$, 512–1024 $\mu g \cdot mL^{-1}$ in agar, respectively. Research showed new interesting information about *P. odorata* and *H. cordata* essential oils and demonstrated that both essential oils could be possibly used in the field of natural medicine or natural food preservation.

Keywords: *Houttuynia cordata*; *Persicaria odorata*; essential oil; distillation; antimicrobial activity; vapor phase; volatile compounds; gas chromatography

1. Introduction

In recent years, many researchers have been focused on finding new antimicrobial agents that could be applied to multi-resistant microorganisms. Medicinal herbs and their products, such as essential oils (EOs), are the main source of natural remedies. They have been used since the time immemorial as the most affordable means of treating diseases. As it has been proven several times, EOs have diverse biological properties. They are bactericidal, viricidal, fungicidal, antiparasitic, antioxidant and insecticidal [1–3]. EOs and their components have activity against a variety of targets, particularly the membrane and cytoplasm, and, in some cases, they can completely change the morphology of the cells [4]. They contain a wide range of complex and structurally different compounds that are biologically, respectively, antimicrobially active. Antimicrobial activity is closely related to the chemical composition of EOs, functional groups and possible synergic interactions among constituents. In addition, due to the combination of these active compounds, bacteria are more difficult to develop a resistance to these compounds in comparison with commercial antimicrobials, which are usually based on one chemical substance [5]. In order to use EOs for treating various diseases caused by bacteria, it is important to understand to the relationship between their chemical composition and the potential antimicrobial activity.

EOs are volatile, and their vapors influence their antimicrobial properties. Therefore, it is necessary to sufficiently investigate the composition and properties of the vapors of essential oil. Currently, several researchers have investigated the vapor phase of the essential oil, but it deserves much wider research and consideration when examining the properties of essential oils [6–12]. It is assumed that active compounds in both phases are synergic and furthermore it has been shown that vapor phase is sometimes even more effective than liquid phase to some microorganisms. If biologically active molecules were present in the vapor phase, there would be, for example, the possibility of using EOs for the inhalation and treatment of respiratory diseases [8,9]. Furthermore, there would be the possibility of using these properties of EOs to produce various food packaging materials that protect food against the spread of microorganisms and extend their shelf life [13–15].

Traditionally, agar dilution, broth dilution or broth microdilution methods have been used to measure the minimal inhibitory concentrations of antimicrobial agents against microorganisms [16,17]. Unfortunately, these methods have been limited to measuring the minimal inhibitory concentrations (MICs) of antimicrobials only in the liquid phase. In the last few years, several methods and the modification-testing MICs of antimicrobials in the vapor phase were developed. Usually, the disc volatilization method with various modifications is used [10–12,17,18].

Two Southeast Asian herbs (*Houttuynia cordata* and *Persicaria odorata*), used mainly in the traditional Chinese medicine and as a spice in Asian cuisine, have been chosen for this study. These two herbs have been chosen for their known health benefits from Chinese medicine and for their geographical occurrence. Moreover, there is a lack of scientific literature about *Houttuynia cordata* [19–24] and *Persicaria odorata* [25,26] EOs. Therefore, they were examined in more detail. *Houttuynia cordata*, also known as a fish mint, belongs to a family *Saururaceae*. It is a flowering perennial herb native to Southeast Asia and it grows in moist shady places [27]. EOs from *H. cordata* have a fishy scent and show a variety of biological activities, such as antimicrobial, antiviral, anti-inflammatory, anticancer and insect repellent [28]. According to the previous reports, the liquid phase of EOs from *H. cordata* has inhibitory effects against *Pseudomonas aeruginosa, Escherichia coli, Bacillus cereus, Bacillus subtilis, Vibrio cholerae* and *Staphylococcus aureus* [21,29]. The dominant volatile compounds are houttuynin, myrcene, decanal, *cis*-ocimene and bornyl acetate [27,28,30]. *Persicaria odorata*, also known as a Vietnamese coriander, belongs to a family *Polygonaceae*. It is a tender perennial herb native to Southeast Asia and it grows in wet environments with a rich, moist soil shady places [25]. EOs isolated from *P. odorata* have a very strong coriander odor and show antimicrobial, anti-inflammatory, antitumor and antioxidant activity. According to the previous reports, the liquid phase of essential oil from *P. odorata* inhibits *Salmonella choleraesuis* [31], *Enterococcus faecalis, Enterococcus faecium, Staphylococcus epidermidis* and *Staphylococcus aureus* [32]. The most abundant volatile compounds are dodecanal, decanal, 3-hexanal, 2-hexanal, β-caryophyllene and α-humulene [25,26,33,34].

Due to the growing demand for natural products and in light of the prevalence of pharmaceuticals, antioxidants or food additives in the food preservation process, it is necessary to find new sources for developing these products and to specify their properties. Plant-derived essential oils have received significant attention in this field. The aim of this study was to find herbs used both in the cuisine and natural medicine and determine their EOs composition and especially their antimicrobial efficiency in both vapor and liquid phase. In this study were two EOs obtained by hydrodistillation and thereafter analyzed by standard techniques GC–MS and GC–FID. Antimicrobial activity was determined by the modern, recently developed method of Houdkova et al. [35] called the broth microdilution volatilization method. It is a simple and rapid simultaneous determination of the antibacterial potential of plant volatile compounds in the liquid and the vapor phase at different concentrations [35].

2. Results and Discussion

2.1. Extraction Yield and Chemical Composition

Hydrodistillation in the Clevenger-type apparatus of *Houttuynia cordata* produced a pale-yellow liquid with a fishy scent. The essential oil content of distilled aerial parts of dried plant was 0.34%. The extraction yield is higher in comparison with those previously published by R. S. Verma et al. [24] who only achieved a yield of 0.06–0.14%. A total of 41 compounds were identified that made up 90.6% of the essential oil composition (Table 1). The essential oil contained a higher amount of terpenoid compounds (75.5%), followed by non-terpenoid compounds (15.1%), such as derivates of phenylpropene, aldehydes, ketones, esters and fatty acids. The major group of substances was monoterpenes with a content of 59.4%, followed by the group of other compounds with a content of 14.8%, oxidized monoterpenes with a content of 7.2% and sesquiterpenes with a content of 6.6%. Other groups were oxidized sesquiterpenes and derivates of phenylpropene. Major compounds of the essential oil were myrcene (51.6%), 2-undecanone (6.7%), tridecan-2-one (6.1%), *cis*-β-ocimene (5.7%), geranyl acetate (3.1%), bornyl acetate (2.9%) and *cis*-caryophyllene (2.6%). The other compounds were present at less than 2%. These results are similar with the results of previous reports [19,22,24]. Only a few fluctuations from other reports were found and are probably attributed to the origin of the plant samples or different extraction method. For the characteristic fishy scent and flavoring of *H. cordata* essential oils is responsible compound houttuynin (decanoyl acetaldehyde). This compound was not identified in our essential oil due to its instability. It is usual that decanoyl acetaldehyde is during the process of distillation easily oxidized into 2-undecanone [24]. This compound had the second highest concentration in our essential oil. Therefore, the amount of 2-undecanone is the primary indicator for the quality of *Houttuynia cordata* essential oil [23,24].

Table 1. Chemical composition of essential oils from *Houttuynia cordata* and *Persicaria odorata*.

Compound	CAS Number	Identification [1]	Retention Index		Peak Area [%]	
			Observed	Published [2]	*H. cordata*	*P. odorata*
Monoterpenes						
α-pinene	80-56-8	MS, RI	930	933	0.53	-
camphene	79-92-5	MS, RI	946	953	0.36	-
β-pinene	127-91-3	MS, RI	974	978	0.45	-
myrcene	123-35-3	MS, RI, Std	993	991	51.64	-
limonene	138-86-3	MS, RI, Std	1027	1030	0.49	-
cis-β-ocimene	3338-55-4	MS, RI	1036	1040	5.72	-
trans-β-ocimene	3779-61-1	MS, RI	1046	1046	0.18	-
7-*epi*-sesquithujene	159407-35-9	MS, RI	1387	1387	-	0.02
Sum [%]					59.37	0.02
Oxidized monoterpenes						
perillene	539-52-6	MS, RI	1097	1098	0.51	-
linalool	78-70-6	MS, RI, Std	1100	1101	0.28	-
myroxide	33281-83-3	MS, RI	1131	1129	0.01	-
isoborneol	10385-78-1	MS, RI	1169	1165	0.08	-
α-terpineol	98-55-5	MS, RI, Std	1193	1195	0.06	-
β-cyclocitral	432-25-7	MS, RI	1218	1219	-	0.04
trans-geraniol	102-24-1	MS, RI	1251	1255	0.2	-
bornyl acetate	92618-89-8	MS, RI	1283	1285	2.85	-
neryl acetate	141-12-8	MS, RI	1358	1365	0.12	-
geranyl acetate	105-87-3	MS, RI, Std	1378	1383	3.11	-
Sum [%]					7.22	0.04

Table 1. *Cont.*

Compound	CAS Number	Identification [1]	Retention Index		Peak Area [%]	
			Observed	Published [2]	*H. cordata*	*P. odorata*
Sesquiterpenes						
cis-caryophyllene	13877-93-5	MS, RI	1419	1424	2.58	3.88
trans-α-bergamotene	13474-59-4	MS, RI	1431	1432	0.1	0.25
isogermacrene D	317819-80-0	MS, RI	1441	1447	-	0.08
trans-caryophyllene	87-44-5	MS, RI	1452	1451	0.99	-
α-humulene	6753-98-6	MS, RI, Std	1455	1454	-	4.50
γ-gurjunene	22567-17- 5	MS, RI	1474	1476	-	0.30
selina-4.11-diene	17627-30-4	MS, RI	1476	1482	1.13	0.23
α-curcumene	644-30-4	MS, RI	1482	1480	-	0.23
β-selinene	17066-67-0	MS, RI	1487	1491	-	0.40
valencene	4630-07-3	MS, RI	1490	1492	0.67	0.04
β-bisabolene	4891-79-6	MS, RI	1508	1508	-	0.07
β-curcumene	72345-84-7	MS, RI	1510	1511	-	0.08
7-*epi*-α-selinene	123123-37-5	MS, RI	1517	1518	0.54	0.52
trans-calamenene	73209-42-4	MS, RI	1519	1527	0.24	-
cis-sesquisabinene hydrate	58319-05-4	MS, RI	1543	1544	-	0.96
Sum [%]					6.25	11.54
Oxidized sesquiterpenes						
β-elemene	33880-83-0	MS, RI	1387	1390	0.13	-
ishwarane	26620-70-2	MS, RI	1465	1468	0.24	-
α-farnesene	502-61-4	MS, RI	1503	1504	0.18	-
β-nerolidol	40716-66-3	MS, RI	1560	1561	0.76	0.53
spathulenol	72203-24-8	MS, RI	1575	1576	0.37	-
caryophyllene oxide	1139-30-6	MS, RI	1580	1587	0.99	1.42
humulene epoxide II	19888-34-7	MS, RI	1608	1613	-	1.09
caryophylla-4(12).8(13)-dien-5-ol	19431-80-2	MS, RI	1632	1636	-	0.69
epi-β-bisabolol	235421-59-7	MS, RI	1669	1675	-	0.34
α-bisabolol	515-69-5	MS, RI	1686	1688	-	0.05
trans-α-bergamotol	88034-74-6	MS, RI	1688	1688	-	0.05
drimenol	19078-37-6	MS, RI	1768	1769	-	1.24
drimenin	2326-89-8	MS, RI	1944	1944	-	0.30
Sum [%]					2.67	5.71
Oxidized diterpenes						
phytone	502-69-2	MS, RI	1840	1841	-	0.38
Sum [%]					0	0.38
Derivates of phenylpropene						
methyl eugenol	93-15-2	MS, RI	1401	1403	0.25	-
Sum [%]					0.25	0
Others						
6-methyl-hept-5-en-2-one	110-93-0	MS, RI	985	986	0.05	-
2-pentyl-furan	3777-69-3	MS, RI	993	991	-	0.06
6-methyl-Hept-5-en-2-ol	1569-60-4	MS, RI	998	995	-	0.03
n-undecane	1120-21-4	MS, RI	1101	1100	-	2.52
n-nonanal	124-19-6	MS, RI	1104	1107	0.12	0.26
1-nonanol	143-08-8	MS, RI	1172	1169	0.33	0.35
n-decanal	112-31-2	MS, RI, Std	1208	1208	0.15	18.4
1-decanol	112-30-1	MS, RI	1276	1278	-	5.37
2-undecanone	112-12-9	MS, RI	1293	1294	6.67	-
n-undecanal	112-44-7	MS, RI	1307	1309	-	1.37
1-undecanol	112-42-5	MS, RI	1377	1379	-	1.16
2-dodecanone	6175-49-1	MS, RI	1393	1393	0.07	-
n-dodecanal	112-54-9	MS, RI, Std	1411	1410	0.02	37.08
1-dodecanol	112-53-8	MS, RI, Std	1477	1476	-	4.81
tridecan-2-one	593-08-8	MS, RI	1496	1495	6.06	-
n-dodecanoic acid	143-07-7	MS, RI	1569	1570	0.70	-
n-tetradecanal	124-5-4	MS, RI	1612	1614	-	0.26
intermedeol	6168-59-8	MS, RI	1661	1668	-	0.13
2-pentadecanone	2345-28-0	MS, RI	1696	1697	0.21	-
n-hexadecanoic acid	57-10-3	MS, RI	1963	1968	0.45	0.44
linoleoyl chloride	7459-33-8	MS, RI	2135	2139	-	0.16
n-dodecenylsuccinic anhydride	1978-11-1	MS, RI	2159	2159	-	0.29
Sum [%]					14.83	72.69
Total peak area [%]					90.59	90.38

[1] MS—mass spectra, RI—retention index, Std—analytical standard; [2] Data published in database ofNational Institute of Standards and Technology (NIST) [36] and Adams [37].

Hydrodistillation in Clevenger-type apparatus of *Persicaria odorata* produced a deep yellow liquid with a strong spicy coriander-like aroma. Due to its aroma, it is also called Vietnamese coriander [25]. The essential oil content of distilled aerial parts of dried plant was 0.41%. The extraction yield is lower in comparison with those previously published by A. A. Almarie et al. [33] who achieved a yield of 0.64%. A total of 41 compounds were identified that made up 90.4% of the essential oil composition. In comparison with other reports, we identified more compounds [38,39]. N. X. Dung et al. [38] used steam distillation for the isolation of essential oils and they identified 28 compounds and the most abundant were β-caryophyllene, dodecanal and caryophyllene oxide. M. V. Hunter et al. [39] only identified 17 compounds using steam distillation as an extraction technique for the isolation of essential oil from *P. odorata*, where the most abundant compounds were α-humulene, decanal and dodecanal. Our essential oil contained a higher amount of non-terpenoid compounds (72.7%) followed by terpenoid compounds (17.7%). Carbonyls and alcohols, especially C10 and C12, made up 68.8% of essential oil composition, followed by the group of sesquiterpenes with a content of 11.5% and oxidized sesquiterpenes with a content of 5.7%. Other groups (monoterpenes, oxidized monoterpenes and oxidized diterpenes) made up less than 1% of essential oil constitution. The major compounds of the essential oil were *n*-dodecanal (37.1%), *n*-decanal (18.1%), 1-decanol (5.4%), 1-dodecanol (4.8%), α-humulene (4.5%), *cis*-caryophyllene (3.9%) and *n*-undecane (2.5%). The other compounds were present at less than 2%. These results in relative percent content are similar with the results of previous reports. It can clearly be seen that the essential oil from *Persicaria odorata* is rich in C10 and C12 carbonyls. Dodecanal and decanal are the main compounds of *Persicaria odorata* essential oil in all previous published reports and ours [25,33,38,39].

When comparing both EOs, it was found that they have only 12 common compounds out of 41 but they vary in the percent content. The essential oil from *Hottuynia cordata* contained much more monoterpenes and monoterpenoids, where the most abundant compound was myrcene (51%) that was not found in the essential oil from the *P. odorata*. On the other hand, the essential oil from *Persicaria odorata* contained more sesquiterpenes, sesquipterpenoids and especially aldehydes, where *n*-dodecanal (37.1%) was the dominant compound compared to *H. cordata* essential oil, where it was only 0.02%. In general, the composition of both essential oils is different. The results are adequate because both herbs are neither from the same genus nor family, so their similarity in composition was not expected. They were selected for this study according to their similar use and same geographical occurrence.

2.2. Antimicrobial Activity

The antimicrobial activity of *H. cordata* and *P. odorata* essential oils is reported in Table 2. Both EOs showed antimicrobial efficiency but in different concentrations. *H. cordata* and *P. odorata* essential oil expressed various antimicrobial activity in the range of 128–1024 $\mu g \cdot mL^{-1}$, 512–1024 $\mu g \cdot mL^{-1}$ in broth and 1024 $\mu g \cdot mL^{-1}$, 512–1024 $\mu g \cdot mL^{-1}$ in agar, respectively. In the liquid phase, the lowest MIC was showed for *H. cordata* (128 $\mu g \cdot mL^{-1}$) against *E. faecalis* and for *P. odorata* (512 $\mu g \cdot mL^{-1}$) against *S. pyogenes*, *E. faecalis* and *B. subtilis*. In the vapor phase, the lowest MIC was observed for *H. cordata* (1024 $\mu g \cdot mL^{-1}$) against *E. faecalis* and *E. coli* and for *P. odorata* (512 $\mu g \cdot mL^{-1}$) against *E. coli*. There are observable differences between the efficiency of the vapor and liquid phases of observed essential oils (EOs). In most cases, the higher MIC reached the liquid phase, except in the case of EOs from *P. odorata* on *E. coli*, where the vapor phase was twice as effective as the liquid phase.

In general, Gram-negative bacteria are more resistant to EOs than Gram-positive bacteria [40]. This is supported by our results because, as it is shown in Table 2, Gram-positive bacteria were more sensitive to tested EOs in comparison with Gram-negative bacteria. It is possible that active compounds in EOs can more easily break important bonds (peptidoglycan) in the cell wall structure of Gram-positive bacteria. The structure of the Gram-positive bacteria cell wall allows hydrophobic molecules to easily penetrate the cells and act on both the cell wall and within the cytoplasm. After the cell wall is broken, the reactive constituents of the essential oil can penetrate the interior of the cell and further damage its DNA. The other fact is that phenolic compounds, which are also present in the EOs, generally show antimicrobial activity against Gram-positive bacteria. On the other hand, the cell wall of Gram-negative bacteria is far more complex, and it is, among other reasons, why they are more resistant to biologically active compounds (EOs) [4].

Table 2. Antimicrobial activity of tested essential oils and antibiotic ampicillin against Gram-negative and Gram-positive bacteria.

Bacterium	Sample/Growth/MIC (μg·mL^{-1})					
	Houttuynia cordata		*Persicaria odorata*		*Ampicillin*	
	Agar	Broth	Agar	Broth	Agar	Broth
Gram negative						
Escherichia coli	1024	512	512	1024	>4	0.50
Pseudomonas aeruginosa	>1024	>1024	>1024	1024	>4	1.00
Klebsiella pneumoniae	>1024	1024	>1024	1024	>4	>4.00
Serratia marcescens	>1024	1024	>1024	>1024	>4	4.00
Gram positive						
Staphylococcus aureus	>1024	1024	>1024	>1024	>4	0.50
Enterococcus faecalis	1024	128	>1024	512	>4	0.25
Streptococcus pyogenes	>1024	512	1024	512	>4	1.00
Bacillus subtilis	>1024	>1024	>1024	512	>4	2.00

The most abundant compound in *H. cordata* essential oil is myrcene. Myrcene has an antimicrobial activity and moreover enhances the activity of antibiotics [41]. EOs with a high content of myrcene have a positive effect on urinary and genital infections [42,43]. These infections may be caused among others by *E. coli* and *E. faecalis*; therefore, we could assume that EOs from *H. cordata* will affect them, which has been confirmed in this study. The most abundant compound in *P. odorata* essential oil was α-humulene, which is known for its anti-inflammatory effect. It is well known that EOs with α-humulene are natural antimicrobial agents [44–46]. Pichette et al. [46] have tested the antimicrobial activity of α-humulene against *E. coli* and *S. aureus* using the microdillution method. α-humulene exhibited an MIC of 2.6 μg·mL^{-1} against *S. aureus* and an MIC of more than 20 μg·mL^{-1} against *E. coli*. Jang et al. [45] have tested the antimicrobial activity of α-humulene against *B. fragilis* and obtained MIC of 0.5 μg·mL^{-1}. The high content of α-humulene could be the reason why *P. odorata* EOs inhibited the growth of six from eight tested bacteria. On the other hand, there is one possible disadvantage when using these oils orally or internally, which is possible irritation or allergy caused by *cis*-caryophyllene, which both EOs contains [47]. It would be necessary to further examine the negative effect of each compound in the essential oil on the human body before using it for treating illness.

As far as authors know, there are no previous reports about *Persicaria odorata* and *Houttuynia cordata* essential oils and its antimicrobial activity in the vapor phase, so it is not possible to further compare those results with other publications. However, there are some reports about testing the liquid phase of EOs from *Houttuynia cordata*. Verma et al. [24] tested the antimicrobial activity of the essential oil from *Houttuynia cordata* against four bacteria (*Staphylococcus aures*, *Streptococcus mutans*, *Mycobacterium smegmatis* and *Enteroccocus faecalis*). Their essential oil exhibited MIC in the range of 0.52–1.04 μL·mL^{-1}. Ji et al. [20] performed a disc diffusion test to determine antimicrobial activity of *H. cordata* essential oil against *Bacillus subtilis*, *Escherichia coli* and *Staphylococcus aureus*; unfortunately, the disc diffusion test is only a screening method, which is not possible to compare with MIC. Lu et al. [22] tested the antimicrobial activity of *H. cordata* EOs against *Staphylococcus aureus* and *Sarcina ureae* using the broth and agar dilution method. Their reached minimal inhibitory concentration was in the range of 0.5–1.0 μL·mL^{-1}.

3. Materials and Methods

3.1. Plant Material

Approximately 120 g of fresh Chinese herbs (*H. cordata* and *P. odorata*) was purchased in a local Vietnamese market (TTTM Sapa, Prague, Czech Republic). Each sample was air dried in a dark room at the laboratory's temperature. Prior to the distillation, both herbs, including leaves and stems, were crushed into smaller pieces.

3.2. Essential Oil Isolation

Essential oils were obtained by hydrodistillation using Clevenger-type apparatus. The EOs was prepared as follows: 26.7 g (*H. cordata*) or 18.4 g (*P. odorata*) of dried herb was weighted into a 2000 mL distillation flask, 1000 mL of water was added and the EOs was distilled for 4 h. The essential oil was then separated from hydrosol and stored in sealed dark-glass vials at 4 °C until the analysis.

3.3. Bacterial Strains and Culture Media

Four Gram-negative and four Gram-positive bacteria that caused respiratory infections, including upper and lower airway diseases, were chosen. Standard strains were used for the experiment: *Escherichia coli* CCM 3954, *Pseudomonas aeruginosa* CCM 3955, *Klebsiella pneumoniae* NPK12, *Serratia marcescens* CCM 303, *Staphylococcus aureus* CCM 4223, *Enterococcus faecalis* CCM 4224, *Streptococcus pyogenes* NPK01 and *Bacillus subtilis* CCM 2215. All standard strains were purchased from Czech Collection of Microorganisms (CCM), Brno, Czech Republic. Bacteria were incubated at 37 °C for 24 h. Prior to the experiment, bacterial suspensions with turbidity according to the McFarland scale were prepared corresponding to the grade 0.5 (~1.5×10^8 CFU·mL^{-1}). The cultivation and assay media were Mueller-Hinton (agar and broth; Himedia, India). The pH of the broth was adjusted to a final value of 7.6 using Trizma base and hydrochloric acid (both Himedia, India). Ampicilin (St. Louis, MO, USA) was used as a positive antibiotic control.

3.4. Antimicrobial Activity Assay

The antimicrobial activity of the liquid and vapor phase of Eos was determined by the broth microdilution volatilization method [35]. The experiments were carried out in 96-well microtiter plates with one well volume of 400 µL. The test is designed for the testing of 6 essential oils in total. For this study, we tested only 2 essential oils, and different samples were in other wells. The plates were covered with wooden plates and clamped to prevent vapor phase leakage. The edge wells were left blank to avoid the edge effect. First, essential oil samples were prepared as follows: approximately 2 µL of EOs was added to corresponding amount of dimethyl sulfoxide (DMSO) at a concentration of 1%, then further diluted in the corresponding broth to initial concentration. Then, the antibiotic was prepared at an initial concentration of 4 µg·mL^{-1}. In the first part of the experiment, 30 µL of agar was pipetted onto the plate lid and inoculated with 5 µL of bacterial suspension for vapor phase testing. In the second part (liquid phase assay), 100 µL of buffered Mueller-Hinton broth was pipetted into the wells. Each well had a final volume of 100 µL. Seven two-fold diluted concentrations of samples starting at a concentration of 1024 µg·mL^{-1} were prepared for each essential oil in one row. A positive and negative control of bacterial growth was prepared in the first two columns. In the last column, 6 two-fold diluted concentrations of antibiotic starting at 4 µg·mL^{-1} were prepared. Finally, all wells except the negative control were inoculated with 5 µL of bacterial suspension. Plates were closed, fixed and incubated at 37 °C for 24 h. After incubation, minimal inhibitory concentrations of EOs were evaluated by the visual assessment of bacterial growth after the coloring of a metabolically active bacterial colony with thiazolyl blue tetrazolium bromide dye (MTT; Sigma Aldrich, Prague, Czech Republic). A total of 25 µL of 600 µg·mL^{-1} dye was applied to the lid and each well of the plate and equilibrated for 10 min. The color changed from yellow (dead cells) to purple (live cells). Thereafter, the MICs were recorded. All experiments were performed in triplicate in three independent experiments. The results were expressed as the median of minimal inhibition concentration of the antimicrobial agent values.

3.5. GC–MS Analysis

The GC–MS analysis of samples was carried out by using a Gas Chromatograph GC 2010 coupled to a Mass Selective Detector GCMS-QP2010 Plus (both Shimadzu, Kyoto, Japan) and Combi Pal Autosampler (CTC Analytics, AG, Zwingen, Swizerland) on a capillary column SLB-5ms Supelco (30 m × 0.25 mm, 2.5 μm film thickness; Bellefonte, PA, USA). The carrier gas was Helium 5.0 (Linde, Prague, Czech Republic) with a constant flow of 30 $cm \cdot s^{-1}$. The oven temperature program was set at an initial temperature of 40 °C for 3 min, then heated up to 250 °C at 2 $°C \cdot min^{-1}$ and held at 250 °C for 10 min. The injector and detector temperatures were set at 200 °C. The mass spectrometry detector was operated under electron ionization mode at ionization energy of 70 eV when ions with m/z 33–500 were scanned. A total of 1 μL of diluted essential oil (200 times, *n*-hexane) was injected with a split ratio 1:50. The experimental results of retention indices were calculated relative to C8–C33 *n*-alkanes in concentrations of 100–200 $\mu g \cdot mL^{-1}$, dissolved in *n*-hexane (Restek, Bellefonte, PA, USA). The calculation was performed according to the van den Dool and Kratz method, and the results were further compared to published data [36,37]. Compounds were identified by comparing their mass spectra with mass spectra of several standards (Table 1) and commercial mass spectral databases NIST'14 Mass Spectral Library and FFNSC 2 GC/MS Library Release 2.0 (Flavor and Fragrance Natural and Synthetic Compounds Library) and further checked out by manual mass spectra evaluation.

3.6. GC–FID Analysis

The GC–FID analysis of samples was carried out by using a Gas Chromatograph GC 2010 with a flame ionization detector (Shimadzu, Kyoto, Japan) and Autosampler Combi Pal (CTC Analytics, AG, Zwingen, Swizerland) on a capillary column SLB-5ms Supelco (30 m × 0.25 mm, 2.5 μm film thickness; Bellefonte, PA, USA). The GC–FID conditions were the same as in case of GC–MS analysis. The injector temperature was set at 200 °C and the detector temperature was set at 260 °C. A total of 1 μL of diluted essential oil (200 times, *n*-hexane) was injected with a split ratio 1:50. As in the case of GC–MS, experimental retention indices were calculated and compared to published data.

4. Conclusions

This study shows new interesting knowledge about EOs distilled from two Asian herbs—*Persicaria odorata* and *Houttuynia cordata*. The chemical composition of EOs corresponds to previous studies with a minor deviation that may be caused by agronomic factors, sample storage or sample preparation and other factors. Both EOs showed antimicrobial activity in different concentrations to different bacteria. Due to great antibacterial activity, along with the composition of Eos, we see a great potential for future usages of these oils, such as natural antimicrobials or food preservatives. As far as we know, we were the first to describe the antimicrobial properties of those EOs in both vapor and liquid phases on eight selected bacteria. Furthermore, it is necessary to study the possible cytotoxicity of these oils. The disadvantage is that both oils contain *cis*-caryophyllene that causes allergic reactions and skin irritation. It would be necessary to find the balance in concentrations of beneficial antimicrobial active compounds and potentially toxic compounds.

Author Contributions: Conceptualization, K.Ř. and T.B.; methodology, T.B. and M.H.; formal analysis, K.Ř.; investigation, K.Ř.; resources, K.Ř.; writing—original draft preparation, K.Ř.; writing—review and editing, P.B.; T.B. and D.Š.; supervision, K.V.; project administration, P.B. All authors have read and agreed to the published version of the manuscript.

Funding: This research received no external funding.

Conflicts of Interest: The authors declare no conflict of interest.

References

1. Caputo, L.; Nazzaro, F.; Souza, L.F.; Aliberti, L.; De Martino, L.; Fratianni, F.; Coppola, R.; De Feo, V. Laurus nobilis: Composition of Essential Oil and Its Biological Activities. *Molecules* **2017**, *22*, 930. [CrossRef] [PubMed]
2. Merghni, A.; Marzouki, H.; Hentati, H.; Aouni, M.; Mastouri, M. Antibacterial and antibiofilm activities of Laurus nobilis L. essential oil against Staphylococcus aureus strains associated with oral infections. *Curr. Res. Transl. Med.* **2016**, *64*, 29–34. [CrossRef] [PubMed]
3. Rafiq, R.; Hayek, S.A.; Anyanwu, U.; Hardy, B.I.; Giddings, V.L.; Ibrahim, S.A.; Tahergorabi, R.; Kang, H.W. Antibacterial and Antioxidant Activities of Essential Oils from Artemisia herba-alba Asso., Pelargonium capitatum x radens and Laurus nobilis L. *Foods* **2016**, *5*, 28. [CrossRef] [PubMed]
4. Nazzaro, F.; Fratianni, F.; De Martino, L.; Coppola, R.; De Feo, V. Effect of essential oils on pathogenic bacteria. *Pharmaceuticals* **2013**, *6*, 1451–1474. [CrossRef]
5. Kon, K.V.; Rai, M.K. Plant essential oils and their constituents in coping with multidrug-resistant bacteria. *Expert Rev. Anti-Infect. Ther.* **2012**, *10*, 775–790. [CrossRef]
6. Feyaerts, A.F.; Mathe, L.; Luyten, W.; Tournu, H.; Van Dyck, K.; Broekx, L.; Van Dijck, P. Assay and recommendations for the detection of vapour-phase-mediated antimicrobial activities. *Flavour Frag. J.* **2017**, *32*, 347–353. [CrossRef]
7. Houdkova, M.; Doskocil, I.; Urbanova, K.; Tulin, E.; Rondevaldova, J.; Tulin, A.B.; Kudera, T.; Tulin, E.E.; Zeleny, V.; Kokoska, L. Evaluation of antipneumonic effect of Philippine essential oils using broth microdilution volatilization method and their lung fibroblasts toxicity. *Nat. Prod. Commun.* **2018**, *13*, 1059–1066. [CrossRef]
8. Acs, K.; Bencsik, T.; Boszormenyi, A.; Kocsis, B.; Horvath, G. Essential oils and their vapors as potential antibacterial agents against respiratory tract pathogens. *Nat. Prod. Commun.* **2016**, *11*, 1709–1712.
9. Amat, S.; Baines, D.; Alexander, T.W. A vapour phase assay for evaluating the antimicrobial activities of essential oils against bovine respiratory bacterial pathogens. *Lett. Appl. Microbiol.* **2017**, *65*, 489–495. [CrossRef]
10. Reyes-Jurado, F.; Cervantes-Rincon, T.; Bach, H.; Lopez-Malo, A.; Palou, E. Antimicrobial activity of Mexican oregano (*Lippia berlandieri*), thyme (*Thymus vulgaris*), and mustard (*Brassica nigra*) essential oils in gaseous phase. *Ind. Crops Prod.* **2019**, *131*, 90–95. [CrossRef]
11. Santomauro, F.; Donato, R.; Pini, G.; Sacco, C.; Ascrizzi, R.; Bilia, A.R. Liquid and vapor-phase activity of *Artemisia annua* essential oil against pathogenic *Malassezia* spp. *Planta Med.* **2018**, *84*, 160–167. [CrossRef] [PubMed]
12. Torpol, K.; Wiriyacharee, P.; Sriwattana, S.; Sangsuwan, J.; Prinyawiwatkul, W. Antimicrobia activity of garlic (*Allium sativum* L.) and holy basil (*Ocimum sanctum* L.) essential oils applied by liquid vs. vapour phases. *Int. J. Food Sci. Technol.* **2018**, *53*, 2119–2128. [CrossRef]
13. Azadbakht, E.; Maghsoudlou, Y.; Khomiri, M.; Kashiri, M. Development and structural characterization of chitosan films containing *Eucalyptus globulus* essential oil: Potential as an antimicrobial carrier for packaging of sliced sausage. *Food Packag. Shelf Life* **2018**, *17*, 65–72. [CrossRef]
14. Chen, C.W.; Xu, Z.W.; Ma, Y.R.; Liu, J.L.; Zhang, Q.J.; Tang, Z.P.; Fu, K.J.; Yang, F.X.; Xie, J. Properties, vapour-phase antimicrobial and antioxidant activities of active poly(vinyl alcohol) packaging films incorporated with clove oil. *Food Control* **2018**, *88*, 105–112. [CrossRef]
15. Silva, F.; Domingues, F.C. Antimicrobial activity of coriander oil and its effectiveness as food preservative. *Crit. Rev. Food Sci. Nutr.* **2017**, *57*, 35–47. [CrossRef]
16. Laird, K.; Phillips, C. Vapour phase: A potential future use for essential oils as antimicrobials? *Lett. Appl. Microbiol.* **2012**, *54*, 169–174. [CrossRef]
17. Ji, H.; Kim, H.; Beuchat, L.R.; Ryu, J.H. Synergistic antimicrobial activities of essential oil vapours against *Penicillium corylophilum* on a laboratory medium and beef jerky. *Int. J. Food Microbiol.* **2019**, *291*, 104–110. [CrossRef]
18. Doran, A.L.; Morden, W.E.; Dunn, K.; Edwards-Jones, V. Vapour-phase activities of essential oils against antibiotic sensitive and resistant bacteria including MRSA. *Lett. Appl. Microbiol.* **2009**, *48*, 387–392. [CrossRef]
19. Asakawa, Y.; Tomiyama, K.; Sakurai, K.; Kawakami, Y.; Yaguchi, Y. Volatile compounds from the different organs of *Houttuynia cordata* and *Litsea cubeba* (*L. citriodora*). *J. Oleo Sci.* **2017**, *66*, 889–895. [CrossRef]

20. Ji, Y.B.; Yang, J.J.; Yu, M.; Cao, Y.; Guo, S.Z.; Qiao, A.N. Study on medicinal plant active substances extraction and antibacterial activity of *Houttuynia cordata*. In Proceedings of the 1st International Global on Renewable Energy and Development, Singapore, 22–25 December 2017; IOP Publishing Ltd: Bristol, UK, 2017; Volume 100.
21. Kwon, H.-D.; Cha, I.-H.; Lee, W.-K.; Song, J.-H.; Park, I.-H. Antibacterial activity of volatile flavor components from *Houttuynia cordata* Thunb. *Prev. Nutr. Food Sci.* **1996**, *1*, 208–213.
22. Lu, H.; Wu, X.; Liang, Y.; Zhang, J. Variation in chemical composition and antibacterial activities of essential oils from two species of *Houttuynia* THUNB. *Chem. Pharm. Bull.* **2006**, *54*, 936–940. [CrossRef] [PubMed]
23. Pang, J.; Dong, W.; Li, Y.; Xia, X.; Liu, Z.; Hao, H.; Jiang, L.; Liu, Y. Purification of *Houttuynia cordata* Thunb. essential oil using macroporous resin followed by microemulsion encapsulation to improve its safety and antiviral activity. *Molecules* **2017**, *22*, 293. [CrossRef] [PubMed]
24. Verma, R.S.; Joshi, N.; Padalia, R.C.; Singh, V.R.; Goswami, P.; Kumar, A.; Iqbal, H.; Verma, R.K.; Chanda, D.; Chauhan, A. Chemical composition and allelopathic, antibacterial, antifungal, and antiacetylcholinesterase activity of fish-mint (*Houttuynia cordata*Thunb.) from India. *Chem. Biodivers.* **2017**, *14*, e1700189. [CrossRef] [PubMed]
25. Shavandi, M.A.; Haddadian, Z.; Ismail, M.H.S. *Eryngium foetidum* L. *Coriandrum sativum* and *Persicaria odorata* L.: A review. *J. Asian Sci. Res.* **2012**, *2*, 410.
26. Starkenmann, C.; Luca, L.; Niclass, Y.; Praz, E.; Roguet, D. Comparison of volatile constituents of *Persicaria odorata* (Lour.) Sojak (*Polygonum odoratum* Lour.) and *Persicaria hydropiper* L. Spach (*Polygonum hydropiper* L.). *J. Agric. Food Chem.* **2006**, *54*, 3067–3071. [CrossRef]
27. Dai, D.N.; Thang, T.D.; Ogunmoye, A.; Eresanya, O.I.; Ogunwande, I.A. Chemical constituents of essential oils from the leaves of *Tithonia diversifolia*, *Houttuynia cordata* and *Asarum glabrum* grown in Vietnam. *Am. J. Essent. Oil.* **2015**, *2*, 17–21.
28. Oh, S. An effective quality control of pharmacologically active volatiles of *Houttuynia cordata* Thunb by fast gas chromatography-surface acoustic wave sensor. *Molecules* **2015**, *20*, 10298–10312. [CrossRef]
29. Zhang, W.; Lu, F.; Pan, S.; Li, S. Extraction of volatile oil from *Houttuynia cordata* and its anti-biotic and anti-virus activities. *Pract. Prev. Med.* **2008**, *15*, 312–316.
30. Jiangang, F.; Ling, D.; Zhang, L.; Hongmei, L. *Houttuynia cordata* Thunb: A review of phytochemistry and pharmacology and quality control. *Chin. Med.* **2013**, *4*, 101–123.
31. Fujita, K.; Chavasiri, W.; Kubo, I. Anti-salmonella activity of volatile compounds of Vietnam coriander. *Phytother. Res.* **2015**, *29*, 1081–1087. [CrossRef]
32. Chansiw, N.; Paradee, N.; Chotinantakul, K.; Srichairattanakool, S. Anti-hemolytic, antibacterial and anti-cancer activities of methanolic extracts from leaves and stems of *Polyg. odoratum*. *Asian Pac. J. Trop. Biomed.* **2018**, *8*, 580–585. [CrossRef]
33. Almarie, A.; Mamat, A.; Wahab, Z.; Rukunudin, I. Chemical composition and phytotoxicity of essential oils isolated from Malaysian plants. *Allelopath. J.* **2016**, *37*, 55–69.
34. Murray, A.F.; Satooka, H.; Shimizu, K.; Chavasiri, W.; Kubo, I. *Polygonum odoratum* essential oil inhibits the activity of mushroom derived tyrosinase. *Heliyon* **2019**, *5*, e02817. [CrossRef] [PubMed]
35. Houdkova, M.; Rondevaldova, J.; Doskocil, I.; Kokoska, L. Evaluation of antibacterial potential and toxicity of plant volatile compounds using new broth microdilution volatilization method and modified MTT assay. *Fitoterapia* **2017**, *118*, 56–62. [CrossRef]
36. Linstrom, P.J.; Mallard, W. NIST Chemistry WebBook, NIST Standard Reference Database Number 69. Available online: https://webbook.nist.gov/chemistry/ (accessed on 30 April 2020).
37. Adams, R.P. *Identification of Essential Oil Components by Gas Chromatography/Mass Spectrometry*, 4th ed.; Allured Publishing Corporation: Carol Stream, IL, USA, 2007.
38. Dũng, N.X.; Van Hac, L.; Leclercq, P.A. Volatile constituents of the aerial parts of Vietnamese *Polygonum odoratum* L. *J. Essent. Oil Res.* **1995**, *7*, 339–340. [CrossRef]
39. Hunter, M.V.; Brophy, J.J.; Ralph, B.J.; Bienvenu, F.E. Composition of *Polygonum odoratum* Lour. from Southern Australia. *J. Essent. Oil Res.* **1997**, *9*, 603–604. [CrossRef]
40. Trombetta, D.; Castelli, F.; Sarpietro, M.G.; Venuti, V.; Cristani, M.; Daniele, C.; Saija, A.; Mazzanti, G.; Bisignano, G. Mechanisms of antibacterial action of three monoterpenes. *Antimicrob. Agents Chemother.* **2005**, *49*, 2474–2478. [CrossRef]

41. Sieniawska, E.; Swatko-Ossor, M.; Sawicki, R.; Skalicka-Woźniak, K.; Ginalska, G. Natural terpenes influence the activity of antibiotics against isolated *Mycobacterium tuberculosis*. *Med. Princ. Pract.* **2017**, *26*, 108–112. [CrossRef]
42. Lis, A.; Banaszczak, P. Chemical composition of the essential oils from flowers and leaves of *Phellodendron chinense* CK Schneid. *J. Essent. Oil-Bear. Plants* **2010**, *13*, 52–58. [CrossRef]
43. Arnal-Schnebelen, B.; Hadji-Minaglou, F.; Peroteau, J.; Ribeyre, F.; De Billerbeck, V. Essential oils in infectious gynaecological disease: A statistical study of 658 cases. *Int. J. Aromather.* **2004**, *14*, 192–197. [CrossRef]
44. Fernandes, E.S.; Passos, G.F.; Medeiros, R.; da Cunha, F.M.; Ferreira, J.; Campos, M.M.; Pianowski, L.F.; Calixto, J.B. Anti-inflammatory effects of compounds alpha-humulene and (−)-*trans*-caryophyllene isolated from the essential oil of *Cordia verbenacea*. *Eur. J. Pharmacol.* **2007**, *569*, 228–236. [CrossRef] [PubMed]
45. Jang, H.-I.; Rhee, K.-J.; Eom, Y.-B. Antibacterial and antibiofilm effects of α-humulene against *Bacteroides fragilis*. *Can. J. Microbiol.* **2020**, *66*, 1–11. [CrossRef] [PubMed]
46. Pichette, A.; Larouche, P.L.; Lebrun, M.; Legault, J. Composition and antibacterial activity of *Abies balsamea* essential oil. *Phytother. Res.* **2006**, *20*, 371–373. [CrossRef] [PubMed]
47. National Center for Biotechnology Information. PubChem Database. Available online: https://pubchem.ncbi.nlm.nih.gov/compound/Caryophyllene (accessed on 28 April 2020).

Sample Availability: Samples of the compounds are not available from the authors.

© 2020 by the authors. Licensee MDPI, Basel, Switzerland. This article is an open access article distributed under the terms and conditions of the Creative Commons Attribution (CC BY) license (http://creativecommons.org/licenses/by/4.0/).

molecules

Article

Assessment of Mint, Basil, and Lavender Essential Oil Vapor-Phase in Antifungal Protection and Lemon Fruit Quality

Renata M. Sumalan [1,*], Raufdzhon Kuganov [2,*], Diana Obistioiu [3], Iuliana Popescu [4,*], Isidora Radulov [4], Ersilia Alexa [5], Monica Negrea [5], Amonullo F. Salimzoda [2], Radu L. Sumalan [1,*] and Ileana Cocan [5]

1 Faculty of Horticulture and Forestry, Banat's University of Agricultural Sciences and Veterinary Medicine "King Michael I of Romania" from Timisoara, 119 C. Aradului, 300645 Timisoara, Romania
2 Department of Food Storing and Processing Technology, Faculty of Agribusiness, Tajik Agrarian University named Shirinsho Shotemur, 146 Rudaki, Dushanbe 73400, Tajikistan; a.faizullozoda@mail.ru
3 Interdisciplinary Research Platform, Banat's University of Agricultural Sciences and Veterinary Medicine "King Michael I of Romania" from Timisoara, 300645 Timisoara, Romania; diana.obistioiu@yahoo.com
4 Faculty of Agriculture, Banat's University of Agricultural Sciences and Veterinary Medicine "King Michael I of Romania" from Timisoara, 300645 Timisoara, Romania; isidoraradulov@yahoo.com
5 Faculty of Food Engineering, Banat's University of Agricultural Sciences and Veterinary Medicine "King Michael I of Romania" from Timisoara, 300645 Timisoara, Romania; alexa.ersilia@yahoo.ro (E.A.); negrea_monica2000@yahoo.com (M.N.); negreaileana@yahoo.com (I.C.)
* Correspondence: srenata_maria@yahoo.com (R.M.S.); kuganovr@mail.ru (R.K.); iuliapopescu2002@yahoo.com (I.P.); sumalanagro@yahoo.com (R.L.S.); Tel.: +40-7233-46537 (R.M.S.); +992-9046-06661 (R.K.); +40-7228-58589 (I.P.); +40-7235-47363 (R.L.S.)

Academic Editor: Ladislav Kokoska
Received: 24 March 2020; Accepted: 14 April 2020; Published: 16 April 2020

Abstract: There is an increasing interest in developing natural methods to replace the current chemicals used for maintaining postharvest quality of citrus fruits. The essential oil antifungal activity of mint (MEO), basil (BEO), and lavender (LEO) acting as the vapor-phases was tested against *Penicillium digitatum*. The minimum doses with fungistatic and fungicidal effect, in vitro, acting as the vapor-phases, were set up. The minimum fungicidal dose was 300 μL for BEO and 350 μL LEO, while for MEO only minimal dose with fungistatic effect was reached. The IC50 values were calculated and used (*v*/*v*) for testing preservation of lemon fruits, in close space enriched in vapor oil. For this purpose, the following two independent in vivo experiments were carried out: experiment 1, inoculated lemons with *P. digitatum* stored without chemical treatments 7 days, at 22 ± 2 °C, at two concentrations (C1—IC50 equivalent; C2—half of C1); and experiment 2, the non-inoculated lemons kept under the same conditions and concentrations of EO vapor served to evaluate the lemon quality properties. The results showed that antifungal protective effect was provided in the order of LEO-C1 > BEO-C1 > MEO-C1 > BEO-C2 > MEO-C2 > LEO-C2. The quality indicators like weight loss, pH, and firmness were not negatively influenced.

Keywords: IC50; fungicide effect; GC/MS analysis; *Penicillium digitatum*; ascorbic acid

1. Introduction

Nowadays, the major concern in postharvest research is keeping the phytonutrients, which assures the nutritional value of fruits and vegetables while minimizing the losses during storage [1]. The fungal decay of lemon fruits (*Citrus limon*) is the main cause of microbiological spoilage during postharvest, leading to high economic declines due to the high water content and to the wounds that often form as a result of harvesting and transportation. The main fungus genus responsible for lemon depreciation is *Penicillium*, among which *Penicillium italicum* is responsible for blue rot, and *Penicillium digitatum* causes green rot [2–5].

Because consumers are more and more concerned about the use of synthetic preservatives, the exploitation of natural compounds has been intensively researched in recent years [6–10]. Innovations in preserving horticultural commodities can be achieved through three directions: (1) introduction of biocontrol agents, such as yeasts and bacteria [11–13]; (2) use of plant essential oils (EOs) extracted from thyme, mint, lemongrass, lemon balm, oregano, or savory [5,14–16]; and (3) by physical methods like sulphur dioxide fumigation, use of ozone, or mixed techniques [7,17,18].

The antifungal activity of EOs has been known and used for centuries, and nowadays efforts to promote natural compounds in post-harvest control of horticultural products have led to an increased interest in their possible applications [19–23]. A particular problem regarding the use of essential oils is related to the decrease in concentration of bioactive compounds due to evaporation [24].

The volatilization properties of essential oils was the idea behind this study. Therefore, it would be particularly interesting to know what amount of essential oil is required to achieve the fungistatic and fungicidal action of the EO vapor-phase against fruit fungal depreciation, and what essential oil is more economically efficient, given their high costs [24]. Considering this background, the aim of this work was to study the effectiveness of the vapor-phases of the essential oils (EOs) of *Mentha piperita* (MEO), basil (*Ocimum basilicum*, BEO), and lavender (*Lavandula angustifolia*, LEO) in lemon preservation for antifungal protection against *P. digitatum* and the effect on fruits quality indicators.

Our research could help to extend the method of storage of freshly cut vegetables or fruit in a modified atmosphere based on the use of natural volatile compounds. To the best of our knowledge, this is the first time that essential oils in the vapor-phase were tested as a preservation method to prevent fungal lemon degradation.

2. Results

2.1. Chemical Composition of Essential Oils by GC/MS Analysis

The MEO, BEO, and LEO composition was determined by the GC/MS method and 23 different components of MEO, 12 of BEO, and 23 of LEO were identified (Table 1). Regarding the extraction of the essential oils, the highest yield was obtained for the dry mass of lavender with 4.85%, followed by the mint with 3.22%, and basil with 0.28%.

About 99.99% of the total constituents were detected in MEO. Of them *p*-menthan-3-one had the highest percent (31.00%), followed by menthol (25.19%), 1,3,12-nonadecatriene (9.76%), eucalyptol (7.44%), and carvone (6.72%). The oxygenated monoterpene compounds predominated in a proportion of 81.28%; sesquiterpene oxygenated compounds were not detected.

BEO had the lowest diversity of compounds compared with MEO and LEO. Only 12 compounds were identified (99.68% proportion), the majority being monoterpene oxygenated compounds (97.02%). The predominant compound was estragole (49.94%), followed by linalool (41.49%) and eucalyptol (3.46%). Other compounds were found in concentrations lower than 1%. LEO showed a chemical composition similar to that of MEO, 23 compounds being identified of which two represented the major constituents linalool (31.44%) and linalyl acetate (31.78%). The next constituents in lower proportion were 4-terpineol (8.43%), caryophyllene (5.39%), and lavandulol (5.24%). LEO was the singular EO that contained the sesquiterpene oxygenated compound, caryophyllene oxide, in a low proportion (0.35%).

Table 1. The chemical composition of essential oils, GC/MS analysis (% from total compounds).

Compounds	Type	LRIc/LRIr	MEO	BEO	LEO
α-Pinene	MH	1021/1015	0.62	-	-
β-Pinene	MH	1106/1096	0.89	0.40	-
Thujene	MH	1118/1122	0.50	-	-
β-Myrcene	MH	1158/1164	0.31	-	0.36
p-Mentha-2,4(8)-diene	MH	1176/1180	0.23	-	-
D-Limonene	MH	1196/1193	3.06	0.39	0.62
Eucalyptol	MO	1204/1209	7.44	3.46	1.48
trans-β-Ocimene	MH	1228/1230	-	-	4.75
Gamma Terpinene	MH	1241/1242	0.43	-	-
cis-β-Ocimene	MH	1245/1250	-	0.62	4.02
p-Cymol	MH	1263/1264	-	-	0.35
o-Cymol	MH	1265/1268	-	0.31	-
6-Methyl hept-5-en-2-one	MO	1325/1325	-	0.30	-
Octen-1-ol acetate	MO	1364/1365	-	-	1.20
p-Menthan-3-one	MO	1457/1458	31.00	-	-
cis-Linaloloxide	MO	1460/1463	-	-	0.14
Menthofurane	MO	1474/1477	1.38	-	-
D-Menthone	MO	1484/1486	3.19	-	-
Camphor	MO	1507/1518	-	-	0.26
Linalool	MO	1533/1537	0.39	41.49	31.44
Linalyl acetate	MO	1541/1543	0.94	-	31.78
Menthyl acetate	MO	1552/1551	2.20	-	-
p-Menth-8-en-3-one	MO	1562/1561	0.19	-	-
1-Terpineol	MO	1565/1562	-	-	0.17
Alfa Santalene	SH	1571/1574	-	-	0.56
4-Terpineol	MO	1593/1592	-	–	8.43
Caryophyllene	SH	1598/1599	2.35	-	5.39
Cyclohexanone,5 methyl-2-(1 methylethyliden)	MO	1635/1633	2.07	-	-
Estragole/Methyl chavicol	MO	1652/1650	-	49.94	-
Beta Farnesene	SH	1653/1652	-	-	1.35
8-*p*-Menthen-2-ol	MO	1655/1656	0.24	-	
Cryptone	MO	1658/1661	-	-	0.47
Cis Citral	MO	1664/1668	-	0.66	-
Alfa Terpineol	MO	1694/1697	-	-	0.13
Germacrene D	SH	1708/1708	0.57	-	0.45
Trans Citral	MO	1712/1714	-	0.80	-
3-Carvomenthenone	MO	1713/1710	0.32	-	-
Carvone	MO	1719/1718	6.72	-	-
Alfa Bisabolene	SH	1734/1736	-	0.94	-
3-Isopropylbenzaldehyde	MO	1765/1765	-	-	0.13
Menthol	MO	1801/1788	25.19	-	-
Anethole	MO	1807/1817	-	-	0.69
Lavandulol	MO	1879/1879	-	-	5.24
Caryophyllene oxide	SO	1998/1989	-	-	0.35
Eugenol	MO	2198/2186	-	0.37	-
1,3,12-Nonadecatriene	SH	2405/2400	9.76	-	-
Total (%)			99.99	99.68	99.76
from which	MH		6.06	1.72	10.10
	MO		81.28	97.02	81.56
	SH		12.65	0.94	7.75
	SO		-	-	0.35

LRIc, calculated linear retention index; LRIr, referred linear retention index [25]; MH, monoterpene hydrocarbonates; MO, monoterpene oxygenated; SH, sesquiterpene hydrocarbonates; SO, sesquiterpene oxygenated.

2.2. *In Vitro Assay of MEO, BEO, and LEO Vapors for Antifungal Performances*

The presence of essential oil vapor led to inhibition of mycelial growth of *P. digitatum* in a dose-dependent manner. The EO dose used in the experiment provided the vapor-enrichment of Petri dish atmosphere for 50 cm^3 of air according to Formula (2). From the first dose of added essential oils (50 μL) we observed decreases of new mycelium growth, but only for BEO and LEO were the differences statistically significant (Figure 1).

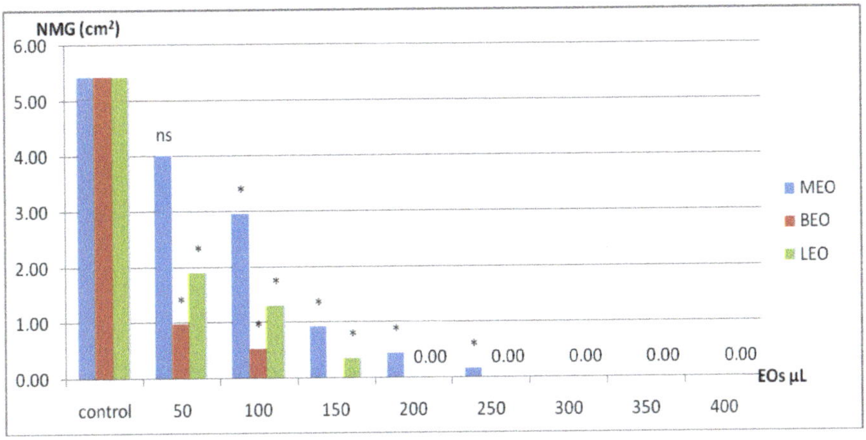

Figure 1. The NMG (new mycelium growth) of *P. digitatum* in EO vapor-phase modified atmospheres in the in vitro assay (cm^2). MEO, mint essential oil; BEO, basil essential oil; LEO, lavender essential oil; * with statistical differences compared to control for $p \leq 0.05$, $n = 4$; ns, without statistical significance.

The mycelial growth of the fungus became null at 150 μL BEO, 200 μL LEO, and 300 μL MEO. Thus, these doses can be attributed to the minimum inhibitory dose which provides a fungistatic effect (MFsD), as seen in Table 2. After passing the mycelium on CYA medium without added EOs, we found that the mycelium restarted the growth after 4 days of incubation. Where the fungal growth was not resumed, the dose of the EO used was considered the minimum dose with fungicidal effect (MFdD).

Table 2. Values of fungistatic and fungicidal doses of essential oils used as the vapor-phases in the in vitro study.

EO Treatment	Effect	EO Doses (μL)							
		50	100	150	200	250	300	350	400
MEO	MFsD [a]	+	+	+	+	+	-	-	-
	MFdD [b]						+	+	+
BEO	MFsD [a]	+	+	-	-	-	-	-	-
	MFdD [b]			+	+	+	-	-	-
LEO	MFsD [a]	+	+	+	-	-	-	-	-
	MFdD [b]				+	+	+	-	-

[a] minimum dose with fungistatic effect; [b] minimum dose with fungicidal effect.

Therefore, by transferring the *P. digitatum* mycelia disks, it was possible to estimate the two different doses for each type of EO, as seen in Table 2. Thus, MFsD for BEO was reached at 150 μL (0.3%), and BEO 300 μL (0.6%) had a fungicidal effect. LEO, in the presence of linalool and linalyl acetate (together totaling 63.22%) achieved the MFsD at 200 μL (0.4%) and MFdD at 350 μL (0.7%), while for MEO only MFsD was reached at 300 μL.

2.3. Experiment 1: In Vivo Assay of Inoculated Lemon Fruits Stored with EO Vapor-Phase

After 7 days of lemon storage in atmosphere enriched with essential oil vapor, the mycelium growth of *P. digitatum* was measured. The non-treated control lemons recorded the largest average diameter of 47 mm Ø followed in order by LEO-C2 > MEO-C2 > BEO-C2 > MEO-C1 > BEO-C1 and finally LEO-C1 with an average of fungal diameter of 16.88 mm (Figure 2).

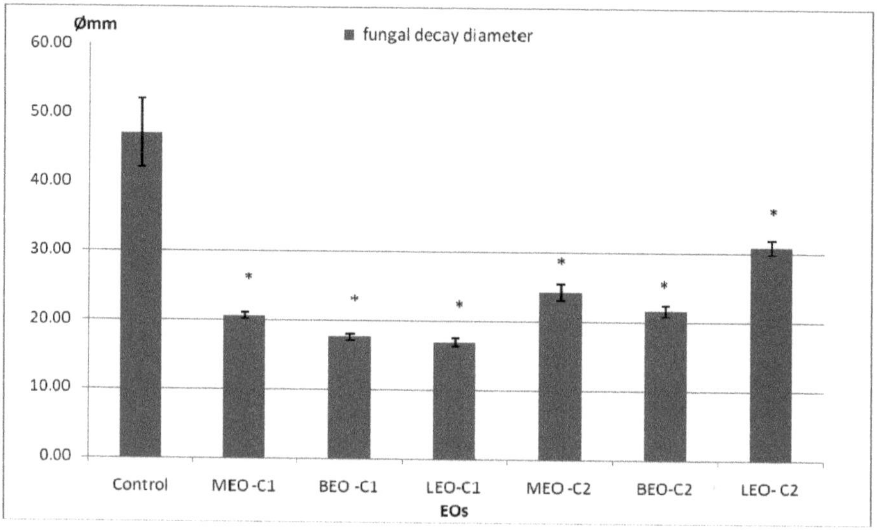

Figure 2. Results of lemon depreciation inoculated with *P. digitatum* stored seven days in the modified atmosphere with EO vapor. Data represent the diameter (mm) affected by the fungal growth with or without reproductive structure presented as mean ± SD. Vertical bars represent standard deviations ($n = 8$); * with statistical differences compared to control at $p < 0.05$; MEO-C1 = 183 μL·L^{-1}, BEO-C1 = 46 μL·L^{-1}, LEO-C1 = 86 μL·L^{-1}, MEO–C2 = 91.5 μL·L^{-1}, BEO-C2 = 23.0 μL·L^{-1}, LEO-C2 = 43.0 μL·L^{-1}.

EO results tested by fumigation of artificially inoculated lemon fruits were compared and classified according to the *t*-test for independent samples. It was observed that between BEO-C1 and LEO-C1 there were no statistically significant differences (Table 3). The same aspect was noted for BEO-C2 and MEO-C2. The lowest effect was recorded for LEO-C2, ranking in second place after the control for the antifungal performance.

Table 3. Ranking the EO treatments acting as vapor in the storage of lemons by *t*-test analysis.

Rank	Treatment	Mean *	LEO C1	BEO C1	MEO C1	BEO C2	MEO C2	LEO C2	Control
1	Control	47.0 ^{a}	30.1	29.4	26.4	25.4	22.8	16.1	
2	LEO-C2	30.9 ^{b}	14.0	13.3	10.3	9.3	6.6		
3	MEO-C2	24.3 ^{c}	7.4	6.6	3.6	2.6			
4	BEO-C2	21.6 ^{d}	4.8	4.0	1.0				
5	MEO-C1	20.6 ^{d}	3.8	3.0					
6	BEO-C1	17.6 ^{e}	0.8						
7	LEO-C1	16.9 ^{e}							

* mean ($n = 8$) with different superscript letters indicating that the differences are statistically significant for $\alpha = 0.05$; C1—concentration corresponding to IC50, C2—half of C1 according to Table 1 (MEO-C1 = 183 μL·L^{-1}, BEO-C1 = 46 μL·L^{-1}, LEO-C1 = 86 μL·L^{-1}, MEO–C2 = 91.5 μL·L^{-1}, BEO-C2 = 23.0 μL·L^{-1}, LEO-C2 = 43.0 μL·L^{-1}).

2.4. Experiment 2: Physiological and Biochemical Indicators of Lemon Quality

Loss of fruit firmness, assessed by the penetration power of lemon fruits, was influenced by type and concentration of EO vapor, as seen in Table 4. In this regard, the treatment LEO-C2 (65.1 N) and MEO–C1 (59.0 N) showed significant differences compared with the control (46.6 N). In contrast, treatments with BEO, at both concentrations (C1 and C2), did not provided significant differences versus the control.

Table 4. Effects of essential oils of mint (MEO), basil (BEO), and lavender (LEO) acting as vapor on firmness (N), weight loss (%), and pH recorded in lemon fruits after storage for 7 days.

Treatment	Firmness [a] (N)	Weight Loss [b] (%)	pH [b]
Initial control	45.0 ± 9.1 ^{A}	-	2.57 ± 0.01
Control	46.6 ± 8.8 ^{A}	0.46 ± 0.1	2.56 ± 0.2 ns
MEO-C1	59.0 ± 15.2 ^{B}	0.44 ± 0.12 *	2.66 ± 0.12 ns
BEO-C1	45.9 ± 10.8 ^{A}	0.51 ± 0.32 *	2.63 ± 0.03 ns
LEO-C1	53.6 ± 15 ^{A,C}	0.33 ± 0.13 ns	2.72 ± 0.06 ns
MEO-C2	51.2 ± 19.5 ^{A,C}	0.47 ± 0.09 *	2.64 ± 0.01 ns
BEO-C2	48.4 ± 9.5 ^{A}	0.42 ± 0.07 *	2.67 ± 0.00 ns
LEO-C2	65.1 ± 7.3 ^{B}	0.35 ± 0.11 *	2.69 ± 0.00 ns

[a] means (n = 10) followed by the same letter do not differ significantly at p = 0.05; [b] means for n = 5; * significant difference according to control; ns without significance; C1-concentration corresponding to IC50; C2—half of C1; MEO-C1 = 183 $\mu L \cdot L^{-1}$, BEO-C1 = 46 $\mu L \cdot L^{-1}$, LEO-C1 = 86 $\mu L \cdot L^{-1}$, MEO–C2 = 91.5 $\mu L \cdot L^{-1}$, BEO-C2 = 23.0 $\mu L \cdot L^{-1}$, LEO-C2 = 43.0 $\mu L \cdot L^{-1}$.

Regarding the weight loss after 7 days of storage in fumigated EO vapor, the smallest decrease was noted for the fruits kept in LEO-C1 treatment. For the rest, the differences were statistically significant (values expressed as a percentage from initial control). Analyzing the pH value, it was found that keeping the lemons in EO vapor had no influence; there were no significant differences compared with the control.

Citrus fruits are widely recognized for rich content in ascorbic acid (AsA). As can be seen from Figure 3a, the AsA content increased in lemon peel in all EO vapor treatments during 7 days of lemon fumigation. The smallest differences compared to the control were noticed for treatments MEO-C1, MEO-C2, and BEO-C1, respectively, differences which proved to be without statistical significance. The highest AsA was determined for LEO-C2. In addition, it can be noted that the C2 concentration of the EO treatments led to a higher AsA content accumulated in lemon peels, compared to C1 concentration. This observation was only valid for BEO and LEO.

In the lemon pulp, the amount of AsA after 7 days of storage in the atmosphere enriched by EOs was higher in all treatments except one, as seen in Figure 3b. For LEO-C1 the amount of AsA determined (17.1 gr 100^{-1}g FW) did not ensure significant differences compared to the control (18.6 gr 100^{-1} g FW). In contrast, LEO-C2 provided the highest amount of AsA in the lemon peel of 31.5 gr 100^{-1} g FW. Similar to AsA content in lemon peel the C2 EO concentration caused a greater amount of AsA accumulation than C1 concentration.

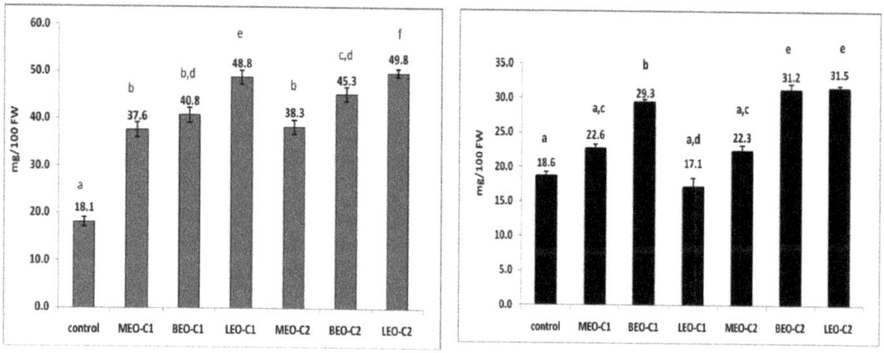

(a) (b)

Figure 3. Levels of ascorbic acid in the peel (**a**) and pulp (**b**) of lemons after 7 days of storage in enriched air-space with EO vapor. Data represent means ± SD (mg AsA·100 g FW). Different letters on top of column indicate differences for $p < 0.05$, *t*-test. Vertical bars represent standard deviations for $n = 10$; MEO-C1 = 183 μL·L^{-1}, BEO-C1 = 46 μL·L^{-1}, LEO-C1 = 86 μL·L^{-1}, MEO–C2 = 91.5 μL·L^{-1}, BEO-C2 = 23.0 μL·L^{-1}, LEO-C2 = 43.0 μL·L^{-1}.

3. Discussion

Essential oils differ from other oils due to the presence of volatile aromatic compounds that plants synthesize in stressful moments [21]. Mostly, the antimicrobial effect of essential oils is due to the predominant compounds of essential oil; based on this observation different chemotypes have been distinguished within aromatic species [26–28].

Obviously, the antifungal effect of EO against *P. digitatum* is mainly due to the presence of monoterpene oxygenated compounds [29]. Within this group, the following compounds represent the majority, as seen in Table 1: menthol in MEO, linalool and methyl chavicol in BEO, and linalool in LEO. This is in accordance with the findings of other researchers; they have shown that linalool and terpinen 4-ol acts against *Aspergillus fumigatus* through vapors and not through agar diffusion contact action [30].

Considering the definition of the minimum inhibitory dose (MID) as the lowest dose that determines 100% inhibition of fungal growth, similar with MIC for testing EO in liquid [29,31], we must consider one important aspect, namely the particularity of fungus to regenerate. Therefore, it appears imperative to define two different aspects: (1) the minimum dose leading to a temporary inhibition of fungal growth, in which case the fungistatic effect was achieved (MFsD); and (2) the minimum dose necessary for fungal irreversible inhibition, when the hyphae do not regenerate, and the fungicidal effect is achieved (MFdD). To be noted, in the case of MEO the establishment of MFdD, the minimal fungicidal dose, was not successful, and future investigations are needed to determine the fungicidal effect of MEO for doses greater than 400 μL (0.8%, *v*/*v*). Taking into account the mycelia growth in the presence of EO vapor valid for 50 cm^3 of Petri dish headspace, the IC50 values for each EO were established using the polynomial equation (Supplementary Materials Figure S1). Accordingly, the doses of EO that provided 50% inhibition of *P. digitatum* growth were 91.49 μL (0.183%, *v*/*v*) for MEO, 23.28 μL (0.046%, *v*/*v*) for BEO, and 43.39 μL (0.086%, *v*/*v*) for LEO. The good antifungal performance of BEO can be due to the synergic action of the two majority compounds like estragole and linalool.

Keeping lemons in the EO enriched atmosphere has the advantage that fruits are not in direct contact with oils because they act on fungi due to its volatilization properties. From what is known so far, only MEO among the EOs tested is recommended for fumigation-based methods due to volatilization properties [32]. However, the action of the EO vapor-phase has proven to be more effective against *Trichophyton mentagrophytes* and *Aspergillus fumigatus* than has been shown in agar essential oil diffusion assays [33]. Similarly, in MEO, BEO, and LEO performed by vapor, the amplitude of EO antifungal effects occurred in a dose-dependent manner (Figure 1). Thus, the presence of MEO vapors assures the antifungal effect due to the menthol, whereas the p-menthan-3-one compound (similar to menthone) is known to have no antifungal action [34,35]. Existence of other compounds, such as eucalyptol (7.44%), carvone (6.72%), and D-limonene (3.06%) recognized for their antimicrobial activity, could interact together or synergic with menthol, thus enhancing the antifungal action. The mode of action of vapor compounds from EO is different, being capable of disrupting the hyphal cells or fungal spore membrane, inhibiting the sporulation, and finally affecting the growth of food spoilage fungi [35]. Identically, the antifungal potential of BEO is due to the presence of the alcoholic compounds linalool, 41.49%, and estragole, 49.94%, (similar to methyl chavicol), which also has proven antifungal activity [28]. The presence of estragole in a high proportion, above the value of linalool, includes the basil plant used in the study into estragole chemotype. Generally, it is widely accepted that the chemical composition of BEO exhibits fluctuations for each component, mainly due to the climatic conditions but influenced by genotype as well.

Similarly, for LEO, linalool represents a major compound of about 31.44% and together with linalyl acetate (31.78%) ensures the antifungal effect of LEO. Additionally, if the ratio value of linalyl acetate-linalool is above one, lavender oil is considered to be of high quality [36]. In the case of the lavender essential oil used in our experiment, the ratio was 1.011, which attests the high quality of the oil. Ranking the EO vapor treatments by *t*-test analysis revealed that the same intensity of the antifungal effect could be reached with different amounts of EOs (Table 3). Evidently, this can be explained by the different proportion of antifungal compounds contained in EOs. The most effective EOs for in vivo antifungal preservation of lemons was significant by LEO-C1 with 86 μLL^{-1}, and BEO-C1 with 46 $\mu L{\cdot}L^{-1}$, respectively. Both essential oils contain linalool as a major compound, but we must make it clear that lemon peel has antifungal compounds such as D-limonene and citral that can potentiate the antifungal effect of the linalool. This argument is in line with Bakkali et al., who affirm that the minority compounds have an important role in the biological effect of EOs by facilitating the penetration of the microbial cell wall or membranes [37]. The antifungal protective effect for lemons was provided in order of LEO-C1 > BEO-C1 > MEO-C1 > BEO-C2 > MEO-C2 > LEO-C2.

Physiological quality indicators like fruit firmness depend on the EO dose used. Firmness is one of the most important indicators used to assess the quality of many fruits. Loss of fruit firmness during storage shows that there are water losses and metabolic changes in fruits [23]. Different polysaccharide, protein, or carboxymethyl cellulose (CMC) coating techniques or edible coatings with chitosan bilayers do not form effective water-vapor barriers [38]. Thus, essential oil vapor-phase treatment is proving to be more efficient during the post-harvest manipulation of fruits. Until now research on the application of cinnamon and eucalypt EO vapor treatments (50 ppm) have been reported with positive effects on maintaining firmness for tomato crop and cherry tomatoes [23]. Additionally, Jiang et al. noted that in shiitake mushrooms after fumigation with cloves, thymus, and cinnamaldehyde EO vapors for 20 days at 4 °C storage, the firmness, texture, and resistance to microorganism attack were augmented. As a result, the mushrooms were better preserved compared to the control sample during storage, but at the same time a higher concentration of EO can have an undesirable effect [39].

Regarding the pH value, it was found that keeping the lemons in EO vapor did not influence the pH values. The result is in agreement with previous research that showed that citrus fruits treated with wax and citral did not change their pH values after 5 days of storage [40].

AsA is one of the most frequently studied and powerful antioxidants. Fruit cells have the ability to protect themselves from oxidative stress by producing low molecular mass antioxidant molecules; in this category ascorbic acid and phenolic compounds are included [41]. Until now, EO was used especially in coating mixtures with chitosan or wax [10]. It appears that certain compounds from the structure of essential oils, such as ethylene, act on lemons in oxidative stress, triggering the enzymatic pathway for the synthesis of AsA molecules [10,39]. Therefore, AsA accumulation in large quantities, especially in the lemon peel, confirms this. However, not all EO vapor can cause the stimulant effect.

MEO vapors in both concentrations did not induce ascorbic acid synthesis. In our experience, BEO and LEO vapor have proven elicitor properties. Additionally, other researchers applied thyme EO vapor in peach treatment and reported increased activities of defense-related enzymes and total phenolic content [42].

The results obtained regarding AsA increases, in lemon pulp and peel, in the presence of EO vapor, are in line with recent studies on treatment with compounds such as methyl jasmonate, nitric oxide, salicylic acid, and essential oil that have proven to elicit and stimulate the inductive-defensive system and extend the shelf life of fruits and vegetables [43–45].

4. Materials and Methods

4.1. Plant Materials, EO Extraction, and Gas Chromatography/Mass Spectrometry Assessment

The research was performed between April and August 2018 in the Interdisciplinary Research Platform of Banat University of Agricultural Sciences and Veterinary Medicine (Romania). Aerial parts of plants, including mint, lavender, and basil, were collected from experimental fields of the Aromatic Plant Department, Faculty of Agriculture Timisoara. The samples were harvested in full flowering stage, in the months of June and July, on sunny days with moderate water deficit when the essential oil content was the highest [21]. Plant samples were cleaned, dried in the shade, milled, and then stored in the dark until use. An amount of 300 g of a homogenous sample was used for steam distillation for 2.0 h, using Clevenger-type equipment for essential oil extraction according to the European Pharmacopeia [46]. The obtained oils were stored at 2–4 °C until GC/MS analysis. The extraction yields of MEO, BEO, and LEO were calculated using the following formula:

$$\text{Yield (\%)} = (\text{amount of EO (g)}/\text{amount of dry plant (g)}) \times 100 \quad (1)$$

The chemical characterization of EO was done using a gas-chromatograph equipment with a Shimadzu QP 2010Plus, Columbia, SC, USA mass spectrometer (GC/MS) with an AT WAX capillary column (characteristics 30 m × 0.32 mm × 1 μm). Helium was used as the carrier gas with a flow rate of 1 mL/min with a column pressure of 42 kPa. Component separation was achieved under the following program: 40 °C for 1 min a rate of 5 °C/min to 210 °C for 5 min. Injector and ion source temperatures were 250 °C and 220 °C, respectively. The injection volume was 1 μL of a hexane solution of EO with a 1:50 split ratio. The NIST 5 Wiley 275 libraries database was used to identify volatile compounds through previously calculation of linear retention index (LRI) [25].

4.2. Antifungal Efficacy Assessment: In Vitro Assay

The fungal isolate, denoted Pd_0318_L, from Microbial Culture Collection, Agricultural Microbiology Department, Faculty of Horticulture and Forestry Timisoara, was used in both in vitro and in vivo assessments. The fungus was previously isolated from spoiled lemons on MEA (malt-extract-agar, Sigma-Aldrich, Chemie, Madrid, Spain) and was identified, by morphological and cultural characteristics according to Pitt and Hocking, as *Penicillium digitatum* [47]. To test the antifungal efficacy of EO vapor, two circular plugs per Petri dish (8 mm Ø) were picked up from the edge of a 4 day-old *P. digitatum* mycelium and were transferred on CYA medium (Czapek-yeast-agar, Sigma-Aldrich). The amount of medium distributed was exactly 20 mL on each Petri dish (Ø = 100 mm). After mycelia plug inoculation (two per Petri dish) to ensure an atmosphere with essential oil vapor, in the lid of the Petri dish a sterile filter paper was placed on which the dose of essential oil was added. The variants were as follows: 0.0 (control), 50, 100, 150, 200, 250, 300, 350, 400 μL for MEO, BEO, and LEO, respectively. To prevent the escape of volatile compounds, the Petri dishes were sealed with adhesive tape. The volume of vapor enriched air in each Petri dish was calculated according to the following equation:

$$V_{HS} = VPD - V_M \quad (2)$$

where V_{HS} is the headspace of the Petri dish with EO vapor; V_{PD} is the volume of the Petri dish (70 cm^3); and V_M is the volume of media in the Petri dish (20 cm^3).

This means that for each Petri dish, 50 cm^3 of air was enriched in EO vapor, resulting in the following concentrations (%, *v*/*v*): 0.1, 0.2, 0.3, 0.4, 0.5, 0.6, 0.7, and 0.8, respectively, valid for each EO used in the experiment. The research was performed twice.

After 5 days, two perpendicular diameters of mycelia were measured, and their average (AD) was calculated. The formula for calculating the new mycelium grown surfaces (NMG) is

$$NMG = ((AD^2 * 3.14/4) - AFI)/100 \quad (3)$$

where NMG is the new mycelium growth surface (cm^2); AD is the average of two perpendicular diameters of the fungal colony (mm); and AFI is the circular plug area of fungal inoculums (50.24 mm^2).

For assessment of fungistatic (MFsD) or fungicidal (MFdD) doses, the fungal discs from treatment samples with no growth recorded were re-inoculated into fresh CYA medium, and after 3 days the fungus revival was checked. The NMG values were then used to determine IC50 (EO concentration with 50% mycelia inhibition).

4.3. *Experiment 1—Antifungal Protection of Lemons, In Vivo Assay*

Lemon (*Citrus limon*, Verna cv) fruits without chemical treatments were used for the evaluation of antifungal protection in the atmosphere enriched in the essential oil vapor-phase. Fourteen desiccant containers of 2 L volume were used; in each of them, 4 lemons were tested as replicates. In total, 56 lemons fruit were washed first in a 9:1 hypochlorite (NaOCl) solution and then rinsed three times in sterile water. After drying in a sterile niche, the lemons were wounded in the peel at the equatorial zone with a cork borer (0.8 cm diameter and 0.5 cm deep); two wounds were produced on each lemon. The wounds were inoculated with 100 μL spore suspension of *P. digitatum* in saline water (0.65%) containing 10^5 spores·mL^{-1}. For each type of essential oil, two doses were tested, in two repetitions each. Thus, EO vapor fumigation treatments were denoted as MEO-C1, MEO-C2, BEO-C1, BEO-C2, LEO-C1, LEO-C2, and the control without EOs. Each treatment was done twice. C1 represents dose of EOs corresponding to the IC50 value, and C2 is the half of it, as seen in Table 5.

Table 5. Doses and concentrations of EOs used for the in vivo study of lemon preservation in vapor-phase modified atmosphere.

EO Doses	IC50 In Vitro *		C1 In Vivo **		C2 In Vivo ***	
	EO Doses μL	EO μL L^{-1} Air Space	EO Doses μL	EO μL L^{-1} Air Space	EO Doses μL	EO μL L^{-1} Air Space
MEO	91.49	183.0	3660	183	1830	91.5
BEO	23.28	46.0	932	46	466	23.0
LEO	43.39	86.0	1736	86	868	43.0

MEO, mint essential oil; BEO, basil essential oil; LEO, lavender essential oil; * IC50 values valid for 50 cm^3 of Petri dish head space; **C1 concentration of EO corresponding to the IC50 value, *** C2 concentration of EO corresponding to half of IC50 value valid for 2 L air from container space.

The essential oil was added on the filter paper at the bottom of the desiccant container along with a Petri glass with 10 mL distilled water for maintaining 90% relative humidity (RH). Four artificially inoculated lemons were placed on the grid. The desiccant containers were hermetically sealed with grease and stored at 24 ± 2 °C in the dark for 7 days. The in vivo experiment was performed twice.

4.4. *Experiment 2: Physiological and Biochemical Indicators of Lemon Fruit Quality, In Vivo Assay*

Forty lemons of the Maglina variety, purchased from the supermarket (conventional culture from Greece), were divided into 8 groups (five lemons per group serving as repetitions). For the initial determination of the indicators, namely firmness, the content of ascorbic acid in peel and pulp, and the pH value, a group of five lemons was used. For the firmness index of the lemons ($n = 5$), a digital penetrometer FR 5120 (tips 8 mm, manufacturer LUTRON ELECTRONIC ENTERPRISE, Taiwan) was used. The rest of the lemons were used to monitor the quality indicators after a 7-day storage period in an atmosphere enriched with EO vapor. The lemon preservation was tested in hermetically desiccant containers (2 L volume each) inside of which 5 lemons (taken as repetitions) were placed. The required volume of each EO, in order to ensure the concentrations C1 and C2 according to Table 5, was added on a filter paper placed on the bottom of the container. Each lemon fruit was weighed, recorded, and labelled prior to storage in EO vapor-enriched air-space. After 7 days of storage in the dark at 24 ± 2 °C, 90% RH, the lemons were weighed again, and the differences were presented as percent of weight loss compared to the initial weight using Formula (4).

$$\text{Weight loss (\%)} = \text{IW} - \text{FW} * 100/\text{IW} \tag{4}$$

where IW is the initial weight; and FW is the final weight.

The pH value was determined using an automatic WTW inoLab pH 720 (Weilheim, Wissenschaftlich- Technische Werkstätten GmbH, Germany). The contents of ascorbic acid (AsA) from peel and lemon pulp were determined with the Tillman method using 2,6-dichloroindophenol (Tillman's reagent, TR) from an average of 5 lemons per treatment [48]. Briefly, after weighing and determining lemon firmness, the peel was scratched until the white lemon skin was exposed. An amount of 4 g was weighed from an average sample to obtain the extract. Therefore, for each gram, 10 mL of 70% ethanol (ethanol 96%, Merck KGaA, Darmstadt, Germany) was added, and the sample was shaken for 120 min at room temperature using a GFL 3005 Analogue Orbital Shaker (Gesellschaft für Labortechnik GMBH, Burgwedel, Germany). Extracts were filtered using Whatman membrane filters (nylon, 0.45 μm, 30 mm diameter; Sigma-Aldrich; Merck, Germany). AsA content was estimated titrimetrically using 5 mL of extract diluted with 5 mL of oxalic acid (CAS Number 144-62-7, Merck, Germany), adding 1 mL acetaldehyde (6.9 g L^{-1}; CAS Number 75-07-0, Merck, Germany) and then titrated with 1 mM solution 2,6-dichloroindophenol sodium salt hydrate (Merck, Germany) to pink color. The experiment was performed twice.

4.5. Statistical Analysis

The results of both in vivo and in vitro assays, regarding the potential of EO treatments, were presented as mean ± standard deviation (SD). ANOVA analyses were applied for data obtained from new mycelial growth area and diameters of fungal growth on lemons preserved with EO vapor treatment. The IC50 for each EO was established by the polynomial equation; previous data were log-transformed. Additionally, *t*-test analysis was performed for ranking the EO treatments acting as vapor in the storage of lemons, and for AsA accumulation in lemon peel and pulp. Data were processed with Statistica 10 (StatSoft, USA).

5. Conclusions

Due to the volatilization properties, essential oils represent a natural method with high potential for use in preserving or storing organic fruits and vegetables. EO utilized in the vapor-phase could be effective against fungal deteriorations. Our data suggest that integration of EO treatments in the vapor-phase could be an alternative to enhance the health-promoting phytochemicals in lemons.

The presence of EO in the vapor-phase stimulates the activity of the defense system in fruits by increasing the ascorbic acid content responsible for protecting against the oxidative stress.

Our study clearly indicated that the vapor activity of BEO contributed considerably to the inhibitory diameter of *P. digitatum*, and suggested, from an economic point of view, that BEO has better efficiency than LEO and MEO since it needs the smallest quantity of EO in the lemon antifungal preservation for the same time period.

For the future, research is needed using higher doses of mint essential oil or research under controlled conditions, taking into account lower temperature, humidity control, and the possibility of airflow to determine the maximum potential that each EO can provide.

Supplementary Materials: Figure S1: The IC50 values determined by polynomial equation for MEO (a), BEO (b), and LEO (c).

Author Contributions: Conceptualization, R.M.S., E.A., and R.K.; methodology, I.P. and D.O.; software, I.C.; validation, A.F.S., E.A., and R.L.S.; formal analysis, D.O. and I.P.; resources, I.R.; data curation, R.K.; writing—original draft preparation, R.K. and I.P.; writing—review and editing, R.M.S. and I.C.; visualization, M.N.; supervision, E.A.; project administration, I.R.; funding acquisition, I.R. All authors have read and agreed to the published version of the manuscript.

Funding: This work was supported by the project "Ensuring excellence in RDI activities within USAMVBT", project number 35PFE/2018, financed by the Ministry of Research and Innovation (MCI) through Program 1, Development of The National Research and Development System; Subprogram 1.2, Institutional Performance, Institutional Development Projects—Projects to fund excellence in RDI.

Acknowledgments: We have been able to carry out this research with the support of the Interdisciplinary Research Platform of Banat University of Agricultural Sciences and Veterinary Medicine, "King Michael I of Romania" from Timisoara, where the chemical analysis and antifungal experiments were made.

Conflicts of Interest: The authors declare no conflict of interest.

References

1. Porat, R.; Lichter, A.; Terry, L.A.; Harker, R.; Buzby, J. Postharvest Losses of Fruit and Vegetables during Retail and in Consumers' Homes: Quantifications, Causes, and Means of Prevention. *Postharvest Biol. Technol.* **2018**, *139*, 135–149. [CrossRef]
2. Pérez-Alfonso, C.O.; Martínez-Romero, D.; Zapata, P.J.; Serrano, M.; Valero, D.; Castillo, S. The Effects of Essential Oils Carvacrol and Thymol on Growth of *Penicillium digitatum* and *P. italicum* Involved in Lemon Decay. *Int. J. Food Microbiol.* **2012**, *158*, 101–106. [CrossRef] [PubMed]
3. He, L.; Liu, Y.; Mustapha, A.; Lin, M. Antifungal Activity of Zinc Oxide Nanoparticles against *Botrytis cinerea* and *Penicillium expansum*. *Microbiol. Res.* **2011**, *166*, 207–215. [CrossRef] [PubMed]
4. Liu, Y.; Wang, W.; Zhou, Y.; Yao, S.; Deng, L.; Zeng, K. Isolation, Identification and in Vitro Screening of Chongqing Orangery Yeasts for the Biocontrol of Penicillium Digitatum on Citrus Fruit. *Biol. Control* **2017**, *110*, 18–24. [CrossRef]
5. Simas, D.L.R.; de Amorim, S.H.B.M.; Goulart, F.R.V.; Alviano, C.S.; Alviano, D.S.; da Silva, A.J.R. Citrus Species Essential Oils and Their Components Can Inhibit or Stimulate Fungal Growth in Fruit. *Ind. Crops Prod.* **2017**, *98*, 108–115. [CrossRef]
6. Palou, L.; Valencia-Chamorro, S.A.; Pérez-Gago, M.B. Antifungal Edible Coatings for Fresh Citrus Fruit: A Review. *Coatings.* **2015**, *5*, 962–986. [CrossRef]
7. Romanazzi, G.; Smilanick, J.L.; Feliziani, E.; Droby, S. Integrated Management of Postharvest Gray Mold on Fruit Crops. *Postharvest Biol. Technol.* **2016**, *113*, 69–76. [CrossRef]
8. Ncama, K.; Magwaza, L.S.; Mditshwa, A.; Tesfay, S.Z. Plant-Based Edible Coatings for Managing Postharvest Quality of Fresh Horticultural Produce: A Review. *Food Packag. Shelf Life.* **2018**, *16*, 157–167. [CrossRef]
9. Bill, M.; Sivakumar, D.; Korsten, L.; Thompson, A.K. The Efficacy of Combined Application of Edible Coatings and Thyme Oil in Inducing Resistance Components in Avocado (Persea Americana Mill.) against Anthracnose during Post-Harvest Storage. *Crop Prot.* **2014**, *64*, 159–167. [CrossRef]
10. Mditshwa, A.; Magwaza, L.S.; Tesfay, S.Z.; Opara, U.L. Postharvest Factors Affecting Vitamin C Content of Citrus Fruits: A Review. *Scientia Horticulturae* **2017**, *218*, 95–104. [CrossRef]
11. Perez, M.F.; Perez Ibarreche, J.; Isas, A.S.; Sepulveda, M.; Ramallo, J.; Dib, J.R. Antagonistic Yeasts for the Biological Control of *Penicillium digitatum* on Lemons Stored under Export Conditions. *Biol. Control* **2017**, *115*, 135–140. [CrossRef]
12. Ferraz, L.P.; Cunha, T.d.; da Silva, A.C.; Kupper, K.C. Biocontrol Ability and Putative Mode of Action of Yeasts against *Geotrichum Citri-Aurantii* in Citrus Fruit. *Microbiol. Res.* **2016**, *188*, 72–79. [CrossRef] [PubMed]

13. Sperandio, E.M.; Martins do Vale, H.M.; Moreira, G.A.M. Yeasts from Native Brazilian Cerrado Plants: Occurrence, Diversity and Use in the Biocontrol of Citrus Green Mould. *Fungal Biol.* **2015**, *119*, 984–993. [CrossRef] [PubMed]
14. Sivakumar, D.; Bautista-Baños, S. A Review on the Use of Essential Oils for Postharvest Decay Control and Maintenance of Fruit Quality during Storage. *Crop Prot.* **2014**, *64*, 27–37. [CrossRef]
15. Elshafie, H.S.; Mancini, E.; Camele, I.; Martino, L.D.; de Feo, V. In Vivo Antifungal Activity of Two Essential Oils from Mediterranean Plants against Postharvest Brown Rot Disease of Peach Fruit. *Ind. Crops Prod.* **2015**, *66*, 11–15. [CrossRef]
16. Acevedo-Fani, A.; Soliva-Fortuny, R.; Martín-Belloso, O. Nanoemulsions as Edible Coatings. *Curr. Opin. Food Sci.* **2017**, *15*, 43–49. [CrossRef]
17. Usall, J.; Ippolito, A.; Sisquella, M.; Neri, F. Physical Treatments to Control Postharvest Diseases of Fresh Fruits and Vegetables. *Postharvest Biol. Technol.* **2016**, *122*, 30–40. [CrossRef]
18. Manganyi, M.C.; Regnier, T.; Olivier, E.I. Antimicrobial Activities of Selected Essential Oils against *Fusarium oxysporum* Isolates and Their Biofilms. *S. Afr. J. Bot.* **2015**, *99*, 115–121. [CrossRef]
19. Da Silva Bomfim, N.; Nakassugi, L.P.; Faggion Pinheiro Oliveira, J.; Kohiyama, C.Y.; Mossini, S.A.G.; Grespan, R.; Nerilo, S.B.; Mallmann, C.A.; Alves Abreu Filho, B.; Machinski, M. Antifungal Activity and Inhibition of Fumonisin Production by *Rosmarinus Officinalis* L. Essential Oil in *Fusarium verticillioides* (Sacc.) Nirenberg. *Food Chem.* **2015**, *166*, 330–336. [CrossRef]
20. Munhuweyi, K.; Caleb, O.J.; Lennox, C.L.; van Reenen, A.J.; Opara, U.L. In Vitro and In Vivo Antifungal Activity of Chitosan-Essential Oils against Pomegranate Fruit Pathogens. *Postharvest Biol. Technol.* **2017**, *129*, 9–22. [CrossRef]
21. Calo, J.R.; Crandall, P.G.; O'Bryan, C.A.; Ricke, S.C. Essential Oils as Antimicrobials in Food Systems—A Review. *Food Control* **2015**, *54*, 111–119. [CrossRef]
22. Frankova, A.; Smid, J.; Bernardos, A.; Finkousova, A.; Marsik, P.; Novotny, D.; Legarová, V.; Pulkrabek, J.; Kloucek, P. The Antifungal Activity of Essential Oils in Combination with Warm Air Flow against Postharvest Phytopathogenic Fungi in Apples. *Food Control* **2016**, *68*, 62–68. [CrossRef]
23. Tzortzakis, N.G. Maintaining Postharvest Quality of Fresh Produce with Volatile Compounds. *Innov. Food Sci. Emerg. Technol.* **2007**, *8*, 111–116. [CrossRef]
24. Sadgrove, N.; Jones, G. A Contemporary Introduction to Essential Oils: Chemistry, Bioactivity and Prospects for Australian Agriculture. *Agriculture* **2015**, *5*, 48–102. [CrossRef]
25. Babushok, V.I.; Linstrom, P.J.; Zenkevich, I.G. Retention indices for frequently reported compounds of plant essential oils. *J. Phys. Chem. Ref. Data* **2011**, *40*, 1–47. [CrossRef]
26. Rota, M.C.; Herrera, A.; Martínez, R.M.; Sotomayor, J.A.; Jordán, M.J. Antimicrobial Activity and Chemical Composition of *Thymus vulgaris, Thymus zygis* and *Thymus hyemalis* Essential Oils. *Food Control* **2008**, *19*, 681–687. [CrossRef]
27. Calvo-Irabien, L.M. Native Mexican Aromatic Flora and Essential Oils: Current Research Status, Gaps in Knowledge and Agro-Industrial Potential. *Ind. Crops Prod.* **2018**, *111*, 807–822. [CrossRef]
28. Oxenham, S.K.; Svoboda, K.P.; Walters, D.R. Antifungal Activity of the Essential Oil of Basil (*Ocimum Basilicum*). *J. Phytopathol.* **2005**, *153*, 174–180. [CrossRef]
29. Laird, K.; Phillips, C. Vapor Phase: A Potential Future Use for Essential Oils as Antimicrobials? *Lett. Appl. Microbiol.* **2012**, *54*, 169–174. [CrossRef]
30. Inouye, S.; Tsuruoka, T.; Watanabe, M.; Takeo, K.; Akao, M.; Nishiyama, Y.; Yamaguchi, H. Inhibitory effect of essential oils on apical growth of *Aspergillus fumigatus* by vapor contact. *Mycoses* **2000**, *43*, 17–23. [CrossRef]
31. Yahyazadeh, M.; Omidbaigi, R.; Zare, R.; Taheri, H. Effect of Some Essential Oils on Mycelial Growth of *Penicillium digitatum* Sacc. *World J. Microbiol. Biotechnol.* **2008**, *24*, 1445–1450. [CrossRef]
32. Kamatou, G.P.P.; Vermaak, I.; Viljoen, A.M.; Lawrence, B.M. Menthol: A Simple Monoterpene with Remarkable Biological Properties. *Phytochemistry* **2013**. [CrossRef] [PubMed]
33. Inouye, S.; Uchida, K.; Maruyama, N.; Yamaguchi, H.; Abe, S. A Novel Method to Estimate the Contribution of the Vapor Activity of Essential Oils in Agar Diffusion Assay. *Jpn. J. Med. Mycol.* **2006**, *47*, 91–98. [CrossRef] [PubMed]
34. Edris, A.E.; Farrag, E.S. Antifungal Activity of Peppermint and Sweet Basil Essential Oils and Their Major Aroma Constituents on Some Plant Pathogenic Fungi from the Vapor Phase. *Nahrung Food* **2003**, *47*, 117–121. [CrossRef]

35. Nazzaro, F.; Fratianni, F.; Coppola, R.; de Feo, V. Essential Oils and Antifungal Activity. *Pharmaceuticals* **2017**, *10*, 86. [CrossRef]
36. Morsy, N.F.S. Chemical structure, quality indices and bioactivity of essential oil constituents. In *Active Ingredients from Aromatic and Medicinal Plants*; El-Shemy, H.A., Ed.; InTech: Rijeka, Croatia, 2017; pp. 189–191. [CrossRef]
37. Bakkali, F.; Averbeck, S.; Averbeck, D.; Idaomar, M. Biological Effects of Essential Oils–a Review. *Food Chem. Toxicol.* **2008**, *46*, 446–475. [CrossRef]
38. Arnon, H.; Zaitsev, Y.; Porat, R.; Poverenov, E. Effects of Carboxymethyl Cellulose and Chitosan Bilayer Edible Coating on Postharvest Quality of Citrus Fruit. *Postharvest Biol. Technol.* **2014**, *87*, 21–26. [CrossRef]
39. Jiang, T.; Luo, Z.; Ying, T. Fumigation with Essential Oils Improves Sensory Quality and Enhanced Antioxidant Ability of Shiitake Mushroom (*Lentinus Edodes*). *Food Chem.* **2015**, *172*, 692–698. [CrossRef]
40. Fan, F.; Tao, N.; Jia, L.; He, X. Use of Citral Incorporated in Postharvest Wax of Citrus Fruit as a Botanical Fungicide against *Penicillium digitatum*. *Postharvest Biol. Technol.* **2014**, *90*. [CrossRef]
41. Léchaudel, M.; Darnaudery, M.; Joët, T.; Fournier, P.; Joas, J. Genotypic and Environmental Effects on the Level of Ascorbic Acid, Phenolic Compounds and Related Gene Expression during Pineapple Fruit Development and Ripening. *Plant Physiol. Biochem.* **2018**, *130*, 127–138. [CrossRef]
42. Cindi, M.D.; Soundy, P.; Romanazzi, G.; Sivakumar, D. Different Defense Responses and Brown Rot Control in Two Prunus Persica Cultivars to Essential Oil Vapors after Storage. *Postharvest Biol. Technol.* **2016**, *119*, 9–17. [CrossRef]
43. Vithana, M.D.K.; Singh, Z.; Johnson, S.K.; Gupta, R. Concentrations of Health-Promoting Phytochemicals in Ripe Mango Fruit Triggered by Postharvest Application of Elicitors. *J. Sci. Food Agric.* **2019**, *99*, 1126–1134. [CrossRef] [PubMed]
44. Romanazzi, G.; Sanzani, S.M.; Bi, Y.; Tian, S.; Gutiérrez Martínez, P.; Alkan, N. Induced Resistance to Control Postharvest Decay of Fruit and Vegetables. *Postharvest Biol. Technol.* **2016**, *122*, 82–94. [CrossRef]
45. Bill, M.; Korsten, L.; Remize, F.; Glowacz, M.; Sivakumar, D. Effect of Thyme Oil Vapors Exposure on Phenylalanine Ammonia-Lyase (PAL) and Lipoxygenase (LOX) Genes Expression, and Control of Anthracnose in 'Hass' and 'Ryan' Avocado Fruit. *Sci. Hortic.* **2017**, *224*, 232–237. [CrossRef]
46. *European Pharmacopoeia*; European Directorate for the Quality of Medicines & Healthcare (EDQM): Strasbourg, France, 1975; Volume 3, pp. 89–99.
47. Pitt, J.I.; Hocking, A.D. *Fungi and Food Spoilage*; Springer: New York, NY, USA, 2009; pp. 53–142.
48. Santos, D.A.; Lima, K.P.; Março, P.H.; Valderrama, P. Vitamin C Determination by Ultraviolet Spectroscopy and Multiproduct Calibration. *J. Braz. Chem. Soc.* **2016**, *27*, 1912–1917. [CrossRef]

Sample Availability: Samples of BEO, MEO, and LEO are available from the authors.

© 2020 by the authors. Licensee MDPI, Basel, Switzerland. This article is an open access article distributed under the terms and conditions of the Creative Commons Attribution (CC BY) license (http://creativecommons.org/licenses/by/4.0/).

MDPI
St. Alban-Anlage 66
4052 Basel
Switzerland
www.mdpi.com

Molecules Editorial Office
E-mail: molecules@mdpi.com
www.mdpi.com/journal/molecules

Disclaimer/Publisher's Note: The statements, opinions and data contained in all publications are solely those of the individual author(s) and contributor(s) and not of MDPI and/or the editor(s). MDPI and/or the editor(s) disclaim responsibility for any injury to people or property resulting from any ideas, methods, instructions or products referred to in the content.

www.ingramcontent.com/pod-product-compliance
Lightning Source LLC
Chambersburg PA
CBHW061556210926
51850CB00009B/2125
* 9 7 8 3 0 3 6 5 8 7 5 1 6 *